T0264820

# LMIs in Control Systems

## Analysis, Design and Applications

# LMIs in Control Systems

## Analysis, Design and Applications

Guang-Ren Duan

Hai-Hua Yu

CRC Press
Taylor & Francis Group
Boca Raton   London   New York

CRC Press is an imprint of the
Taylor & Francis Group, an **informa** business

MATLAB® is a trademark of The MathWorks, Inc. and is used with permission. The MathWorks does not warrant the accuracy of the text or exercises in this book. This book's use or discussion of MATLAB® software or related products does not constitute endorsement or sponsorship by The MathWorks of a particular pedagogical approach or particular use of the MATLAB® software.

CRC Press
Taylor & Francis Group
6000 Broken Sound Parkway NW, Suite 300
Boca Raton, FL 33487-2742

© 2013 by Taylor & Francis Group, LLC
CRC Press is an imprint of Taylor & Francis Group, an Informa business

No claim to original U.S. Government works

International Standard Book Number-13: 978-1-4665-8299-6 (Hardback)

This book contains information obtained from authentic and highly regarded sources. Reasonable efforts have been made to publish reliable data and information, but the author and publisher cannot assume responsibility for the validity of all materials or the consequences of their use. The authors and publishers have attempted to trace the copyright holders of all material reproduced in this publication and apologize to copyright holders if permission to publish in this form has not been obtained. If any copyright material has not been acknowledged please write and let us know so we may rectify in any future reprint.

Except as permitted under U.S. Copyright Law, no part of this book may be reprinted, reproduced, transmitted, or utilized in any form by any electronic, mechanical, or other means, now known or hereafter invented, including photocopying, microfilming, and recording, or in any information storage or retrieval system, without written permission from the publishers.

For permission to photocopy or use material electronically from this work, please access www.copyright.com (http://www.copyright.com/) or contact the Copyright Clearance Center, Inc. (CCC), 222 Rosewood Drive, Danvers, MA 01923, 978-750-8400. CCC is a not-for-profit organization that provides licenses and registration for a variety of users. For organizations that have been granted a photocopy license by the CCC, a separate system of payment has been arranged.

**Trademark Notice:** Product or corporate names may be trademarks or registered trademarks, and are used only for identification and explanation without intent to infringe.

**Library of Congress Cataloging-in-Publication Data**

Duan, Guangren.
  LMIs in control systems : analysis, design and applications / author, Guang-Ren Duan.
    pages cm
  Includes bibliographical references and index.
  ISBN 978-1-4665-8299-6 (hardback)
  1. Control theory. 2. Matrix inequalities. 3. Mathematical optimization. I. Title.

QA402.3.D7973 2013
629.8--dc23                                                                    2013008019

**Visit the Taylor & Francis Web site at**
**http://www.taylorandfrancis.com**

**and the CRC Press Web site at**
**http://www.crcpress.com**

To
Shichao and Jiefu
Degang and Ruike

# Contents

# List of Figures

# List of Tables

# Preface

In the last two decades, linear matrix inequalities (LMIs) have emerged as a powerful tool in the field of control systems analysis and design. Many problems, such as state feedback synthesis, robustness analysis and design, and $H_2$ and $H_\infty$ control, can all be reduced to convex or quasi-convex problems that involve LMIs. Due to successful developments of efficient and reliable algorithms for solving LMIs, these problems can now be solved both efficiently and numerically reliably, thereby making this the most attractive feature of LMIs.

Today, LMIs have become a real technique. By using LMIs in control systems analysis and design, and with the help of the mature MATLAB® LMI toolbox, or the open-source software packages YALMIP and CVX, more and more theoretical and practical control applications can be solved, which might not have been otherwise possible using traditional methods.

## Goal of the Book

The goal of this book is to provide a textbook for graduate and senior undergraduate courses in the field of control systems analysis and design. The book contains not only the basic but also systematic knowledge about LMIs and LMI-based control systems analysis and design.

Boyd et al. (1994), which is the first book on LMIs, certainly has performed a pioneering function in the field, but due to time constraints it does not contain the numerous new theories and techniques in LMIs that emerged in the last two decades. Yu (2002) is a book on LMI approach to robust control, which is suitable to be used as a textbook, but the Chinese language limits its usage worldwide.

There are some other books available that are related to LMIs, but none of them provide a simple and systematic introduction to LMIs in control systems analysis and design. They are all research-oriented and focus on special research topics related to LMI approaches such as analysis and synthesis of multidimensional systems (Paszke (2006)), analysis and control of fuzzy systems (Lin et al. (2007), Tanaka and Wang (2001), Lam and Leung (2011)), and dynamic surface control of uncertain nonlinear systems (Song and Hedrick (2011)). Different from these, Cychowski (2009) and Guo and Wang (2010) respectively address convex optimization approaches to robust model predictive control and stochastic distribution control, both involving LMIs.

Besides the ones mentioned earlier, there are two other books related to LMIs. The first is Ghaoui and Niculescu (2000), which is a volume on advances in LMI methods in control, consisting of 17 different topics from 33 contributors. The second is Ostertag (2011), which treats mono- and multivariable control and estimation. Although the title of Ostertag (2011) partly refers to "LMI methods," only a small portion of the text is devoted to LMIs.

Although some books on LMIs are available now, a textbook on this topic suitable for graduate and senior undergraduate students in the discipline of systems and control is still lacking. Since LMIs today have become a real technique, and as they have also become an advanced, powerful tool for control systems analysis and design, a comprehensive textbook on LMIs in control systems analysis and design is now in great demand. The goal of this book is to fill this gap.

## Coverage and Supplements

The book consists of 12 chapters and focuses on the analysis and design problems on both continuous- and discrete-time linear systems based on LMI methods. Besides an introductory chapter, there are totally five parts including two appendices. Part I provides some technical lemmas to be used in the sequential chapters and also gives a brief review of optimization theory, while Parts II and III deal with the analysis and design problems, respectively. Specifically, Part II discusses the problems of stability analysis, performance analysis, and property analysis using LMI techniques, and Part III studies the problems of feedback stabilization, $H_2$ and $H_\infty$ control, state observation and filtering, and multiple objective designs. Part IV deals with two applications using LMI techniques: missile attitude control and satellite attitude control. Part V consists of two appendices, which provide lengthy proofs of certain results presented in the chapters and a brief introduction of the MATLAB LMI Toolbox.

It has to be mentioned that many researchers worldwide have made great contributions to LMI-based control systems analysis and design. However, due to length limitation and the structural arrangement of the book, many of their published results could not be included or even cited. We extend our apologies to these researchers.

Exercise problems are given at the end of each chapter, and all the solutions to these problems have also been worked out. Interested readers are welcome to contact Professor Duan (g.r.duan@hit.edu.cn) or Dr. Yu (yuhaihua@hit.edu.cn) to enquire about the following:

- Electronic version of the *Solutions Manual* (hardcopies can be obtained from the publisher for lecturing purposes)
- MATLAB codes for all the computational exercise problems in the form of a set of M files
- MATLAB codes for all the computational examples in the form of a set of M files

## Audience

This book mainly has two types of audience.

First, since the book gives a general and systematic introduction to LMI-based control systems analysis and design, it can be used as a primary or alternate textbook for some LMI-related courses for senior undergraduate and postgraduate students majoring in the fields of control systems theory and applications, and possibly for those majoring in applied mathematics, mechanical engineering, electrical engineering, and aerospace engineering as well.

Second, since the book also contains some advanced materials and materials related to aerospace control applications, it can be used as a reference book by postgraduates, researchers, scientists, engineers, and university lecturers in the fields of control systems and applications, applied mathematics, mechanical engineering, electrical engineering, and aerospace engineering.

MATLAB® is a registered trademark of The MathWorks, Inc. For product information, please contact:

The MathWorks, Inc.
3 Apple Hill Drive
Natick, MA 01760-2098 USA
Tel: 508-647-7000
Fax: 508-647-7001
E-mail: info@mathworks.com
Web: www.mathworks.com

# Acknowledgments

The students who took the graduate course "LMIs in control systems" at Harbin Institute of Technology during 2004–2012 found several errors and typos in the manuscript. Their help has greatly improved the quality of the book and is indeed very much appreciated.

Many other persons have also helped with the writing of this book. Dr. Ming-Zhe Hou, Dr. Ling-Ling Lv, Dr. Hong-Liang Liu, Dr. Yan-Jiang Li, Dr. Guang-Bin Cai, Dr. Shi-Jie Zhang, and Dr. Da-Ke Gu, who were all PhD students of Professor Duan years ago, have helped in revising certain parts of the book. Long Zhang, Feng Zhang, and Shi Li, who are present PhD students of Professor Duan, helped with the examples and exercises in the book. Our thanks also go to Professor Bin Zhou and Dr. Feng Tan, former PhD students of Professor Duan and now his colleagues, who helped in completing the proofs of Theorems 5.1 and 5.2 and in preparing the manuscript for Chapter 11, respectively.

Several persons have proofread the manuscripts. These include Professor Chen-xiao Cai of Nanjing University of Science and Technology; Professor Xiaohua Li, a visiting scholar from University of Science and Technology Liaoning; and Professor Aiguo Wu at the Shenzhen Graduate School of Harbin Institute of Technology, who was also Professor Duan's PhD student. In particular, we would like to thank Professor Stephen Boyd from Stanford University, who is also the chief academic advisor for the faculty of control science and engineering at Harbin Institute of Technology, for his helpful comments and suggestions.

Professor Duan would like to acknowledge the financial support provided by the many sponsors, including NSFC, the National Natural Science Foundation of China, the Chinese Ministry of Science and Technology, and the Ministry of Education of China, for projects funded by the Program of the National Science Fund for Distinguished Young Scholars, the Innovative Scientific Research Team Program, the National Key Basic Research and Development Program (973 Program), and the Program of ChangJiang Scholars. Thanks also to the Aerospace Industry Companies of China for funded projects that have provided the background and data for the applications in Chapters 11 and 12.

Last but not the least, we would like to thank in advance all the readers for choosing to read and use this textbook. We would be grateful if they provide via email at g.r.duan@hit.edu.cn or yuhaihua@hit.edu.cn feedback about any problems found. Their help will certainly make any future editions of the book much better.

**Guang-Ren Duan**
*Harbin Institute of Technology*

**Hai-Hua Yu**
*Heilongjiang University*

# Authors

**Guang-Ren Duan, BSc, MSc, PhD,** received his BSc in applied mathematics and his MSc and PhD in control systems theory. From 1989 to 1991, he was a post-doctoral researcher at the Harbin Institute of Technology, Harbin, Heilongjiang, People's Republic of China, where he became full professor of control systems theory in 1991. He is the founder and currently the director of the Center for Control Theory and Guidance Technology at the Harbin Institute of Technology.

Professor Duan visited the University of Hull, United Kingdom, and the University of Sheffield, United Kingdom, from December 1996 to October 1998 and worked as a lecturer at the Queen's University of Belfast, United Kingdom, from October 1998 to October 2002. He was elected IEEE senior member in 1994 and also won the 4th Chinese National Youth Award of Science and Technology in the same year. In 1997, he won the fund for Over-century Talents by the Chinese Ministry of Education, and in 1999, he was awarded the National Science Fund for Distinguished Young Scholars by Natural Science Foundation of China (NSFC). In August 2000, he was selected by the Cheung Kong Scholars Program of the Chinese government, and in 2001, he became fellow of IEE (IET) and an IET chartered engineer. His research group was selected in 2005 by the Cheung Kong Scholar Innovative Team Program sponsored by the Chinese Ministry of Education and in 2009 by the Innovative Research Group Program sponsored by NSFC. As a principal investigator, he has won several science and technology awards, including the Chinese National Award of Natural Sciences in 2008 on "Parametric approaches for robust control systems design with applications."

Professor Duan has served at many domestic and international conferences and symposiums as general chair, associate general chair, chair and co-chair of IPCs, and member of IPCs and has also been invited as a keynote speaker at several international conferences. Besides, he is a member of the Science and Technology Committee of the Chinese Ministry of Education, the Information Branch; associate director of the Control Theory and Applications Committee, Chinese Association of Automation; and associate director of the Foundation Committee for the "Zhao-Zhi Guan" Award. He is associate editor of many international journals, including *European Journal of Control* and *International Journal of Computing and Automation* and the editor in chief of *Journal of Harbin Institute of Technology* and director of the Academic Journal Center at the Harbin Institute of Technology.

Professor Duan's main research interests include parametric robust control systems design, LMI-based control systems, descriptor systems, spacecraft control, and magnetic bearing control. He is the author and coauthor of five books and has won two Chinese national best book awards. Besides this text book, he has written a monograph entitled *Analysis and Design of Descriptor Linear Systems* (published by Springer). He has also published over 200 SCI indexed papers, particularly publishing over 40 papers in *IEEE Transactions*.

He has supervised 53 master students and 54 PhD students. Among them, two students won the Chinese National 100 Best PhD Theses Award, two students won the Fund for New Century Talents by the Chinese Ministry of Education, and one student won the National Science Fund for Distinguished Young Scholars by NSFC and also the Cheung Kong Scholars Program of the Chinese government.

**Hai-Hua Yu, BS, MS, PhD,** received her BS in applied mathematics and MS in control theory and control engineering from Heilongjiang University and PhD in control science and engineering from the Harbin Institute of Technology, Harbin, Heilongjiang, People's Republic of China. She was a lecturer at Heilongjiang University from 2008 to 2010 and is now associate professor in the Department of Automation at Heilongjiang University and postdoctoral researcher at the Harbin Institute of Technology. She has taught the following courses: computer control, digital signal processing, computer simulation, C programming language, probability and statistics, nonlinear systems, and matrix analysis.

# List of Notations

## Notations Related to Subspaces

| | |
|---|---|
| $\mathbb{R}$ | set of all real numbers |
| $\mathbb{R}^+$ | set of all positive real numbers |
| $\mathbb{R}^-$ | set of all negative real numbers |
| $\mathbb{C}$ | set of all complex numbers |
| $\mathbb{C}^+$ | right-half complex plane |
| $\mathbb{C}^-$ | left-half complex plane |
| $\mathbb{R}^n$ | set of all real vectors of dimension $n$ |
| $\mathbb{C}^n$ | set of all complex vectors of dimension $n$ |
| $\mathbb{R}^{m \times n}$ | set of all real matrices of dimension $m \times n$ |
| $\mathbb{C}^{m \times n}$ | set of all complex matrices of dimension $m \times n$ |
| $\mathbb{R}_r^{m \times n}$ | set of $m \times n$ real matrices with rank $r$ |
| $\mathbb{C}_r^{m \times n}$ | set of $m \times n$ complex matrices with rank $r$ |
| $\bar{\mathbb{C}}^+$ | closed right-half complex plane, $\bar{\mathbb{C}}^+ = \{s \mid s \in \mathbb{C},\ \mathrm{Re}(s) \geq 0\}$ |
| $\ker(T)$ | kernel of transformation or matrix $T$ |
| $\mathrm{Image}(T)$ | image of transformation or matrix $T$ |
| $\mathrm{conv}(\mathbb{F})$ | convex hull of set $\mathbb{F}$ |
| $\mathbb{S}^n$ | set of symmetric matrix in $\mathbb{R}^{n \times n}$, $\mathbb{S}^n = \left\{M \mid M = M^{\mathrm{T}}, M \in \mathbb{R}^{n \times n}\right\}$ |
| $\mathbb{F}_{\mathrm{B}}$ | boundary set of $\mathbb{F}$ |
| $\mathbb{F}_{\mathrm{E}}$ | set of all extreme points of $\mathbb{F}$ |

## Notations Related to Vectors and Matrices

| | |
|---|---|
| $0_n$ | zero vector in $\mathbb{R}^n$ |
| $0_{m \times n}$ | zero matrix in $\mathbb{R}^{m \times n}$ |
| $I_n$ | identity matrix of order $n$ |
| $A^{-1}$ | inverse matrix of matrix $A$ |
| $A^{\mathrm{T}}$ | transpose of matrix $A$ |
| $\bar{A}$ | complex conjugate of matrix $A$ |
| $A^{\mathrm{H}}$ | transposed complex conjugate of matrix $A$ |
| $\mathrm{Re}(A)$ | real part of matrix $A$ |

| | |
|---|---|
| $\text{Im}(A)$ | imaginary part of matrix $A$ |
| $\det(A)$ | determinant of matrix $A$ |
| $\text{Adj}(A)$ | adjoint of matrix $A$ |
| $\text{trace}(A)$ | trace of matrix $A$ |
| $\text{rank}(A)$ | rank of matrix $A$ |
| $\kappa\,(A)$ | condition number of matrix $A$ |
| $\rho\,(A)$ | spectral radius of matrix $A$ |
| $A > 0$ | $A$ is Hermite (symmetric) positive definite |
| $A \geq 0$ | $A$ is Hermite (symmetric) semi-positive definite |
| $A > B$ | $A - B > 0$ |
| $A \geq B$ | $A - B \geq 0$ |
| $A < 0$ | $A$ is Hermite (symmetric) negative definite |
| $A \leq 0$ | $A$ is Hermite (symmetric) semi-negative definite |
| $A < B$ | $A - B < 0$ |
| $A \leq B$ | $A - B \leq 0$ |
| $A^{\frac{1}{2}}$ | matrix $Z$ satisfying $Z^{\mathrm{T}}Z = A > 0$ |
| $\lambda\,(A)$ | set of all eigenvalues of matrix $A$ |
| $\lambda_i\,(A)$ | $i$th eigenvalue of matrix $A$ |
| $\lambda_{\max}\,(A)$ | maximum eigenvalue of matrix $A$ |
| $\lambda_{\min}\,(A)$ | minimum eigenvalue of matrix $A$ |
| $\sigma_i\,(A)$ | $i$th singular value of matrix $A$ |
| $\sigma_{\max}\,(A)$ | maximum singular of matrix $A$ |
| $\sigma_{\min}\,(A)$ | minimum singular value of matrix $A$ |
| $\langle A \rangle_s$ | sum of matrix $A$ and its transpose, $\langle A \rangle_s = A + A^{\mathrm{T}}$ |
| $\|A\|_2$ | spectral norm of matrix $A$ |
| $\|A\|_F$ | Frobenius norm of matrix $A$ |
| $\|A\|_1$ | row–sum norm of matrix $A$ |
| $\|A\|_\infty$ | column–sum norm of matrix $A$ |
| $\dot{x}(t)$ | first-order derivative of vector $x$ with respective to $t$ |
| $\ddot{x}(t)$ | second-order derivative of vector $x$ with respective to $t$ |
| $x \prec 0$ | elements of vector $x$ are less than 0 |
| $x \preceq 0$ | elements of vector $x$ are not greater than 0 |
| $x \prec y$ | $x - y \prec 0$ |
| $x \preceq y$ | $x - y \preceq 0$ |
| $A \overset{\leftrightsquigarrow}{=} B$ | the equivalent relation between matrices $A$ and $B$ |

## Notations of Relations and Manipulations

| | |
|---|---|
| $\Longrightarrow$ | imply |
| $\Longleftrightarrow$ | if and only if |
| $\in$ | belong to |
| $\subset$ | subset |
| $\cap$ | intersection |

| | |
|---|---|
| $\cup$ | union |
| $\forall$ | arbitrarily chosen |
| $\exists$ | there exist(s) |
| s.t. | subject to |
| $A \otimes B$ | Kronecker product of matrices $A$ and $B$ |

## Other Notations

| | |
|---|---|
| diag | $\mathrm{diag}(d_1, d_2, \ldots, d_n)$ represents the diagonal matrix with diagonal elements $d_i,\ i = 1, 2, \ldots, n$ |
| $F_{\mathbf{D}}(s)$ | characteristic function of the LMI region $\mathbb{D}$ |
| $S_{ch}(\bullet)$ | $S_{ch}(A_{11}) = A_{22} - A_{12}^{\mathrm{T}} A_{11}^{-1} A_{12}$ and $S_{ch}(A_{22}) = A_{11} - A_{12} A_{22}^{-1} A_{12}^{\mathrm{T}}$ are the Schur complements of $A_{11}$ and $A_{22}$ in symmetric matrix $A = \begin{bmatrix} A_{11} & A_{12} \\ A_{12}^{\mathrm{T}} & A_{22} \end{bmatrix}$, respectively. |

# Chapter 1

# Introduction

In this chapter, we first introduce the definition of linear matrix inequalities (LMIs) and then present to the readers some simple examples that are related to LMIs. Following a brief introduction of the histories and advantages of LMI theory and its applications in control systems analysis and design, the structure of the book is finally briefly stated.

## 1.1 What Are Linear Matrix Inequalities (LMIs)?

Just as the name implies, linear matrix inequalities are matrix inequalities that are linear in the matrix variables. The abbreviation, LMI, has been widely recognized and accepted today.

This book is about LMIs and their applications in control systems analysis and design. Thus as the first thing to be introduced, let us present in this section, in a relatively strict way, the definition of LMIs.

### 1.1.1 General Form

To most readers in the control community, may be the best known LMI is the following continuous-time Lyapunov matrix inequality

$$F(P) = A^\mathrm{T} P + PA + Q < 0, \tag{1.1}$$

where $A \in \mathbb{R}^{m \times m}$,

$$Q \in \mathbb{S}^m = \{M \mid M = M^\mathrm{T} \in \mathbb{R}^{m \times m}\}$$

are given matrices, and $P \in \mathbb{S}^m$ is the unknown matrix. Here "$<$" represents symmetric negative definiteness.

1

Let $Q \in \mathbb{S}^n$, $D, E_i \in \mathbb{R}^{m \times n}$, $F_i \in \mathbb{R}^{n \times n}$, $i = 1, 2, \ldots, l$, then it is easy to verify that

$$L(X) = D^\mathrm{T} X + X^\mathrm{T} D + \sum_{i=1}^{l} \left( E_i^\mathrm{T} X F_i + F_i^\mathrm{T} X^\mathrm{T} E_i \right) + Q$$

is linear in the matrix $X \in \mathbb{R}^{m \times n}$. Note that the matrix function $L(X)$ is symmetric, we can define the inequality

$$L(X) < 0, \tag{1.2}$$

which holds in the sense that the matrix $L(X)$ is symmetric negative definite. This inequality is a linear matrix inequality in the parameter matrix $X \in \mathbb{R}^{m \times n}$. Furthermore, (1.2) is called the general form of LMIs.

When $m = n$, and the unknown parameter matrix $X$ is symmetric, we have $X^\mathrm{T} = X$, and the aforementioned LMI in (1.2) becomes

$$L(X) = D^\mathrm{T} X + XD + \sum_{i=1}^{l} \left( E_i^\mathrm{T} X F_i + F_i^\mathrm{T} X E_i \right) + Q < 0. \tag{1.3}$$

Further, when $D = A$, $F_i = \pm E_i$, $i = 1, 2, \ldots, l$, the aforementioned LMI becomes a special type of LMI in a parameter matrix $X \in \mathbb{S}^m$, which takes the following form:

$$L(X) = A^\mathrm{T} X + XA \pm 2 \sum_{i=1}^{l} E_i^\mathrm{T} X E_i + Q < 0. \tag{1.4}$$

Obviously, the continuous-time Lyapunov matrix inequality (1.1) is a special form of (1.4). Moreover, the following discrete-time Lyapunov matrix inequality

$$A^\mathrm{T} P A - P + Q < 0 \tag{1.5}$$

is also a special form of (1.4).

## 1.1.2 Standard Form

The standard form for an LMI appears as follows:

$$A(x) = A_0 + x_1 A_1 + \cdots + x_n A_n < 0, \tag{1.6}$$

where $A_i \in \mathbb{S}^m$, $i = 0, 1, 2, \ldots, n$, are known, $x_i$, $i = 1, 2, \ldots, n$, are unknown scalars and are called the decision variables.

It can be clearly observed that $A(x)$ is a mapping from $\mathbb{R}^n$ to $\mathbb{S}^m$.

**Example 1.1**

Let $x_1, x_2 \in \mathbb{R}$, and

$$A(x) = A_0 + A_1 x_1 + A_2 x_2,$$

where

$$A_0 = \begin{bmatrix} 1 & 0 \\ 0 & -1 \end{bmatrix}, \quad A_1 = \begin{bmatrix} -1 & -1 \\ -1 & 4 \end{bmatrix}, \quad A_2 = \begin{bmatrix} -1 & 1 \\ 1 & -2 \end{bmatrix}.$$

Since

$$A(x) = \begin{bmatrix} 1 - x_1 - x_2 & -x_1 + x_2 \\ -x_1 + x_2 & -1 + 4x_1 - 2x_2 \end{bmatrix},$$

then $A(x) < 0$ is equivalent to

$$1 - x_1 - x_2 < 0, \tag{1.7}$$

and

$$\det \begin{bmatrix} 1 - x_1 - x_2 & -x_1 + x_2 \\ -x_1 + x_2 & -1 + 4x_1 - 2x_2 \end{bmatrix} = -5x_1^2 + 5x_1 + x_2^2 - x_2 - 1 > 0. \tag{1.8}$$

Thus, all $x_1, x_2 \in \mathbb{R}$ satisfying $A(x) < 0$ are determined by (1.7) and (1.8), which can be compactly written as

$$\begin{cases} -1 + x_1 + x_2 > 0 \\ \left( x_2 - \sqrt{5}x_1 + \frac{\sqrt{5}-1}{2} \right) \left( x_2 + \sqrt{5}x_1 - \frac{\sqrt{5}+1}{2} \right) > 0. \end{cases}$$

This is the double-shaded region in the $(x_1, x_2)$ plane shown in Figure 1.1.

**Example 1.2**

The continuous-time Lyapunov matrix inequality (1.1) with $A \in \mathbb{R}^{m \times m}$, $Q \in \mathbb{S}^m$ can be converted into a standard form. In fact, let $E_i$, $i = 1, 2, \ldots, d$, where $d = m(m+1)/2$, be a set of basis in $\mathbb{S}^m$, then there exist $x_i$, $i = 1, 2, \ldots, d$, such that

$$P = x_1 E_1 + x_2 E_2 + \cdots + x_d E_d.$$

Thus,

$$\begin{aligned} F(P) &= A^\mathsf{T} P + PA + Q \\ &= x_1 \left( A^\mathsf{T} E_1 + E_1 A \right) + x_2 \left( A^\mathsf{T} E_2 + E_2 A \right) + \cdots \\ &\quad + x_d \left( A^\mathsf{T} E_d + E_d A \right) + Q. \end{aligned}$$

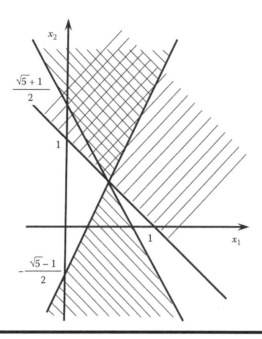

**Figure 1.1** **Region in the ($x_1, x_2$) plane.**

Therefore, the Lyapunov matrix inequality (1.1) becomes the following LMI:

$$F(x) = F_0 + x_1 F_1 + \cdots + x_d F_d < 0$$

with

$$F_0 = Q, \quad F_i = A^T E_i + E_i A, \quad i = 1, 2, \ldots, d.$$

The aforementioned example has presented a way of converting continuous-time Lyapunov matrix inequalities into standard LMIs. Similarly, the discrete-time Lyapunov matrix inequality (1.5) can also be converted into a standard LMI form. More generally, we have the following proposition.

**Proposition 1.1** The general LMI in (1.2) can be converted into an LMI standard form.

*Proof* Without loss of generality, here we only consider the case of $l = 1$.

Let $R_{ij}$, $i = 1, 2, \ldots, m, j = 1, 2, \ldots, n$, be a set of basis in $\mathbb{R}^{m \times n}$, then there exist $x_{ij}$, $i = 1, 2, \ldots, m, j = 1, 2, \ldots, n$, such that

$$X = \sum_{i=1}^{m} \sum_{j=1}^{n} x_{ij} R_{ij}.$$

Thus,

$$D^{\mathrm{T}} X + X^{\mathrm{T}} D = \sum_{i=1}^{m} \sum_{j=1}^{n} x_{ij} \left( D^{\mathrm{T}} R_{ij} + R_{ij}^{\mathrm{T}} D \right),$$

and

$$E^{\mathrm{T}} X F + F^{\mathrm{T}} X^{\mathrm{T}} E = \sum_{i=1}^{m} \sum_{j=1}^{n} x_{ij} \left( E^{\mathrm{T}} R_{ij} F + F^{\mathrm{T}} R_{ij}^{\mathrm{T}} E \right).$$

Denote

$$Q_{ij} = D^{\mathrm{T}} R_{ij} + R_{ij}^{\mathrm{T}} D + E^{\mathrm{T}} R_{ij} F + F^{\mathrm{T}} R_{ij}^{\mathrm{T}} E,$$
$$i = 1, 2, \ldots, m, \quad j = 1, 2, \ldots, n.$$

Then, $Q_{ij} \in \mathbb{S}^n$, $i = 1, 2, \ldots, m$, $j = 1, 2, \ldots, n$, and

$$D^{\mathrm{T}} X + X^{\mathrm{T}} D + E^{\mathrm{T}} X F + F^{\mathrm{T}} X^{\mathrm{T}} E = \sum_{i=1}^{m} \sum_{j=1}^{n} x_{ij} Q_{ij}.$$

Therefore, the inequality (1.2) can be arranged into the form of

$$L(x_{ij}, \ i = 1, 2, \ldots, m, \ j = 1, 2, \ldots, n) = Q + \sum_{i=1}^{m} \sum_{j=1}^{n} x_{ij} Q_{ij},$$

which is clearly in a standard form of LMIs. ■

**Remark 1.1** When the strictly less symbol "<" in the LMI (1.6) is replaced by the not greater symbol "≤", the LMI (1.6) becomes a nonstrict one. We emphasize that many conclusions for strict LMIs also hold for nonstrict ones.

### Example 1.3

Consider the *combined constraints* (in the unknown x) of the form

$$\begin{cases} F(x) < 0 \\ x = Bu + \alpha, \end{cases} \tag{1.9}$$

where the affine function $F : \mathbb{R}^n \rightarrow \mathbb{S}^m$, matrix $B \in \mathbb{R}^{n \times r}$, and vector $\alpha \in \mathbb{R}^n$ are given, with $r \leq n$. Then, we can show that (1.9) is in fact an LMI.

Denote

$$
B = \begin{bmatrix} b_{11} & b_{12} & \cdots & b_{1r} \\ b_{21} & b_{22} & \cdots & b_{2r} \\ \vdots & \vdots & \ddots & \vdots \\ b_{n1} & b_{n2} & \cdots & b_{nr} \end{bmatrix}, \quad \alpha = \begin{bmatrix} \alpha_1 \\ \alpha_2 \\ \vdots \\ \alpha_n \end{bmatrix}, \quad u = \begin{bmatrix} u_1 \\ u_2 \\ \vdots \\ u_r \end{bmatrix}.
$$

Thus, we can represent $x$ as

$$
x_i = \alpha_i + \sum_{k=1}^{r} b_{ik} u_k, \quad i = 1, 2, \ldots, n.
$$

Then,

$$
F(x) = F_0 + \sum_{l=1}^{n} x_l F_l
$$

$$
= F_0 + \sum_{l=1}^{n} \left( \alpha_l + \sum_{k=1}^{r} b_{lk} u_k \right) F_l
$$

$$
= F_0 + \sum_{l=1}^{n} \alpha_l F_l + \sum_{k=1}^{r} u_k \left( \sum_{l=1}^{n} b_{lk} F_l \right)
$$

$$
= \tilde{F}_0 + u_1 \tilde{F}_1 + \cdots + u_r \tilde{F}_r
$$

$$
= \tilde{F}(u),
$$

where

$$
\tilde{F}_0 = F_0 + \sum_{l=1}^{n} \alpha_l F_l,
$$

$$
\tilde{F}_k = \sum_{l=1}^{n} b_{lk} F_l, \quad k = 1, 2, \ldots, r.
$$

This implies that (1.9) is satisfied for $x \in \mathbb{R}^n$ if and only if the LMI $\tilde{F}(u) < 0$ is satisfied for $u \in \mathbb{R}^r$.

**Example 1.4**

(Yu 2002, p. 7) Consider the *combined constraints* (in the unknown $x$) of the form

$$
\begin{cases} F(x) < 0 \\ x \in \mathbb{M}, \end{cases} \tag{1.10}
$$

where $\mathbb{M}$ is an *affine set* in $\mathbb{R}^n$. Recall that affine set $\mathbb{M}$ can be written as

$$\mathbb{M} = \{x \,|\, x = x_0 + y, \ y \in \mathbb{M}_0\}$$

with $x_0 \in \mathbb{R}^n$ and $\mathbb{M}_0$ a linear subspace in $\mathbb{R}^n$. Suppose that $r = \dim(\mathbb{M}_0)$ and let $e_1, e_2, \ldots, e_r$ be a basis of $\mathbb{M}_0$, then an arbitrary $y \in \mathbb{M}_0$ can be expressed in the following form:

$$y = \sum_{j=1}^{r} c_j e_j,$$

where $c_i$, $i = 1, 2, \ldots, r$, are a series of scalars. Therefore, any $x \in \mathbb{M}$ can be expressed as

$$x = x_0 + \sum_{j=1}^{r} c_j e_j.$$

Let

$$x = \begin{bmatrix} x_1 & x_2 & \cdots & x_n \end{bmatrix}^\mathsf{T},$$

$$x_0 = \begin{bmatrix} x_1^0 & x_2^0 & \cdots & x_n^0 \end{bmatrix}^\mathsf{T},$$

$$e_i = \begin{bmatrix} e_{1i} & e_{2i} & \cdots & e_{ni} \end{bmatrix}^\mathsf{T}, \quad i = 1, 2, \ldots, r,$$

then the components of vector $x$ can be written as

$$x_j = x_j^0 + c_1 e_{j1} + c_2 e_{j2} + \cdots + c_r e_{jr}, \quad j = 1, 2, \ldots, n.$$

Substituting the aforementioned equation into

$$F(x) = F_0 + x_1 F_1 + x_2 F_2 + \cdots + x_n F_n,$$

yields

$$
\begin{aligned}
F(x) = F_0 &+ \left( x_1^0 + c_1 e_{11} + c_2 e_{12} + \cdots + c_r e_{1r} \right) F_1 \\
&+ \left( x_2^0 + c_1 e_{21} + c_2 e_{22} + \cdots + c_r e_{2r} \right) F_2 \\
&+ \cdots + \left( x_n^0 + c_1 e_{n1} + c_2 e_{n2} + \cdots + c_r e_{nr} \right) F_n \\
= F_0 &+ x_1^0 F_1 + \cdots + x_n^0 F_n \\
&+ c_1 \left( e_{11} F_1 + e_{21} F_2 + \cdots + e_{n1} F_n \right) \\
&+ \cdots + c_r \left( e_{1r} F_1 + e_{2r} F_2 + \cdots + e_{nr} F_n \right).
\end{aligned}
$$

Putting

$$\tilde{F}_0 = F_0 + x_1^0 F_1 + \cdots + x_n^0 F_n,$$
$$\tilde{F}_i = e_{1i} F_1 + e_{2i} F_2 + \cdots + e_{ni} F_n, \quad i = 1, 2, \ldots, r,$$
$$c = \begin{bmatrix} c_1 & c_2 & \cdots & c_r \end{bmatrix}^{\mathsf{T}},$$

we finally have

$$F(x) = \tilde{F}_0 + c_1 \tilde{F}_1 + \cdots + c_r \tilde{F}_r \overset{\triangle}{=} \tilde{F}(c).$$

This implies that $x \in \mathbb{R}^n$ satisfies (1.10) if and only if $\tilde{F}(c) < 0$, $c \in \mathbb{R}^r$.

### *1.1.3 Manipulations*

To finish the section, we finally mention some obvious manipulations related to LMIs.

**Proposition 1.2**  Let $A(x) = \begin{bmatrix} A_{ij}(x) \end{bmatrix}_{q \times p}$. Then, $A(x) < 0$ is an LMI in vector $x$ only if

$$A_{ij}(x), \quad i = 1, 2, \ldots, q, \quad j = 1, 2, \ldots, p$$

are all linear in $x$.

**Proposition 1.3**  Let

$$A_i(x) < 0, \quad i = 1, 2, \ldots, l$$

be a set of linear matrix functions in $x$. Then,

1.  $A_i(x) < 0$, $i = 1, 2, \ldots, l$, are a set of LMIs in $x$ if and only if

    $$\mathrm{diag}\,(A_1(x), \ldots, A_l(x)) < 0$$

    is an LMI in $x$;
2.  $A_i(x) < 0$, $i = 1, 2, \ldots, l$, are a set of LMIs in $x$ implies that

    $$\sum_{i=1}^{l} \alpha_i A_i(x) < 0$$

    is also an LMI in $x$, where $\alpha_i \geq 0$, $i = 1, 2, \ldots, l$, are a set of real scalars that are not simultaneously zero.

**Proposition 1.4**  Let

$$A_i(y_i), \quad i = 1, 2, \ldots, l$$

be a set of linear matrix functions in $y_i$, $i = 1, 2, \ldots, l$, and define

$$z = \left[ y_1^{\mathrm{T}}, y_2^{\mathrm{T}}, \ldots, y_l^{\mathrm{T}} \right]^{\mathrm{T}}.$$

Then,

1. $A_i(y_i) < 0$, $i = 1, 2, \ldots, l$, are a set of LMIs in $y_i$, $i = 1, 2, \ldots, l$, if and only if

$$\mathrm{diag}\left( A_1(y_1), A_2(y_2), \ldots, A_l(y_l) \right) < 0$$

   is an LMI in $z$:
2. $A_i(y_i) < 0$, $i = 1, 2, \ldots, l$, are a set of LMIs in $y_i$, $i = 1, 2, \ldots, l$, implies that

$$\sum_{i=1}^{l} \alpha_i A_i(y_i) < 0$$

   is an LMI in $z$, where $\alpha_i \geq 0$, $i = 1, 2, \ldots, l$, are a set of real scalars that are not simultaneously zero.

## 1.2  A Few Examples Involving LMIs

In this section, a few simple examples are presented. The purpose is to let the readers understand that many problems in linear algebra can be converted into problems of solving some LMIs.

### 1.2.1  Eigenvalue Minimization

First let us consider a problem of minimizing the maximal eigenvalue of a linear matrix function, which is stated as follows.

**Problem 1.1**  Let $A_i \in \mathbb{R}^{m \times m}$, $i = 0, 1, 2, \ldots, n$, be symmetric matrices, and define the matrix function

$$A(x) = A_0 + A_1 x_1 + \cdots + A_n x_n. \tag{1.11}$$

Find $x = \begin{bmatrix} x_1 & x_2 & \cdots & x_n \end{bmatrix}$ to minimize

$$J(x) = \lambda_{\max}(A(x)).$$

In order to convert the aforementioned problem into one related to LMIs, we first present the following lemma.

**Lemma 1.1**   Let $M$ be a symmetric matrix. Then,

$$\lambda_{\max}(M) \le t \Longleftrightarrow M - tI \le 0. \tag{1.12}$$

*Proof*   Note that for an arbitrary matrix $M$ with an eigenvalue $s$ and a corresponding eigenvector $x$, there holds

$$(M - tI)x = Mx - tx = (s - t)x.$$

This states that for an arbitrary matrix $M$ there holds

$$\lambda(M - tI) = \lambda(M) - t.$$

Thus, when $M$ is symmetric, we have

$$\lambda_{\max}(M) \le t \Longleftrightarrow \lambda_{\max}(M - tI) \le 0$$
$$\Longleftrightarrow M - tI \le 0.$$

∎

Applying the aforementioned lemma to the matrix defined in (1.11), we immediately have

$$\lambda_{\max}(A(x)) \le t \Longleftrightarrow A(x) - tI \le 0.$$

This clearly gives us the following conclusion.

**Conclusion 1.1**   Problem 1.1 is equivalent to the following minimization problem

$$\begin{cases} \min & t \\ \text{s.t.} & A(x) - tI \le 0, \end{cases} \tag{1.13}$$

where $x_i$, $i = 1, 2, \ldots, n$, and $t > 0$, are the parameters to be optimized.

Obviously, (1.13) is a minimization with a matrix inequality constraint that is linear with respect to the optimizing parameters $x_i$, $i = 1, 2, \ldots, n$, and $t > 0$. Therefore, the eigenvalue minimization problem has been in fact converted into an LMI problem.

## 1.2.2 Matrix Norm Minimization

The matrix norm minimization problem is a slight generalization of the aforementioned eigenvalue minimization problem.

**Problem 1.2**   Let $A_i \in \mathbb{R}^{p \times q}$, $i = 0, 1, 2, \ldots, n$, be given matrices, again define the matrix function $A(x)$ as in (1.11). Find $x = \begin{bmatrix} x_1 & x_2 & \cdots & x_n \end{bmatrix}$ to minimize

$$J(x) = ||A(x)||_2.$$

   To convert this problem into an LMI problem, the following result is needed, which is actually a special case of the Schur complement lemma to be introduced later.

**Lemma 1.2**   Let $A$ be a matrix of appropriate dimensions, and $t$ a positive scalar. Then,

$$A^{\mathrm{T}}A - t^2 I \leq 0 \iff \begin{bmatrix} -tI & A \\ A^{\mathrm{T}} & -tI \end{bmatrix} \leq 0. \tag{1.14}$$

*Proof*   Put

$$Q = \begin{bmatrix} I & A \\ 0 & tI \end{bmatrix},$$

then $Q$ is nonsingular since $t > 0$. Note that

$$Q^{\mathrm{T}} \begin{bmatrix} -tI & A \\ A^{\mathrm{T}} & -tI \end{bmatrix} Q = \begin{bmatrix} -tI & 0 \\ 0 & t\left(A^{\mathrm{T}}A - t^2 I\right) \end{bmatrix},$$

again in view of the fact that $t > 0$ it is clearly observed from the aforementioned relation that the equivalence in (1.14) holds.   ∎

   Based on the aforementioned lemma, we can derive the following conclusion.

**Conclusion 1.2**   Problem 1.2 is equivalent to the following minimization problem

$$\begin{cases} \min & t \\ \text{s.t.} & \begin{bmatrix} -tI & A(x) \\ A^{\mathrm{T}}(x) & -tI \end{bmatrix} \leq 0, \end{cases} \tag{1.15}$$

with $x_i$, $i = 1, 2, \ldots, n$, and $t$ being parameters to be optimized.

*Proof* According to the definition of $||A||_2$, we have

$$||A||_2 = \left(\lambda_{\max}\left(A^T A\right)\right)^{\frac{1}{2}}. \tag{1.16}$$

Thus, using Lemmas 1.1 and 1.2, we have, for $t > 0$,

$$||A||_2 \leq t$$
$$\Longleftrightarrow \lambda_{\max}\left(A^T A\right) \leq t^2$$
$$\Longleftrightarrow A^T A - t^2 I \leq 0$$
$$\Longleftrightarrow \begin{bmatrix} -tI & A \\ A^T & -tI \end{bmatrix} \leq 0.$$

The proof is thus completed. ■

Again, like the solution to the eigenvalue minimization, the matrix norm minimization problem is converted into a minimization with a linear matrix inequality constraint with respect to the optimizing parameters $x_i$, $i = 1, 2, \ldots, n$, and $t$.

### 1.2.3 Key Step in μ-Analysis

In the well-known $\mu$-analysis for control systems, we need to treat the following problem.

**Problem 1.3** Given a constant real matrix $A$, find a nonsingular matrix $D$ such that

$$||DAD^{-1}||_2 < 1.$$

Using relations (1.12), (1.16), and Lemma 1.1, we have

$$||DAD^{-1}||_2 < 1$$
$$\Longleftrightarrow \lambda_{\max}\left(D^{-T}A^T D^T DAD^{-1}\right) < 1$$
$$\Longleftrightarrow D^{-T}A^T D^T DAD^{-1} < I$$
$$\Longleftrightarrow A^T D^T DA < D^T D.$$

Denote

$$X = D^T D,$$

then the aforementioned deduction yields

$$||DAD^{-1}||_2 < 1 \Longleftrightarrow A^T X A < X.$$

This gives the following conclusion.

**Conclusion 1.3**   Problem 1.3 has a solution if and only if there exists a symmetric positive definite matrix $X$ satisfying the following LMI:

$$A^{\mathrm{T}}XA - X < 0. \tag{1.17}$$

In this case, the solution to the problem is given by $D = X^{\frac{1}{2}}$.

Condition (1.17) is clearly a linear inequality with respect to the variable $X$. Obviously, the solution to this problem is again converted into one of solving an LMI.

Recall the Lyapunov stability theory for discrete-time linear systems, the LMI (1.17) is a necessary and sufficient condition for the matrix $A$ to be stable in the discrete-time sense, or to be Schur stable. Therefore, there exists a solution to the aforementioned problem if and only if the matrix $A$ is stable in the discrete-time sense. In certain cases, the matrix $D$ can be restricted to be diagonal.

## 1.2.4 Schur Stabilization

In this section, we consider the control of the following linear system

$$x(k+1) = Ax(k) + Bu(k),$$

where $x \in \mathbb{R}^n$ and $u \in \mathbb{R}^r$ are the state vector and the input vector, respectively. With the following state feedback control law

$$u(k) = Kx(k),$$

the closed-loop system is given by

$$x(k+1) = (A + BK)x(k).$$

Note that for an arbitrary matrix $L \in \mathbb{R}^{n \times n}$ there holds

$$|\lambda_i(L)| \leq ||L||_2, \quad i = 1, 2, \ldots, n,$$

we have

$$|\lambda_i(A + BK)| \leq ||A + BK||_2, \quad i = 1, 2, \ldots, n.$$

Thus, the closed-loop system is stable in the discrete-time sense, or the matrix $(A + BK)$ is Schur stable, when

$$||A + BK||_2 < 1.$$

Inspired by this fact, we propose the following problem.

**Problem 1.4**   Given matrices $A \in \mathbb{R}^{n \times n}$, $B \in \mathbb{R}^{n \times r}$, and a scalar $0 < \gamma \leq 1$, find a matrix $K \in \mathbb{R}^{r \times n}$ such that

$$||A + BK||_2 < \gamma. \tag{1.18}$$

It again follows from the definition of the spectral norms of matrices, the condition (1.18) is clearly equivalent to

$$(A + BK)^{\mathrm{T}} (A + BK) < \gamma^2 I.$$

Using Lemma 1.2 we can equivalently convert the aforementioned inequality into

$$\begin{bmatrix} -\gamma I & (A + BK) \\ (A + BK)^{\mathrm{T}} & -\gamma I \end{bmatrix} < 0.$$

Thus, the problem is converted into one of finding the feasible solution $K$ of the aforementioned LMI. For the case of finding a minimal $\gamma$ satisfying (1.18), we can turn to solve the following optimization problem with LMI constraints:

$$\begin{cases} \min & \gamma \\ \text{s.t.} & \begin{bmatrix} -\gamma I & (A + BK) \\ (A + BK)^{\mathrm{T}} & -\gamma I \end{bmatrix} < 0. \end{cases}$$

**Remark 1.2**   This problem is a special case of the problem treated in Section 7.5.3. It is reasoned there that this problem may not have a solution even when the system is stabilizable, but once there exists a solution, the solution is robust in the sense that the closed-loop eigenvalues are not easy to get out of a circle region within the unit circle when there are parameter perturbations.

## 1.3 Brief History

In this section, we give an overview of the brief history of LMIs. For a more detailed description of the development process of LMIs before 1990, please refer to the first chapter of the book by Boyd et al. (1994). We point out that some of the points given in this section coincide with some of those in the first chapter of Boyd et al. (1994).

### 1.3.1 Seed Planted (1890)

In 1890, Lyapunov published his work and introduced the celebrated Lyapunov stability theory. Analytical solution of Lyapunov LMI was proposed via Lyapunov equation. Concretely he showed that the linear system

$$\dot{x}(t) = Ax(t) \tag{1.19}$$

is asymptotically stable if and only if there exists a positive definite matrix $P$ such that

$$A^\mathrm{T} P + PA < 0. \tag{1.20}$$

This requirement is what we now call a Lyapunov inequality on $P$, which is a simple LMI in $P$, and usually the variable $P$ is restricted to be symmetric positive or negative definite.

Lyapunov also showed that this LMI could be explicitly solved. Indeed, we can pick any $Q = Q^\mathrm{T} > 0$ and then solve the linear equation $A^\mathrm{T} P + PA = -Q$, for the matrix $P$, which is guaranteed to be positive definite if the system (1.19) is stable.

In conclusion, the first LMI used to analyze stability of a dynamical system was the Lyapunov inequality (1.20), which can be solved analytically by solving a set of linear equations, and this is the seed planted in the field of LMIs.

### 1.3.2 Rooting Period (1940–1970)

One of the pioneering researchers in the field of LMIs in the period of 1940–1970 was Lur'e.

In the 1940s, Lur'e, Postnikov, and some other scholars applied the Laypunov LMI to some specific practical problems in control engineering. They characterized the stability condition of a control system with a nonlinearity in the actuator in the form of LMIs, and through some reduction, they also succeeded in checking these LMI conditions through finding their analytical solutions by hand. This was indeed a major step in the history toward the application of Lyapunov stability theory in control engineering.

At the beginning of the 1950s, Lur'e summarized the earlier results in a book entitled *Some Nonlinear Problems in the Theory of Automatic Control* (Lur'e 1951), in which we can find the following insightful comment:

> This book represents the first attempt to demonstrate that the ideas expressed 60 years ago by Lyapunov, which even comparatively recently appeared to be remote from practical application, are now about to become a real medium for the examination of the urgent problems of contemporary engineering.

The major breakthrough in the 1960s was the development of simple graphical criteria for the type of Lyapunov LMI conditions arose in the problem of Lur'e, and this was due to Yakubovich, Popov, Kalman, and some other researchers. In this development, the well-known positive-real (PR) lemma was proposed, as well as the celebrated Popov criterion, circle criterion, Tsypkin criterion, and many other variations. These criteria could be applied to higher-order systems, but did not gracefully or usefully extend to systems containing more than one non-linearity. The contribution was to show how to solve a certain family of LMIs by graphical methods.

The PR lemma and its extensions were intensively studied in the latter half of the 1960s and were found to be related to the ideas of passivity, the small-gain criteria introduced by Zames and Sandberg, and also quadratic optimal control.

As early as the 1960s, the important role of LMIs in control theory was well recognized and utilized, especially by Yakubovich (see, Yakubovich 1962, 1964, 1967).

### 1.3.3 Growing Period (1970–2000)

#### 1.3.3.1 PR Lemma and ARE

By 1970, it was known that the LMI appearing in the positive-real lemma could be solved not only by graphical means but also by solving a certain algebraic Riccati equation (ARE). Such a relation was established by J. C. Willems. In his work, which is on quadratic optimal control (Willems 1971b), he led to the following LMI:

$$\begin{bmatrix} A^{\mathrm{T}}P + PA + Q & PB + C^{\mathrm{T}} \\ B^{\mathrm{T}}P + C & R \end{bmatrix} \geq 0,$$

and found that this LMI can be solved by studying the symmetric solutions of an ARE in the form of

$$A^{\mathrm{T}}P + PA - (PB + C^{\mathrm{T}})R^{-1}(B^{\mathrm{T}}P + C) + Q = 0,$$

which in turn can be found by an eigendecomposition of a related Hamiltonian matrix.

So by 1971, researchers knew several methods for solving special types of LMIs, including direct methods (for small systems), graphical methods, and by solving Lyapunov or Riccati equations. The latter obviously lies in the category of analytical solutions. What lacked at that time were computational algorithms for solving LMIs. Regarding this, Willems (1971b) also commented as follows:

> The basic importance of the LMI (Lyapunov type) seems to be largely unappreciated. It would be interesting to see whether or not it can be exploited in computational algorithm, for example.

#### 1.3.3.2 Convex Optimization and Computational Algorithms

As expected by J. C. Willems, with the development of computational technologies and optimization theory, researchers began to formulate the stability problems related to LMIs into convex optimizations and solved them via optimization algorithms. For instance, Horisberger and Belanger (1976) showed that the existence of a quadratic Lyapunov function that guarantees simultaneous stability of a set of linear systems is a convex problem involving LMIs, and Pyatnitskii and Skorodin-skii (1982) reduced the original problem of Lur'e (extended to the case of multiple nonlinearity) to a convex optimization problem involving LMIs, which can then be solved using the ellipsoid algorithm for convex optimization.

The convex optimization formulation of LMI problems has then opened a prosperous way for the research in the field.

*Ellipsoid algorithms:*
The ellipsoid algorithm was developed for convex optimization by Nemirovskii and Yudin (1983) and Shor (1985). Based on this algorithm, some researchers proved the important fact that linear programs can be solved in polynomial-time (Khachiyan 1979, Gács and Lovász 1981, Grötschel et al. 1988). For more discussions on the ellipsoid method as well as its extensions and variations, one can refer to Bland et al. (1981) and Boyd and Barratt (1991). Furthermore, a detailed history of the development of ellipsoid algorithm, including English and Russian references, can be found in Chapter 3 of Akgül (1984).

*Interior-Point algorithms:*
General interior-point methods were really proposed in the 1960s (Lieu and Huard 1966, Dikin 1967, Huard 1967, Fiacco and McCormick 1968). While for interior-point algorithms, solving various LMI problems have been developed in the 1990s by several researchers. The first were Nesterov and Nemirovsky (1988, 1990a,b, 1991, 1993, 1994); others include Alizadeh (1991, 1992a,b), Jarre (1993), Vandenberghe and Boyd (1995), Rendl et al. (1993), and Yoshise (1994).

Particularly, interior-point methods for generalized eigenvalue problems (GEVP) are described in Boyd and Ghaoui (1993), Nesterov and Nemirovsky (1995), and Section 4.4 of the book by Nesterov and Nemirovsky (1994). Since GEVPs are not convex problems, devising a reliable stopping criterion is more challenging than for convex problems. For a detailed complexity analysis, please refer to Freund and Jarre (1994) and Nesterov and Nemirovsky (1994, 1995).

Also, in the middle of the 1990s, the MATLAB®-based LMI toolbox was produced (Gahinet et al. 1994, 1995). Ever since then, this toolbox has found so many applications and has performed a great role in the development of LMI-based control theories and applications. Without the efficient software packages such as LMI Toolbox, LMIs certainly would not have become so popular in the control community.

## 1.3.3.3 LMIs in Control Systems

As algorithms for LMIs are getting more and more mature, more and more problems arising in control systems analysis and design are reduced by researchers to a few standard convex or quasi-convex optimization problems involving LMIs. By the end of the 1980s, the importance of LMIs was fully recognized. The following quotes from Doyle et al. (1991) may probably highlight the importance of LMIs:

> LMIs play the same central role in the postmodern theory as Lyapunov and Riccati equations played in the modern, and in turn various graphical techniques such as Bode, Nyquist and Nichols plots played in the classical.

Instead of citing too many papers in the LMI field, let us here mention again the first book in the LMI field, Boyd et al. (1994), which covers a great deal of the developments in this aspect by studying 11 types of LMI problems, ranging from matrix scaling to various controller synthesis problems.

Besides Boyd et al. (1994), there is a paper that needs to be mentioned, namely, Skelton and Iwasaki (1995), which is entitled "Increased roles of linear algebra in control education." It lists 17 different control problems for both the continuous-time and the discrete-time system cases, which are solvable via LMI technique. These include stabilization, positive-realness, robust $H_2/H_\infty$ control, etc. LMIs, as an optimization technique, have found great significance in system and control theories and applications by the end of the 1990s.

## 1.3.4 Flourishing Period (2000–Present)

Since 2000, more and more applications of LMIs, both theoretical and practical, have been appearing in a very diverse and dramatically fast way.

Entering into the new millennium, more books on this topic have appeared. The first book on this topic in the new millennium may be Ghaoui and Niculescu (2000), which is in fact a volume on advances in LMI methods in control, which collects altogether 17 separated topics from 33 contributors. Yu (2002) is a book in Chinese on LMI approach to robust control, which has been widely used in China. The new book Ostertag (2011) treats mono- and multivariable control and estimation and also introduces in its Chapter 6 some basic concepts and results of LMIs. There are also some other ones, which are research-oriented, and focus on special research topics related to LMI approaches, such as analysis and synthesis of multidimensional systems (Paszke 2006), analysis and control of fuzzy systems (Tanaka and Wang 2001, Lin et al. 2007, Lam and Leung 2011), dynamic surface control of uncertain nonlinear systems (Song and Hedrick 2011), and convex optimization approaches to robust model predictive control (Cychowski 2009) and stochastic distribution control (Guo and Wang 2010), both involving LMIs in certain contents.

The number of papers published on LMIs after 2000 has really increased explosively. Therefore, instead of commenting on any individual paper, we here share some of our search records.

## 1.3.4.1 General Search Record

We have searched using the keyword "LMI" or "Linear matrix inequality" in the library of Science Citation Index Expanded (SCI-Expanded), and the results are shown in Table 1.1, where

- $x$ is the total number of papers found
- $y$ is the partial number of papers on automatic control systems while
- $z$ is the percentage of $y$ in $x$

It can be observed from Table 1.1 that, since 2000, there have been nearly 250 SCI papers published on LMIs each year. Compared with the number of papers published on LMIs in the period from 1985 to 2000, the number of papers published since 2000 has increased dramatically. This increasement is shown intuitively in Figure 1.2.

## 1.3.4.2 Record Related to Control

It can be further observed from Table 1.1 that, since 1995, nearly half of the papers published on LMIs are concerned with automatic control systems. In order to have a clear classification, we have also searched in the library of Science Citation Index Expanded (SCI-Expanded), by the following: "LMI" or "Linear matrix inequality" plus "a subject term."

All together we have chosen 13 subject terms, as shown in Table 1.2, and the total number of papers found corresponding to each subject term is also given in Table 1.2. Meanwhile, we have also refined each search by the category of automatic control systems, and the refined results are also listed in the table. As seen in the table, corresponding to each subject term, the number of papers published in the category of automatic control system occupies nearly half of those published

**Table 1.1  Searched Results in SCI**

| Time Periods | x | y | z |
|---|---|---|---|
| 1985.1.1–1989.12.31 | 6 | 0 | 0 |
| 1990.1.1–1994.12.31 | 38 | 3 | 7.89 |
| 1995.1.1–1999.12.31 | 238 | 133 | 55.88 |
| 2000.1.1–2004.12.31 | 1425 | 706 | 49.54 |
| 2005.1.1–2009.12.31 | 3880 | 1638 | 42.22 |
| 2010.1.1–2012.6.30 | 2611 | 1279 | 48.99 |

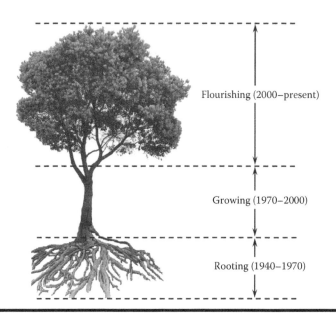

Flourishing (2000–present)

Growing (1970–2000)

Rooting (1940–1970)

**Figure 1.2   Developing process of LMIs.**

**Table 1.2   Searched Results by Subjects**

| Subject Term | Total Numbers | Refined Numbers |
|---|---|---|
| Linear system | 8013 | 3904 |
| Complex system | 4992 | 2266 |
| Fuzzy system | 5156 | 2359 |
| Nonlinear system | 5664 | 2634 |
| Network control system | 5254 | 2453 |
| Stability | 6985 | 3378 |
| H infinity | 6053 | 2990 |
| H2 | 4838 | 2229 |
| Passivity | 4910 | 2254 |
| Nonexpansivity | 4827 | 2219 |
| Observer | 5092 | 2380 |
| Filtering | 5212 | 2419 |
| Multiobjective control | 4849 | 2232 |

in the field of LMIs. LMI in control, as a subdiscipline in systems and control, now has been very rich in content. LMI today has really become a technique for control.

# 1.4  Advantages

Having reviewed the development process of LMIs, we can now give a brief summarization of the advantages of LMI approaches for control systems analysis and design.

## 1.4.1  Global Optimal Solution and Numerical Reliability

As we have seen from Section 1.3, and will learn from the next chapter, that LMIs form convex constraints, and therefore, many optimization problems involving LMIs

1. are convex, and hence have solvable global optimal solutions
2. can be solved numerically efficiently and reliably
3. are always solvable even when their sizes are very large.

So, for an LMI problem, a solution can always be obtained with acceptable precision due to existence of efficient algorithms as long as the problem has a solution.

In Chapter 3, we have presented some basic theories and results for convex optimization. The aforementioned points will be understood better after going over the content of that chapter.

## 1.4.2  Multiple Objective Design

Another extreme advantage of the LMI approaches for control systems analysis and design is that it is very convenient to handle multiple objective control systems design. General multiobjective control problems are difficult and remain mostly open to this date. By multiobjective control, we refer to synthesis problems with a mix of certain design specifications ranging from $H_2$ and $H_\infty$ performance to regional pole placement, asymptotic tracking or regulation, and settling time or saturation constraints. With the LMI approach, as long as the design objectives can be converted into certain LMI constraints, the solution to the multiobjective design problem is straightforward.

Besides the design objectives mentioned earlier, we may have other design objectives as passivity, asymptotic disturbance rejection, time-domain constraints, and constraints on the closed-loop pole location. In addition, these objectives can be specified on different channels of the closed-loop system. When all the objectives are formulated in terms of a common Lyapunov function, controller design amounts to solving a system of linear matrix inequalities.

### 1.4.3 Mature Software Packages

For solving problems related to LMIs, there have been a few mature software packages, for example, YALMIP, CVX, and the MATLAB LMI toolbox. These packages have all been implemented and tested on specific families of LMIs that arise in control theory and have been found to be very efficient. With the help of these standard and open-source software packages, an application of the LMI approach to a control system analysis and design today has been only a problem of converting an analysis or design problem into an LMI problem. Then, the problem can be handed over to some standard software package, and the next step for us to do is to analyze the results provided by the package. This is why we are saying that LMI today has been a technique and has been widely used in the world.

## 1.5 About the Book

### 1.5.1 Structure

Besides an introduction chapter, this whole book contains the following five parts:

- Part I. Preliminaries
- Part II. Control systems analysis
- Part III. Control systems design
- Part IV. Applications
- Appendices

Part I comprises two chapters: "Technical lemmas" (Chapter 2) and "Review of optimization theory" (Chapter 3).

Chapter 2 presents some useful technical lemmas which are used in the sequential chapters of the book. These include some matrix inequalities, the Schur complement lemma, and some lemmas for elimination of variables.

Chapter 3 gives a brief review of optimization theory. It first presents some basic concepts related to mathematical optimization then introduces a few types of solvable optimization problems, and finally leads to the problem of convex optimization and reveals two important advantages of the LMI approaches, that is, numerical reliability and global optimality.

Part II examines some control system analysis problems using LMI Techniques. It consists of three chapters: "Stability analysis" (Chapter 4), "$H_\infty$/$H_2$ performance" (Chapter 5), and "Property analysis" (Chapter 6).

Chapter 4 treats the problems of stability analysis. First, the LMI conditions for stability of both continuous- and discrete-time systems and that of the general $\mathbb{D}$-stability are presented. The quadratic stability for uncertain system and stability for time-delay systems are also studied based on LMI techniques.

Chapter 5 looks at the LMI conditions for the $H_\infty$ and $H_2$ indices of a linear system to be bounded by certain given levels.

Chapter 6 examines certain properties of control systems, which include the stabilizability and detectability, dissipativity, passivity and positive-realness, non-expansivity, and bounded-realness. This chapter shows the equivalence relation between passivity and positive-realness, and that between nonexpansivity and bounded-realness.

Part III deals with several types of control systems design problems. The structure of this part is as follows: "Feedback stabilization" (Chapter 7), "$H_\infty/H_2$ control" (Chapter 8), "State observation and filtering" (Chapter 9), and "Multiple objective designs" (Chapter 10).

From Chapters 7 through 10, certain typical control systems design problems are considered. Chapter 7 treats several types of stabilization problems, Chapter 8 deals with the $H_\infty$ and $H_2$ state feedback control, Chapter 9 treats the problems of full-order observer, reduced-order observer, $H_\infty$ and $H_2$ state observer and filtering, while Chapter 10 handles a few multiple objective design problems based on the results given in previous chapters.

We must mention that, as a basic introduction to LMIs in control systems analysis and design, here all the control problems are limited to state feedback case.

Part IV introduces some practical applications of LMI-based control system designs. It contains two chapters: "Missile attitude control" (Chapter 11) and "Satellite control" (Chapter 12).

These respectively look into the problems of the attitude control of a type of nonrotating missiles and a type of BTT missiles and the problem of attitude stabilization of a type of satellites.

In the last part, some appendices are given, which involve proofs of some theorems and a brief introduction of the LMI toolbox. The structure diagram is shown in Figure 1.3.

### *1.5.2 Features*

A great part of audience of this book is university graduates and senior undergraduates, hence we have paid great attention to the readability and easy understanding in the writing of the book. Toward this goal, we have taken various measures including the following:

1. It is well structured, and the contents are reasonably arranged into four parts: Preliminaries, Control systems analysis, Control systems design, and Applications. Readers can master the contents step by step with ease.
2. It summarizes most of the technical lemmas used in the book in Chapter 2 ("Technical lemmas") of the preliminary part, and these lemmas are systematically classified into different groups. This certainly provides the reader with great convenience.
3. It provides many examples, and some of them possess practical application backgrounds. It also provides exercises at the end of each chapter, and the

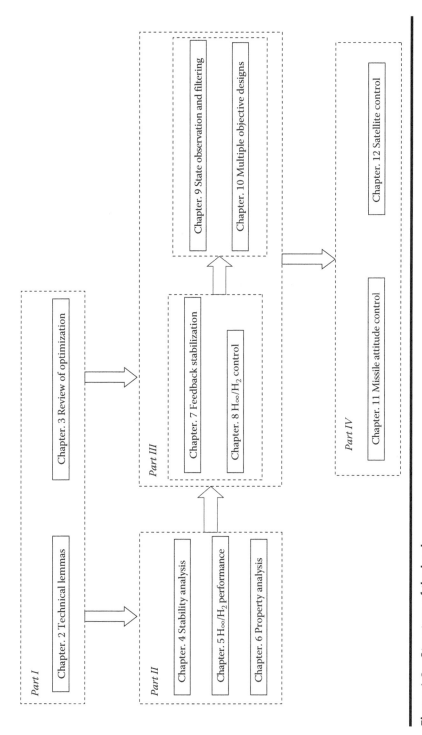

**Figure 1.3   Structure of the book.**

solutions to all the exercises are also available and can be obtained on request. These help the reader in understanding the materials.

4. It summarizes most of the important results in each chapter in the last section of that chapter, usually in clear table form. This provides convenience for readers to look up or review the results.

5. It contains an application part composed of two chapters that respectively deal with missile and satellite control using LMI techniques. From these, readers can indeed see the important roles of the LMI methods and learn to use LMIs in practical problems.

6. It provides, for convenience of readers, a brief and clear introduction to the use of the MATLAB LMI toolbox in an appendix, which will help the reader to get to use LMI techniques quickly.

7. It provides detailed proofs for all main results, lengthy ones are clearly divided into different subsections or steps, and also elementary mathematics is used whenever possible, for instance, in the proof of an important lemma (Lemma 2.5), the solution theory of second-order linear algebraic equation with a single variable is used, which makes the proof much easier to understand.

8. It shows, in contrast to the numerical reliability of the LMI problems, with a pole assignment Benchmark problem that a computational solution to an ill-condition problem could have gone far away from the true one, and this makes the reader get well aware of the importance of numerical reliability and hence the LMI approaches.

## 1.5.3 Using the Book in Courses

Most of the materials of the book have been lectured by the first author of the book in the Fall semester of 2004 and 2005 and also in the Spring semester of 2006–2012 in a postgraduate course entitled "LMIs in control systems" at the Harbin Institute of Technology. The prerequisite knowledge that the students should have includes linear algebra, basic calculus, and linear systems theory.

This book may be useful at other universities in the world as the primary or alternate textbook for a similar course for graduate and senior undergraduate students majoring in control systems and theory, or in applied mathematics, mechanical, electrical and aerospace engineering.

In the authors' opinion, this book is suitable for such a master course with about 60 lecturing hours and 10 tutorials. In the case that the lecturing hours are fewer than this, certain materials may be skipped. Table 1.3 gives our suggestions. Certain sections in Appendix A may be lectured according to the preference of the lecturer.

At the end of each chapter, we have given certain exercise problems. We have also worked out all the solutions to these problems. They can be asked for, together with the MATLAB codes for the computational examples and exercises problems, through emails g.r.duan@hit.edu.cn and yuhaihua@hit.edu.cn, or by contacting the publisher.

# Exercises

**1.1** Let $A \in \mathbb{S}^m$. Show that for arbitrary $M \in \mathbb{R}^{m \times n}$, $A \leq 0$ implies $M^{\mathrm{T}} A M \leq 0$.

**1.2** (Duan and Patton 1998, Zhang and Yang 2003, p. 175) Let $A \in \mathbb{C}^{n \times n}$. Show that $A$ is Hurwitz stable if $A + A^{\mathrm{H}} < 0$.

**1.3** (Duan and Patton 1998) Let $A \in \mathbb{R}^{n \times n}$. Show that $A$ is Hurwitz stable if and only if

$$A = PQ, \qquad (1.21)$$

with $P > 0$ and $Q$ being some matrix satisfying $Q + Q^{\mathrm{T}} < 0$.

**1.4** Give an example to show that certain set of nonlinear inequalities can be converted into LMIs.

**1.5** Verify for which integer $i$ the following inequality is true:

$$\begin{bmatrix} 1 & i \\ i & 1 \end{bmatrix} > \begin{bmatrix} 0 & 1 \\ 1 & 0 \end{bmatrix}.$$

**1.6** Consider the combined constraints (in the unknown $x$) of the form

$$\begin{cases} F(x) < 0 \\ Ax = a, \end{cases} \qquad (1.22)$$

where the affine function $F : \mathbb{R}^n \rightarrow \mathbb{S}^m$, matrix $A \in \mathbb{R}^{m \times n}$, and vector $a \in \mathbb{R}^m$ are given, and the equation $Ax = a$ has a solution. Show that (1.22) is in fact an LMI.

**1.7** Write the Hermite matrix $A \in \mathbb{C}^{n \times n}$ as $X + iY$ with real $X$ and $Y$. Show that $A < 0$ only if $X < 0$.

**1.8** Let $A, B$ be symmetric matrices of the same dimension. Show
   i. $A > B$ implies $\lambda_{\max}(A) > \lambda_{\max}(B)$,
   ii. $\lambda_{\max}(A + B) \leq \lambda_{\max}(A) + \lambda_{\max}(B)$.

# PRELIMINARIES

# Chapter 2

# Technical Lemmas

As the first chapter in this preliminary part, some technical lemmas are presented here, which are to be used in the following parts.

## 2.1 Generalized Square Inequalities

For arbitrary scalars $x$, $y$, and $\delta > 0$, we have

$$\left(\sqrt{\delta}x - \frac{1}{\sqrt{\delta}}y\right)^2 = \delta x^2 + \frac{1}{\delta}y^2 - 2xy \geq 0,$$

from which we obtain

$$2xy \leq \delta x^2 + \frac{1}{\delta}y^2.$$

This is the well-known square inequality. In this section, we give some generalized versions of this inequality, which will play very fundamental roles in the LMI techniques.

### 2.1.1 Restriction-Free Inequality

**Lemma 2.1** Let $X, Y \in \mathbb{R}^{m \times n}$, $F \in \mathbb{S}^m$, $F > 0$, and $\delta > 0$ be a scalar, then

$$X^{\mathrm{T}}FY + Y^{\mathrm{T}}FX \leq \delta X^{\mathrm{T}}FX + \delta^{-1}Y^{\mathrm{T}}FY. \tag{2.1}$$

Particularly, when $X = x$ and $Y = y$ are vectors, the aforementioned inequality reduces to

$$2x^{\mathrm{T}}Fy \leq \delta x^{\mathrm{T}}Fx + \delta^{-1}y^{\mathrm{T}}Fy. \tag{2.2}$$

*Proof*   Inequality (2.1) clearly follows from

$$\delta X^{\mathrm{T}} FX + \delta^{-1} Y^{\mathrm{T}} FY - \left( X^{\mathrm{T}} FY + Y^{\mathrm{T}} FX \right)$$
$$= \left( \sqrt{\delta} X - \sqrt{\delta^{-1}} Y \right)^{\mathrm{T}} F \left( \sqrt{\delta} X - \sqrt{\delta^{-1}} Y \right)$$
$$\geq 0.$$

When $X = x$ and $Y = y$ are vectors, $x^{\mathrm{T}} Fy$ reduces to a scalar. In this case, (2.1) clearly reduces to (2.2).   ■

## 2.1.2  Inequalities with Restriction

The following several results in this section are associated with the following restriction set:

$$\mathbb{F} = \{ F \mid F \in \mathbb{R}^{n \times n}, \; F^{\mathrm{T}} F \leq I \}. \tag{2.3}$$

**Lemma 2.2**   Let $X \in \mathbb{R}^{m \times n}$, $Y \in \mathbb{R}^{n \times m}$. Then for arbitrary scalar $\delta > 0$, there holds

$$XFY + Y^{\mathrm{T}} F^{\mathrm{T}} X^{\mathrm{T}} \leq \delta XX^{\mathrm{T}} + \delta^{-1} Y^{\mathrm{T}} Y, \quad \forall F \in \mathbb{F}. \tag{2.4}$$

This lemma can be similarly proven by completing the square technique. It can also be immediately obtained from the following result.

**Lemma 2.3**   Let $X \in \mathbb{R}^{m \times n}$, $Y \in \mathbb{R}^{n \times m}$. Then for arbitrary scalar $\delta > 0$, arbitrary nonzero vectors $x$ and $y$, there holds,

$$2x^{\mathrm{T}} XFYy \leq \delta x^{\mathrm{T}} XX^{\mathrm{T}} x + \delta^{-1} y^{\mathrm{T}} Y^{\mathrm{T}} Yy, \quad \forall F \in \mathbb{F}. \tag{2.5}$$

*Proof*   Using the property of set $\mathbb{F}$, we can obtain

$$0 \leq \left( \sqrt{\delta} X^{\mathrm{T}} x - \sqrt{\delta^{-1}} FYy \right)^{\mathrm{T}} \left( \sqrt{\delta} X^{\mathrm{T}} x - \sqrt{\delta^{-1}} FYy \right)$$
$$= \delta x^{\mathrm{T}} XX^{\mathrm{T}} x + \delta^{-1} y^{\mathrm{T}} Y^{\mathrm{T}} F^{\mathrm{T}} FYy - 2x^{\mathrm{T}} XFYy$$
$$\leq \delta x^{\mathrm{T}} XX^{\mathrm{T}} x + \delta^{-1} y^{\mathrm{T}} Y^{\mathrm{T}} Yy - 2x^{\mathrm{T}} XFYy.$$

This gives the inequality (2.5).   ■

We can rewrite (2.5) as follows:

$$x^{\mathrm{T}} XFYy + y^{\mathrm{T}} Y^{\mathrm{T}} F^{\mathrm{T}} X^{\mathrm{T}} x \leq \delta x^{\mathrm{T}} XX^{\mathrm{T}} x + \delta^{-1} y^{\mathrm{T}} Y^{\mathrm{T}} Yy, \tag{2.6}$$

and taking $x = y$ in the aforementioned relation, gives

$$x^\mathrm{T} \left( XFY + X^\mathrm{T} F^\mathrm{T} Y^\mathrm{T} - \delta X X^\mathrm{T} - \delta^{-1} Y^\mathrm{T} Y \right) x \leq 0. \tag{2.7}$$

In view of the arbitrariness of $x$, we know that the aforementioned inequality is equivalent to (2.4).

### 2.1.3 Variable Elimination Lemma

In order to give the last result in this subsection, here we first introduce a property of the set $\mathbb{F}$ defined in (2.3).

**Lemma 2.4**  For arbitrary nonzero vectors $x, y \in \mathbb{R}^n$, there holds

$$\max_{F \in \mathbb{F}} \left( x^\mathrm{T} F y \right)^2 = \left( x^\mathrm{T} x \right) \left( y^\mathrm{T} y \right). \tag{2.8}$$

*Proof*  First, using the well-known Schwarz inequality, we have

$$|x^\mathrm{T} F y| \leq \sqrt{x^\mathrm{T} x} \sqrt{y^\mathrm{T} F^\mathrm{T} F y}$$

$$\leq \sqrt{x^\mathrm{T} x} \sqrt{y^\mathrm{T} y}.$$

Therefore, we have for arbitrary nonzero vectors $x, y \in \mathbb{R}^n$,

$$\left( x^\mathrm{T} F y \right)^2 \leq \left( x^\mathrm{T} x \right) \left( y^\mathrm{T} y \right). \tag{2.9}$$

In the following, it suffices to show that the equality in (2.9) does hold with some $F_0 \in \mathbb{F}$.

Taking

$$F_0 = \frac{x y^\mathrm{T}}{\sqrt{x^\mathrm{T} x} \sqrt{y^\mathrm{T} y}}.$$

Let us first show $F_0 \in \mathbb{F}$. To do this, let us consider

$$F_0^\mathrm{T} F_0 = \frac{y x^\mathrm{T} x y^\mathrm{T}}{\left( x^\mathrm{T} x \right) \left( y^\mathrm{T} y \right)} = \frac{y y^\mathrm{T}}{y^\mathrm{T} y}.$$

Since

$$\sigma_{\max}(F_0^\mathrm{T} F_0) = \sigma_{\max}(F_0 F_0^\mathrm{T}) = 1,$$

we know that $F_0^\mathrm{T} F_0 \leq I$, thus $F_0 \in \mathbb{F}$.

Next, we show that, with this matrix $F_0$, the equality in (2.9) holds. This can be clearly seen from

$$\left(x^\mathsf{T} F_0 y\right)^2 = \left(x^\mathsf{T} \frac{xy^\mathsf{T}}{\sqrt{x^\mathsf{T} x}\sqrt{y^\mathsf{T} y}} y\right)^2 = \left(x^\mathsf{T} x\right)\left(y^\mathsf{T} y\right).$$

Thus the proof is done.  ■

The following variable elimination lemma often plays an important role in certain analysis problems involving LMIs.

**Lemma 2.5** Let $X \in \mathbb{R}^{m \times n}$, $Y \in \mathbb{R}^{n \times m}$, $Q \in \mathbb{R}^{m \times m}$. Then

$$Q + XFY + Y^\mathsf{T} F^\mathsf{T} X^\mathsf{T} < 0, \quad \forall F \in \mathbb{F}, \tag{2.10}$$

if and only if there exists a scalar $\delta > 0$, such that

$$Q + \delta XX^\mathsf{T} + \delta^{-1} Y^\mathsf{T} Y < 0. \tag{2.11}$$

*Proof*  Sufficiency: Let (2.11) hold. By adding both sides of (2.4) the matrix $Q$, we obtain

$$Q + XFY + Y^\mathsf{T} F^\mathsf{T} X^\mathsf{T}$$
$$\leq Q + \delta XX^\mathsf{T} + \delta^{-1} Y^\mathsf{T} Y$$
$$< 0.$$

That is, (2.10) holds.

Necessity: Suppose (2.10) holds. Then, for arbitrary nonzero vector $x$, we have

$$x^\mathsf{T} \left(Q + XFY + Y^\mathsf{T} F^\mathsf{T} X^\mathsf{T}\right) x < 0,$$

that is,

$$x^\mathsf{T} Qx + 2x^\mathsf{T} XFYx < 0. \tag{2.12}$$

On the other hand, using Lemma 2.4 we have

$$\max_{F \in \mathbb{F}} \left(x^\mathsf{T} XFYx\right) = \sqrt{x^\mathsf{T} XX^\mathsf{T} x}\sqrt{x^\mathsf{T} Y^\mathsf{T} Yx}.$$

Therefore, in order that (2.12) holds for arbitrary $F \in \mathbb{F}$, the following must hold:

$$x^\mathsf{T} Qx + 2\sqrt{x^\mathsf{T} XX^\mathsf{T} x}\sqrt{x^\mathsf{T} Y^\mathsf{T} Yx} < 0, \tag{2.13}$$

and from this we can get

$$\left(x^{\mathrm{T}} Q x\right)^2 - 4\left(x^{\mathrm{T}} X X^{\mathrm{T}} x\right)\left(x^{\mathrm{T}} Y^{\mathrm{T}} Y x\right) > 0. \tag{2.14}$$

Denote

$$\begin{cases} a = x^{\mathrm{T}} X X^{\mathrm{T}} x \\ b = x^{\mathrm{T}} Q x \\ c = x^{\mathrm{T}} Y^{\mathrm{T}} Y x, \end{cases} \tag{2.15}$$

then (2.14) can be written briefly as

$$b^2 - 4ac > 0.$$

Further, note from (2.13) that $b = x^{\mathrm{T}} Q x < 0$, which gives

$$\delta_0 = -\frac{b}{2a} > 0,$$

then the following equation

$$a\delta^2 + b\delta + c = 0$$

has at least one positive root (see Figure 2.1), and thus there exists a $\delta > 0$ such that

$$a\delta^2 + b\delta + c < 0.$$

Dividing both sides of the aforementioned inequality by $\delta$ gives

$$b + a\delta + \frac{1}{\delta}c < 0. \tag{2.16}$$

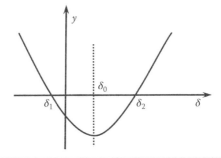

**Figure 2.1  Existence of $\delta$.**

Substituting the expressions for $a, b,$ and $c$ in (2.15) back into the aforementioned inequality immediately gives

$$x^T Q x + \delta x^T X X^T x + \frac{1}{\delta} x^T Y^T Y x < 0,$$

that is,

$$x^T \left( Q + \delta X X^T + \frac{1}{\delta} Y^T Y \right) x < 0.$$

This is clearly equivalent to (2.11) in view of the arbitrariness of the vector $x$. The proof is then completed. ■

## 2.2 Schur Complement Lemma

### 2.2.1 Schur Complements

First let us define the so-called Schur complements.

**Definition 2.1**    Consider the partitioned matrix

$$A = \begin{bmatrix} A_{11} & A_{12} \\ A_{21} & A_{22} \end{bmatrix}. \qquad (2.17)$$

1. When $A_{11}$ is nonsingular, $A_{22} - A_{21} A_{11}^{-1} A_{12}$ is called the Schur complement of $A_{11}$ in $A$, denoted by $S_{ch}(A_{11})$.
2. When $A_{22}$ is nonsingular, $A_{11} - A_{12} A_{22}^{-1} A_{21}$ is called the Schur complement of $A_{22}$ in $A$, denoted by $S_{ch}(A_{22})$.

The meaning of Schur complements lies in the following result.

**Lemma 2.6**    Let $\overset{\leftrightsquigarrow}{=}$ represent the equivalence relation between two matrices. Then for the partitioned matrix (2.17), the following conclusions hold.

1. When $A_{11}$ is nonsingular,

$$A \overset{\leftrightsquigarrow}{=} \begin{bmatrix} A_{11} & 0 \\ 0 & A_{22} - A_{21} A_{11}^{-1} A_{12} \end{bmatrix} = \begin{bmatrix} A_{11} & 0 \\ 0 & S_{ch}(A_{11}) \end{bmatrix}, \qquad (2.18)$$

and hence $A$ is nonsingular if and only if $S_{ch}(A_{11})$ is nonsingular, and

$$\det A = \det A_{11} \det S_{ch}(A_{11}). \tag{2.19}$$

2. When $A_{22}$ is nonsingular,

$$A \overset{\sim}{=} \begin{bmatrix} A_{11} - A_{12}A_{22}^{-1}A_{21} & 0 \\ 0 & A_{22} \end{bmatrix} = \begin{bmatrix} S_{ch}(A_{22}) & 0 \\ 0 & A_{22} \end{bmatrix}, \tag{2.20}$$

hence $A$ is nonsingular if and only if $S_{ch}(A_{22})$ is nonsingular, and

$$\det A = \det A_{22} \det S_{ch}(A_{22}). \tag{2.21}$$

*Proof*   When $A_{11}$ is nonsingular, we can define the nonsingular matrices

$$T_1 = \begin{bmatrix} I & 0 \\ -A_{21}A_{11}^{-1} & I \end{bmatrix}, \qquad T_2 = \begin{bmatrix} I & -A_{11}^{-1}A_{12} \\ 0 & I \end{bmatrix}, \tag{2.22}$$

and have

$$
\begin{aligned}
A \overset{\sim}{=} \; & T_1 A T_2 \\
= \; & \begin{bmatrix} I & 0 \\ -A_{21}A_{11}^{-1} & I \end{bmatrix} \begin{bmatrix} A_{11} & A_{12} \\ A_{21} & A_{22} \end{bmatrix} \begin{bmatrix} I & -A_{11}^{-1}A_{12} \\ 0 & I \end{bmatrix} \\
= \; & \begin{bmatrix} A_{11} & A_{12} \\ 0 & A_{22} - A_{21}A_{11}^{-1}A_{12} \end{bmatrix} \begin{bmatrix} I & -A_{11}^{-1}A_{12} \\ 0 & I \end{bmatrix} \\
= \; & \begin{bmatrix} A_{11} & 0 \\ 0 & A_{22} - A_{21}A_{11}^{-1}A_{12} \end{bmatrix} \\
= \; & \begin{bmatrix} A_{11} & 0 \\ 0 & S_{ch}(A_{11}) \end{bmatrix}.
\end{aligned}
$$

That is, the relation (2.18) holds. Further, since

$$\det (T_1) = \det (T_2) = 1,$$

equation (2.19) clearly follows. Thus, the first conclusion is proven. The second conclusion can be shown similarly.   ■

## 2.2.2 Matrix Inversion Lemma

Based on Lemma 2.6, we can easily obtain the following well-known matrix inversion lemma.

**Lemma 2.7** For the partitioned matrix (2.17), the following conclusions hold.

1. When $A_{11}$ is nonsingular, $A$ is nonsingular if and only if $S_{ch}(A_{11})$ is nonsingular, and

$$A^{-1} = \begin{bmatrix} A_{11}^{-1} + A_{11}^{-1}A_{12}S_{ch}^{-1}(A_{11})A_{21}A_{11}^{-1} & -A_{11}^{-1}A_{12}S_{ch}^{-1}(A_{11}) \\ -S_{ch}^{-1}(A_{11})A_{21}A_{11}^{-1} & S_{ch}^{-1}(A_{11}) \end{bmatrix}. \quad (2.23)$$

2. When $A_{22}$ is nonsingular, $A$ is nonsingular if and only if $S_{ch}(A_{22})$ is nonsingular, and

$$A^{-1} = \begin{bmatrix} S_{ch}^{-1}(A_{22}) & -S_{ch}^{-1}(A_{22})A_{12}A_{22}^{-1} \\ -A_{22}^{-1}A_{12}S_{ch}^{-1}(A_{22}) & A_{22}^{-1} + A_{22}^{-1}A_{21}S_{ch}^{-1}(A_{22})A_{12}A_{22}^{-1} \end{bmatrix}. \quad (2.24)$$

*Proof* Let

$$S \triangleq \begin{bmatrix} A_{11}^{-1} + A_{11}^{-1}A_{12}S_{ch}^{-1}(A_{11})A_{21}A_{11}^{-1} & -A_{11}^{-1}A_{12}S_{ch}^{-1}(A_{11}) \\ -S_{ch}^{-1}(A_{11})A_{21}A_{11}^{-1} & S_{ch}^{-1}(A_{11}) \end{bmatrix} \begin{bmatrix} A_{11} & A_{12} \\ A_{21} & A_{22} \end{bmatrix}$$

$$\triangleq \begin{bmatrix} S_{11} & S_{12} \\ S_{21} & S_{22} \end{bmatrix},$$

we are going to show that the matrix $S$ is an identity matrix.

By direct multiplication, we clearly have

$$S_{11} = \left( A_{11}^{-1} + A_{11}^{-1}A_{12}S_{ch}^{-1}(A_{11})A_{21}A_{11}^{-1} \right) A_{11} - A_{11}^{-1}A_{12}S_{ch}^{-1}(A_{11})A_{21} = I,$$

and

$$S_{21} = -S_{ch}^{-1}(A_{11})A_{21}A_{11}^{-1}A_{11} - S_{ch}^{-1}(A_{11})A_{21} = 0.$$

Recalling the definition of $S_{ch}(A_{11})$, we further have

$$S_{12} = A_{11}^{-1}A_{12} + A_{11}^{-1}A_{12}S_{ch}^{-1}(A_{11})A_{21}A_{11}^{-1}A_{12} - A_{11}^{-1}A_{12}S_{ch}^{-1}(A_{11})A_{22}$$

$$= A_{11}^{-1}A_{12} - A_{11}^{-1}A_{12}S_{ch}^{-1}(A_{11}) \left( A_{22} - A_{21}A_{11}^{-1}A_{12} \right)$$

$$= A_{11}^{-1}A_{12} - A_{11}^{-1}A_{12}S_{ch}^{-1}(A_{11})S_{ch}(A_{11})$$

$$= 0,$$

and

$$S_{22} = -S_{ch}^{-1}(A_{11})A_{21}A_{11}^{-1}A_{12} + S_{ch}^{-1}(A_{11})A_{22}$$

$$= S_{ch}^{-1}(A_{11})\left(A_{22} - A_{21}A_{11}^{-1}A_{12}\right)$$

$$= S_{ch}^{-1}(A_{11})S_{ch}(A_{11})$$

$$= I.$$

Thus, the first conclusion is true. Similarly, we can prove the second one. ■

Based on the aforementioned lemma, we obtain immediately the following relations.

**Corollary 2.1**   Let $A_{11}$ and $A_{22}$ be nonsingular matrices of appropriate dimensions. Then,

$$\left(A_{11} - A_{12}A_{22}^{-1}A_{21}\right)^{-1} = A_{11}^{-1} + A_{11}^{-1}A_{12}\left(A_{22} - A_{21}A_{11}^{-1}A_{12}\right)^{-1}A_{21}A_{11}^{-1},$$
$$(2.25)$$

$$\left(A_{22} - A_{21}A_{11}^{-1}A_{12}\right)^{-1} = A_{22}^{-1} + A_{22}^{-1}A_{21}\left(A_{11} - A_{12}A_{22}^{-1}A_{21}\right)^{-1}A_{12}A_{22}^{-1}.$$
$$(2.26)$$

*Proof*   It follows from Lemma 2.7 that, under the given conditions, the inverse of the $A$ matrix has both forms (2.23) and (2.24). Equating the (1,1) element in (2.23) and (2.24), yields

$$S_{ch}^{-1}(A_{22}) = A_{11}^{-1} + A_{11}^{-1}A_{12}S_{ch}^{-1}(A_{11})A_{21}A_{11}^{-1}, \qquad (2.27)$$

while the (2,2) element in (2.23) and (2.24) gives

$$S_{ch}^{-1}(A_{11}) = A_{22}^{-1} + A_{22}^{-1}A_{21}S_{ch}^{-1}(A_{22})A_{12}A_{22}^{-1}. \qquad (2.28)$$

Further, note that

$$S_{ch}(A_{22}) = A_{11} - A_{12}A_{22}^{-1}A_{21},$$

$$S_{ch}(A_{11}) = A_{22} - A_{21}A_{11}^{-1}A_{12},$$

substituting the aforementioned equations into (2.27) and (2.28) gives (2.25) and (2.26). ■

## 2.2.3 Schur Complement Lemma

The following Schur complement lemma is well known and plays a fundamental role in converting nonlinear matrix inequalities into LMIs.

**Lemma 2.8**   (Schur complement lemma) Let the partitioned matrix

$$A = \begin{bmatrix} A_{11} & A_{12} \\ A_{12}^{\mathrm{T}} & A_{22} \end{bmatrix} \tag{2.29}$$

be symmetric. Then,

$$A < 0 \iff A_{11} < 0, \ S_{ch}(A_{11}) < 0 \iff A_{22} < 0, \ S_{ch}(A_{22}) < 0, \tag{2.30}$$

or

$$A > 0 \iff A_{11} > 0, \ S_{ch}(A_{11}) > 0 \iff A_{22} > 0, \ S_{ch}(A_{22}) > 0. \tag{2.31}$$

*Proof*   When the partitioned matrix (2.17) is restricted to be symmetric, the two matrices $T_1$ and $T_2$ in (2.22) satisfy $T_1^{\mathrm{T}} = T_2$. In this case, it is easy to see that the equivalence relations (2.18) and (2.20) both become the congruent relation. Thus, the conclusion holds.                                                                          ■

According to the Schur complement lemma, we can easily obtain the following conclusion.

**Corollary 2.2**   Let

$$A = \begin{bmatrix} A_{11} & A_{12} & \cdots & A_{1r} \\ A_{12}^{\mathrm{T}} & A_{22} & \cdots & A_{2r} \\ \vdots & \vdots & \ddots & \vdots \\ A_{1r}^{\mathrm{T}} & A_{2r}^{\mathrm{T}} & \cdots & A_{rr} \end{bmatrix}.$$

Then, $A < 0$ implies $A_{ii} < 0$, $i = 1, 2, \ldots, r$.

*Proof*   First, we prove that the conclusion holds in the case of $r = 2$. It is obvious that

$$A < 0 \implies A_{11} < 0$$
$$\implies A_{11}^{-1} < 0$$
$$\implies A_{12}^{\mathrm{T}} A_{11}^{-1} A_{12} \leq 0.$$

Applying Schur complement lemma to matrix $A$, we get

$$S_{ch}(A_{11}) = A_{22} - A_{12}^{\mathrm{T}} A_{11}^{-1} A_{12} < 0,$$

which implies

$$A_{22} < A_{12}^{\mathrm{T}} A_{11}^{-1} A_{12} \leq 0.$$

Thus the conclusion holds when $r = 2$.

If $r > 2$, we partition the matrix $A$ as follows:

$$A = \begin{bmatrix} A_{r-1} & g_{r-1} \\ g_{r-1}^{\mathrm{T}} & A_{rr} \end{bmatrix} < 0,$$

then by the conclusion we have just shown, we have

$$A_{rr} < 0, \quad A_{r-1} < 0.$$

Partition the matrix $A_{r-1}$ as

$$A_{r-1} = \begin{bmatrix} A_{r-2} & g_{r-2} \\ g_{r-2}^{\mathrm{T}} & A_{r-1,r-1} \end{bmatrix} < 0,$$

and applying the conclusion again, we obtain

$$A_{r-1,r-1} < 0, \quad A_{r-2} < 0.$$

Continue this process, we can derive

$$A_{ii} < 0, \quad i = 1, 2, \ldots, r.$$

The proof is then done. ■

As an application of the aforementioned Schur complement lemma, we here show the following proposition that establishes an equivalent LMI for a type of Riccati inequalities.

**Proposition 2.1**  Let $A \in \mathbb{R}^{n \times n}$, $B \in \mathbb{R}^{n \times r}$, $C \in \mathbb{R}^{r \times n}$, $Q \in \mathbb{S}^n$, $R \in \mathbb{S}^r$, and denote

$$\Phi(P) = A^{\mathrm{T}} P + PA + (PB + C^{\mathrm{T}}) R^{-1} (B^{\mathrm{T}} P + C) + Q, \qquad (2.32)$$

and

$$\Psi(P) = \begin{bmatrix} A^{\mathrm{T}} P + PA + Q & PB + C^{\mathrm{T}} \\ B^{\mathrm{T}} P + C & -R \end{bmatrix}. \qquad (2.33)$$

Then,

$$\Phi(P) < 0 \iff \Psi(P) < 0, \quad \text{if } R > 0, \tag{2.34}$$

and

$$\Phi(P) > 0 \iff \Psi(P) > 0, \quad \text{if } R < 0. \tag{2.35}$$

*Proof*  It is easy to check that the Schur complement of $R$ in $\Psi(P)$ is given as follows:

$$
\begin{aligned}
S_{ch}(-R) &= A^{\mathrm{T}}P + PA + Q + (PB + C^{\mathrm{T}})R^{-1}(B^{\mathrm{T}}P + C) \\
&= A^{\mathrm{T}}P + PA + (PB + C^{\mathrm{T}})R^{-1}(B^{\mathrm{T}}P + C) + Q \\
&= \Phi(P).
\end{aligned}
$$

Therefore, following from the Schur complement lemma, we have

$$\Psi(P) < 0 \iff \begin{cases} R > 0 \\ \Phi(P) < 0, \end{cases}$$

and

$$\Psi(P) > 0 \iff \begin{cases} R < 0 \\ \Phi(P) > 0. \end{cases}$$

These two relations are clearly equivalent to the relations in (2.34) and (2.35), respectively. Thus the proof is done.  ■

The aforementioned proposition is very important since it realizes the conversion from quadratic to linear inequalities. We remark that this conversion technique based on Schur complement lemma will be used frequently in this book.

## 2.3 Elimination of Variables

In many deduction processes involving LMIs, often the techniques of elimination of variables are used. This section introduces some basic lemmas that serve as the basic tools for these techniques.

### 2.3.1 Variable Elimination in a Partitioned Matrix

The two lemmas in this section both treat the elimination of variables in a partitioned matrix.

**Lemma 2.9**   Let

$$Z = \begin{bmatrix} Z_{11} & Z_{12} \\ Z_{12}^T & Z_{22} \end{bmatrix}, \quad Z_{11} \in \mathbb{R}^{n \times n},$$

be symmetric. Then, there exists a symmetric matrix $X$ such that

$$\begin{bmatrix} Z_{11} - X & Z_{12} & X \\ Z_{12}^T & Z_{22} & 0 \\ X & 0 & -X \end{bmatrix} < 0 \tag{2.36}$$

if and only if

$$Z = \begin{bmatrix} Z_{11} & Z_{12} \\ Z_{12}^T & Z_{22} \end{bmatrix} < 0. \tag{2.37}$$

*Proof*   According to Schur complement lemma, we know that (2.36) holds if and only if $X > 0$ and

$$\begin{aligned} 0 &> S_{ch}(-X) \\ &= \begin{bmatrix} Z_{11} - X & Z_{12} \\ Z_{12}^T & Z_{22} \end{bmatrix} - \begin{bmatrix} X \\ 0 \end{bmatrix} (-X)^{-1} \begin{bmatrix} X & 0 \end{bmatrix} \\ &= \begin{bmatrix} Z_{11} - X & Z_{12} \\ Z_{12}^T & Z_{22} \end{bmatrix} + \begin{bmatrix} X & 0 \\ 0 & 0 \end{bmatrix} \\ &= \begin{bmatrix} Z_{11} & Z_{12} \\ Z_{12}^T & Z_{22} \end{bmatrix}. \end{aligned}$$

The proof is done.   ■

**Lemma 2.10**   Let $Z_{ij}$, $i = 1, 2, 3$, $j = i, \ldots, 3$, be given matrices of appropriate dimensions. Then, there exists a matrix $X$ such that

$$\begin{bmatrix} Z_{11} & Z_{12} & Z_{13} \\ Z_{12}^T & Z_{22} & Z_{23} + X^T \\ Z_{13}^T & Z_{23}^T + X & Z_{33} \end{bmatrix} < 0 \tag{2.38}$$

if and only if

$$\begin{bmatrix} Z_{11} & Z_{12} \\ Z_{12}^T & Z_{22} \end{bmatrix} < 0, \quad \begin{bmatrix} Z_{11} & Z_{13} \\ Z_{13}^T & Z_{33} \end{bmatrix} < 0. \tag{2.39}$$

In this case, such a matrix $X$ is given by

$$X = Z_{13}^{\mathrm{T}} Z_{11}^{-1} Z_{12} - Z_{23}^{\mathrm{T}}. \tag{2.40}$$

*Proof*  Necessity: Let (2.38) hold, then the submatrices corresponding every principle minors are also negative definite. Therefore, (2.39) holds.

Sufficiency: Let (2.39) hold, then $Z_{11} < 0$, and by applying Schur complement lemma to the two inequalities in (2.39), we have

$$Z_{22} - Z_{12}^{\mathrm{T}} Z_{11}^{-1} Z_{12} < 0, \quad Z_{33} - Z_{13}^{\mathrm{T}} Z_{11}^{-1} Z_{13} < 0. \tag{2.41}$$

In this case, by the Schur complement lemma, we need only to show that there exists an $X$ satisfying

$$\begin{bmatrix} Z_{22} & Z_{23} + X^{\mathrm{T}} \\ Z_{23}^{\mathrm{T}} + X & Z_{33} \end{bmatrix} - \begin{bmatrix} Z_{12}^{\mathrm{T}} \\ Z_{13}^{\mathrm{T}} \end{bmatrix} Z_{11}^{-1} \begin{bmatrix} Z_{12} & Z_{13} \end{bmatrix} < 0. \tag{2.42}$$

Note

$$\begin{bmatrix} Z_{22} & Z_{23} + X^{\mathrm{T}} \\ Z_{23}^{\mathrm{T}} + X & Z_{33} \end{bmatrix} - \begin{bmatrix} Z_{12}^{\mathrm{T}} \\ Z_{13}^{\mathrm{T}} \end{bmatrix} Z_{11}^{-1} \begin{bmatrix} Z_{12} & Z_{13} \end{bmatrix}$$

$$= \begin{bmatrix} Z_{22} - Z_{12}^{\mathrm{T}} Z_{11}^{-1} Z_{12} & Z_{23} + X^{\mathrm{T}} - Z_{12}^{\mathrm{T}} Z_{11}^{-1} Z_{13} \\ Z_{23}^{\mathrm{T}} + X - Z_{13}^{\mathrm{T}} Z_{11}^{-1} Z_{12} & Z_{33} - Z_{13}^{\mathrm{T}} Z_{11}^{-1} Z_{13} \end{bmatrix},$$

by simply choosing $X$ as in (2.40), and using the aforementioned relation and (2.41), we have

$$\begin{bmatrix} Z_{22} & Z_{23} + X^{\mathrm{T}} \\ Z_{23}^{\mathrm{T}} + X & Z_{33} \end{bmatrix} - \begin{bmatrix} Z_{12}^{\mathrm{T}} \\ Z_{13}^{\mathrm{T}} \end{bmatrix} Z_{11}^{-1} \begin{bmatrix} Z_{12} & Z_{13} \end{bmatrix}$$

$$= \begin{bmatrix} Z_{22} - Z_{12}^{\mathrm{T}} Z_{11}^{-1} Z_{12} & 0 \\ 0 & Z_{33} - Z_{13}^{\mathrm{T}} Z_{11}^{-1} Z_{13} \end{bmatrix} < 0.$$

Thus, the proof is completed. ■

## 2.3.2 Projection Lemma

In this section, we present a condition for eliminating a variable in an LMI using orthogonal complements. Thus, we first should provide the definition of orthogonal complements.

**Definition 2.2** Let $A \in \mathbb{R}^{m \times n}$. Then, $M_a$ is called a left orthogonal complement of $A$ if it satisfies

$$M_a A = 0, \quad \text{rank}\,(M_a) = m - \text{rank}(A);$$

and $N_a$ is called a right orthogonal complement of $A$ if it satisfies

$$A N_a = 0, \quad \text{rank}\,(N_a) = n - \text{rank}(A).$$

The following projection lemma makes use of the aforementioned concept of orthogonal complement.

**Lemma 2.11** Let $P$, $Q$ and $H = H^{\mathrm{T}}$ be given matrices of appropriate dimensions, $N_p$ and $N_q$ be the right orthogonal complements of $P$ and $Q$, respectively. Then, there exists a matrix $X$ such that

$$H + P^{\mathrm{T}} X^{\mathrm{T}} Q + Q^{\mathrm{T}} X P < 0 \tag{2.43}$$

if and only if

$$N_p^{\mathrm{T}} H N_p < 0, \quad N_q^{\mathrm{T}} H N_q < 0. \tag{2.44}$$

*Proof* By the definitions of matrices $N_p$ and $N_q$, we have

$$P N_p = 0, \quad Q N_q = 0,$$

and hence

$$N_p^{\mathrm{T}} P^{\mathrm{T}} = 0, \quad N_q^{\mathrm{T}} Q^{\mathrm{T}} = 0.$$

Necessity: Premultiplying (2.43) by $N_p^{\mathrm{T}}$ (alternatively, $N_q^{\mathrm{T}}$) and postmultiplying by $N_p$ (alternatively, $N_q$) and using these relations gives the first (alternatively, the second) inequality in (2.44).

Sufficiency: Let $V_1$ be a matrix, whose column vectors are an arbitrary set of base vectors of $\ker(P) \cap \ker(Q)$, then there exist matrices $V_2$ and $V_3$, such that

$$\text{Image}\begin{bmatrix} V_1 & V_2 \end{bmatrix} = \ker(P), \quad \text{Image}\begin{bmatrix} V_1 & V_3 \end{bmatrix} = \ker(Q).$$

Without loss of generality, assume that $V_2$ and $V_3$ are full-rank. Then, the column vectors of $V_1$, $V_2$ and $V_3$ construct the base of $\ker(P) \oplus \ker(Q)$. Therefore, it can be expanded to the base of $\mathbb{R}^n$, that is, there exists a matrix $V_4$, such that

$$V = \begin{bmatrix} V_1 & V_2 & V_3 & V_4 \end{bmatrix}$$

is square and nonsingular. So, (2.43) holds if and only if

$$V^{\mathrm{T}}HV + V^{\mathrm{T}}P^{\mathrm{T}}X^{\mathrm{T}}QV + V^{\mathrm{T}}Q^{\mathrm{T}}XPV < 0. \tag{2.45}$$

Noting the structure of $V$, we have

$$PV = \begin{bmatrix} 0 & 0 & P_1 & P_2 \end{bmatrix}, \quad QV = \begin{bmatrix} 0 & Q_1 & 0 & Q_2 \end{bmatrix}.$$

Denote

$$V^{\mathrm{T}}HV = \begin{bmatrix} H_{11} & H_{12} & H_{13} & H_{14} \\ H_{12}^{\mathrm{T}} & H_{22} & H_{23} & H_{24} \\ H_{13}^{\mathrm{T}} & H_{23}^{\mathrm{T}} & H_{33} & H_{34} \\ H_{14}^{\mathrm{T}} & H_{24}^{\mathrm{T}} & H_{34}^{\mathrm{T}} & H_{44} \end{bmatrix},$$

and

$$Y = \begin{bmatrix} Y_{11} & Y_{12} \\ Y_{21} & Y_{22} \end{bmatrix} = \begin{bmatrix} P_1^{\mathrm{T}} \\ P_2^{\mathrm{T}} \end{bmatrix} X^{\mathrm{T}} \begin{bmatrix} Q_1 & Q_2 \end{bmatrix}, \tag{2.46}$$

then according to the definitions of matrices $P_1$, $P_2$, $Q_1$, and $Q_2$, it follows that

$$\ker \begin{bmatrix} P_1 & P_2 \end{bmatrix} = 0, \quad \ker \begin{bmatrix} Q_1 & Q_2 \end{bmatrix} = 0.$$

Therefore, for arbitrary $Y$, there exists a matrix $X$ satisfying (2.46).

According to the partitioned matrices, (2.45) is equal to

$$\begin{bmatrix} H_{11} & H_{12} & H_{13} & H_{14} \\ H_{12}^{\mathrm{T}} & H_{22} & H_{23} + Y_{11}^{\mathrm{T}} & H_{24} + Y_{21}^{\mathrm{T}} \\ H_{13}^{\mathrm{T}} & H_{23}^{\mathrm{T}} + Y_{11} & H_{33} & H_{34} + Y_{12} \\ H_{14}^{\mathrm{T}} & H_{24}^{\mathrm{T}} + Y_{21} & H_{34}^{\mathrm{T}} + Y_{12}^{\mathrm{T}} & H_{44} + Y_{22} + Y_{22}^{\mathrm{T}} \end{bmatrix} < 0. \tag{2.47}$$

In this case, by the Schur complement lemma, (2.47) holds if and only if

$$\overline{H} = \begin{bmatrix} H_{11} & H_{12} & H_{13} \\ H_{12}^{\mathrm{T}} & H_{22} & H_{23} + Y_{11}^{\mathrm{T}} \\ H_{13}^{\mathrm{T}} & H_{23}^{\mathrm{T}} + Y_{11} & H_{33} \end{bmatrix} < 0, \tag{2.48}$$

$$H_{44} + Y_{22} + Y_{22}^{\mathrm{T}} - \begin{bmatrix} H_{14} \\ H_{24} + Y_{21}^{\mathrm{T}} \\ H_{34} + Y_{12} \end{bmatrix}^{\mathrm{T}} \overline{H}^{-1} \begin{bmatrix} H_{14} \\ H_{24} + Y_{21}^{\mathrm{T}} \\ H_{34} + Y_{12} \end{bmatrix} < 0 \tag{2.49}$$

hold. If (2.48) holds by choosing $Y_{11}$ properly, (2.49) also holds by choosing $Y_{12}$, $Y_{21}$, and $Y_{22}$ properly. Therefore, (2.48) and (2.49) hold if and only if there exists a defined matrix $Y_{11}$ properly, such that $\overline{H} < 0$.

On the other hand, since

$$
\ker(PV) = \text{Image}
\begin{bmatrix}
I & 0 \\
0 & I \\
0 & 0 \\
0 & 0
\end{bmatrix},
\quad
\ker(QV) = \text{Image}
\begin{bmatrix}
I & 0 \\
0 & 0 \\
0 & I \\
0 & 0
\end{bmatrix},
$$

combining the aforementioned equations with (2.44), it follows that

$$
\begin{bmatrix}
H_{11} & H_{12} \\
H_{12}^{\mathrm{T}} & H_{22}
\end{bmatrix} < 0,
\quad
\begin{bmatrix}
H_{11} & H_{13} \\
H_{13}^{\mathrm{T}} & H_{33}
\end{bmatrix} < 0.
$$

Then, according to Lemma 2.10, under the condition of (2.44), there exists a matrix $Y_{11}$, such that $\overline{H} < 0$, which leads to that there exists matrix $Y$ such that (2.47) holds. Noting that, for an arbitrary matrix $Y$, there always exists a matrix $X$, such that (2.46) holds, which is also the solution of (2.43). Then, the desired conclusion is obtained. The proof is completed. ■

### 2.3.3 Reciprocal Projection Lemma

In this section, we further introduce the so-called reciprocal projection lemma, which is due to Apkarian et al. (2000). This will be used in Chapter 5.

**Lemma 2.12**    For a given symmetric matrix $\Psi \in \mathbb{S}^n$, there exists a matrix $S \in \mathbb{R}^{n \times n}$ satisfying

$$
\Psi + S^{\mathrm{T}} + S < 0 \tag{2.50}
$$

if and only if, for an arbitrarily fixed symmetric matrix $P \in \mathbb{S}^n$, there exists a matrix $W \in \mathbb{R}^{n \times n}$ satisfying

$$
\begin{bmatrix}
\Psi + P - (W^{\mathrm{T}} + W) & S^{\mathrm{T}} + W^{\mathrm{T}} \\
S + W & -P
\end{bmatrix} < 0. \tag{2.51}
$$

*Proof*    Let

$$
\Theta =
\begin{bmatrix}
\Psi + P & S^{\mathrm{T}} \\
S & -P
\end{bmatrix},
$$

$$
\Phi = \begin{bmatrix} -I_n & 0 \end{bmatrix},
\quad
\Psi = \begin{bmatrix} I_n & -I_n \end{bmatrix},
$$

then we can easily verify that

$$N_\Phi = \begin{bmatrix} 0 \\ P^{-1} \end{bmatrix} \quad \text{and} \quad N_\Psi = \begin{bmatrix} I_n \\ I_n \end{bmatrix},$$

are the right orthogonal complements of $\Phi$ and $\Psi$, respectively. Furthermore, we can show that

$$N_\Phi^T \Theta N_\Phi = -P^{-1},$$

$$N_\Psi^T \Theta N_\Psi = \Psi + S^T + S,$$

and

$$\Theta + \Phi^T W^T \Psi + \Psi^T W \Phi = \begin{bmatrix} \Psi + P - (W^T + W) & S^T + W^T \\ S + W & -P \end{bmatrix}.$$

Thus, the conclusion holds according to the projection lemma. ■

## 2.4 Some Other Useful Results

In this section, we further present some other useful results involving LMIs.

### 2.4.1 Trace of an LMI

In certain applications, we often encounter a type of constraints which requires the trace of a matrix function to be less than a given level. The following lemma provides a way of dealing with this type of constraints.

**Lemma 2.13**  Let $A(x) \in \mathbb{S}^m$ be a matrix function in $\mathbb{R}^n$, and $\gamma$ be a positive scalar. Then, the following statements are equivalent:

(1) $\exists x \in \mathbb{R}^n$, such that trace $(A(x)) < \gamma$
(2) $\exists x \in \mathbb{R}^n$, $Z \in \mathbb{S}^m$, such that $A(x) < Z$, while trace$(Z) < \gamma$

*Proof*  Note that for arbitrary square matrices $X$ and $Y$ of the same dimensions, there holds

$$\text{trace}(X + Y) = \text{trace}(X) + \text{trace}(Y).$$

Also, for any symmetric positive definite matrix $P$ we have trace$(P) > 0$. Based on these facts, it is easy to show that, for arbitrary $P, Q \in \mathbb{S}^m$,

$$Q - P > 0 \implies \text{trace}(Q) - \text{trace}(P) > 0,$$

that is,

$$P < Q \implies \text{trace}(P) < \text{trace}(Q).$$

When the second conclusion holds, with the aforementioned relation, it is clearly seen that

$$\text{trace}\,(A(x)) < \text{trace}(Z) < \gamma.$$

Thus, the second conclusion implies the first one. Now let us show the converse.

Let the first conclusion hold, then there exist a vector $x \in \mathbb{R}^n$ and a scalar $\beta$ satisfying

$$\text{trace}\,(A(x)) < \beta < \gamma. \tag{2.52}$$

Taking

$$Z = \alpha I_n + A(x),$$

with

$$\alpha = \frac{\beta - \text{trace}\,(A(x))}{n} > 0,$$

then we have

$$Z - A(x) > 0, \tag{2.53}$$

and

$$\begin{aligned} \text{trace}\,(Z) &= \text{trace}\,(\alpha I_n + A(x)) \\ &= \beta - \text{trace}\,(A(x)) + \text{trace}\,(A(x)) \\ &= \beta < \gamma. \end{aligned} \tag{2.54}$$

These two relations obviously give the second conclusion.  ■

## 2.4.2 Maximum Modulus Principle

**Lemma 2.14** Let $\mathcal{D}$ be a connected set and $\partial\mathcal{D}$ be its boundary. Assume that $f(z) : \mathbb{C} \to \mathbb{C}$ is analytic in $\mathcal{D} \cup \partial\mathcal{D}$. Then,

$$\sup_{z \in (\mathcal{D} \cup \partial\mathcal{D})} \{|f(z)|\} = \sup_{z \in \partial\mathcal{D}} \{|f(z)|\}. \tag{2.55}$$

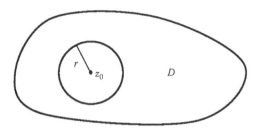

**Figure 2.2   Set $\mathcal{D}$ and the inside ball $\mathcal{B}(z_0, r)$.**

*Proof*   Since $f(z)$ is analytic in $\mathcal{D} \cup \partial\mathcal{D}$, we know that $\sup_{z \in (\mathcal{D} \cup \partial\mathcal{D})} \{|f(z)|\}$ exists, that is,

$$M \triangleq \sup_{z \in (\mathcal{D} \cup \partial\mathcal{D})} \{|f(z)|\} < \infty.$$

Clearly, if $f(z)$ is a constant function, the result is clearly true. Thus, in the following we assume that $f(z)$ is not constant. We prove the conclusion by contradiction.

Assume that there exists a point $z_0 \in \mathcal{D}$ but $z_0 \notin \partial\mathcal{D}$ such that $|f(z_0)| = M$. Then, there exists a circle $\mathcal{B}(z_0, r)$ centered at $z_0$ with radius $r$ such that $\mathcal{B}(z_0, r) \subset \mathcal{D}$ (see Figure 2.2). Using the mean value theorem, we know that

$$f(z_0) = \frac{1}{2\pi} \int_0^{2\pi} f(z_0 + re^{i\theta})d\theta.$$

According to the assumption, we have

$$\left|f(z_0 + re^{j\theta})\right| \le M, \quad \theta \in [0, 2\pi].$$

Note that $\left|f(z_0 + re^{j\theta})\right|$ is continuous with respect to $\theta$ and is not constant, we know that there exist a $\theta_0$ and a $\delta > 0$ such that

$$\left|f(z_0 + re^{j\theta})\right| < M, \quad \theta \in [\theta_0 - \delta, \theta_0 + \delta].$$

Consequently, we have

$$M = |f(z_0)|$$
$$= \frac{1}{2\pi} \left| \int_0^{2\pi} f(z_0 + re^{i\theta})\, d\theta \right|$$

$$\leq \frac{1}{2\pi} \int_0^{2\pi} \left| f\left(z_0 + re^{i\theta}\right) \right| d\theta$$

$$= \frac{1}{2\pi} \int_0^{\theta_0 - \delta} \left| f\left(z_0 + re^{i\theta}\right) \right| d\theta + \frac{1}{2\pi} \int_{\theta_0 - \delta}^{\theta_0 + \delta} \left| f\left(z_0 + re^{i\theta}\right) \right| d\theta$$

$$+ \frac{1}{2\pi} \int_{\theta_0 + \delta}^{2\pi} \left| f\left(z_0 + re^{i\theta}\right) \right| d\theta$$

$$< \frac{1}{2\pi} \left( M \left( \theta_0 - \delta + 2\delta + 2\pi - (\theta_0 + \delta) \right) \right)$$

$$= M,$$

which is obviously a contradiction. That is to say, there does not exist $z_0 \in \mathcal{D}$ but $z_0 \notin \partial \mathcal{D}$ such that $\left| f(z_0) \right| = M = \sup_{z \in (\mathcal{D} \cup \partial \mathcal{D})} \left\{ \left| f(z) \right| \right\}$. Thus (2.55) holds. ∎

### 2.4.3 Parseval Lemma

In this section, we introduce the well-known Parseval lemma.

**Lemma 2.15** Let $u(t) \in \mathbb{L}_2^n$, $U(s) = \mathcal{L}(u(t))$. Then,

$$\| u(t) \|_2 = \| U(s) \|_2, \tag{2.56}$$

holds with

$$\| u(t) \|_2 = \left( \int_0^{+\infty} u^{\mathrm{H}}(t) u(t) dt \right)^{\frac{1}{2}}, \tag{2.57}$$

and

$$\| U(s) \|_2 = \left( \frac{1}{2\pi} \int_{-\infty}^{+\infty} U^{\mathrm{H}}(j\omega) U(j\omega) d\omega \right)^{\frac{1}{2}}. \tag{2.58}$$

The Parseval lemma reveals the fact that Laplace transform maintains the magnitude of the signal being transformed. In the following, we give a generalized version of the Parseval lemma.

Let $Q \in \mathbb{S}^n$, and define

$$
\mathbb{L}_Q^n = \left\{ u(t) \, \middle| \, u(t) \in \mathbb{R}^n, \, \left| \int_0^{+\infty} u^H(t) Q u(t) dt \right| < \infty \right\}. \tag{2.59}
$$

Then, it is easy to see that $\mathbb{L}_Q^n$ reduces to $\mathbb{L}_2^n$ when $Q \in \mathbb{S}^n$ is an identity matrix. Further, define the following quantities

$$
\lfloor u(t) \rfloor_Q = \int_0^{+\infty} u^H(t) Q u(t) dt, \tag{2.60}
$$

and

$$
\lfloor U(s) \rfloor_Q = \frac{1}{2\pi} \int_{-\infty}^{+\infty} U^H(j\omega) Q U(j\omega) dw. \tag{2.61}
$$

Then it is easy to see that

$$
\lfloor u(t) \rfloor_{I_n} = \| u(t) \|_2^2, \quad \lfloor U(s) \rfloor_{I_n} = \| U(s) \|_2^2.
$$

With the aforementioned preparations, we can now state the following generalized Parseval lemma.

**Lemma 2.16**   Let $Q \in \mathbb{S}^n$, $u(t) \in \mathbb{L}_Q^n$, $U(s) = \mathcal{L}(u(t))$. Then,

$$
\lfloor u(t) \rfloor_Q = \lfloor U(s) \rfloor_Q. \tag{2.62}
$$

*Proof*   Since $U(s) = \mathcal{L}(u(t))$, by Laplace transform we have

$$
U(s) = \int_0^{\infty} u(t) e^{-st} dt,
$$

and

$$
u(t) = \frac{1}{2\pi} \int_{-\infty}^{\infty} U(j\omega) e^{j\omega t} dw.
$$

Then, using the two relations and also (2.60) and (2.61), we get

$$
\lfloor u(t) \rfloor_Q = \int_0^{+\infty} u^{\mathrm{H}}(t) Q u(t) \mathrm{d}t
$$

$$
= \int_0^{+\infty} u^{\mathrm{H}}(t) Q \left( \frac{1}{2\pi} \int_{-\infty}^{\infty} U(j\omega) e^{j\omega t} \mathrm{d}\omega \right) \mathrm{d}t
$$

$$
= \frac{1}{2\pi} \int_0^{+\infty} u^{\mathrm{H}}(t) Q \left( \int_{-\infty}^{\infty} U(j\omega) e^{j\omega t} \mathrm{d}\omega \right) \mathrm{d}t
$$

$$
= \frac{1}{2\pi} \left[ \int_0^{+\infty} u^{\mathrm{H}}(t) Q \left( \int_{-\infty}^{\infty} U(j\omega) e^{j\omega t} \mathrm{d}\omega \right) \mathrm{d}t \right]^{\mathrm{H}}
$$

$$
= \frac{1}{2\pi} \int_{-\infty}^{+\infty} U^{\mathrm{H}}(j\omega) Q^{\mathrm{T}} \int_0^{\infty} u(t) e^{-j\omega t} \mathrm{d}t \mathrm{d}\omega
$$

$$
= \frac{1}{2\pi} \int_{-\infty}^{+\infty} U^{\mathrm{H}}(j\omega) Q U(j\omega) \mathrm{d}\omega
$$

$$
= \lfloor U(s) \rfloor_Q,
$$

thus the conclusion holds. ■

To end this section, we point out that the matrix $Q$ usually is taken positive definite. However, assumption of positive definiteness on the matrix $Q$ is not made in the aforementioned generalized Parseval lemma. When the matrix $Q$ is taken to be the identity matrix, the generalized Parseval lemma reduces to the normal one, that is, Lemma 2.15.

The generalized Parseval lemma is used in the proof of Theorem 6.11.

## 2.5 Notes and References

This chapter summarizes systematically some of the technical lemmas to be used in the later chapters.

Section 2.1 presents a set of generalized square inequalities that are often used in the deduction processes involving LMIs. For Lemma 2.5, we have provided a proof which uses elementary mathematics—the theory of second-order linear algebraic equations with a single variable.

Section 2.2 introduces the well-known Schur complement lemma. It is not exaggerating to say that, without this powerful tool, at least one-third of the results in the LMI fields would not have existed.

In control systems theory, many analysis and design problems are closely related to Riccati algebraic equations or inequalities. Although this class of results based on Riccati algebraic equations or inequalities are theoretically neat and elegant, they are not quite easy to implement numerically, especially when the problem size is large. Riccati algebraic inequalities are quadratic in the unknown decision matrix and is generally more complicated than linear matrix inequalities in both forms and numerical implementation. While with the Schur complement lemma the Riccati algebraic inequalities can be easily turned into linear matrix inequalities.

Section 2.3 introduces three basic lemmas which serve as basic tools for the techniques of variable elimination. The projection lemma gives a condition for elimination a variable in an LMI using orthogonal complements, while the reciprocal projection lemma is proven using the projection lemma and is used in Chapter 5.

Section 2.4 includes some other useful results. Lemma 2.13 provides a way of dealing with a type of constraints that requires the trace of a matrix function to be less than a given level. The well-known maximum modulus principle and the Parseval lemma are also represented in Lemmas 2.14 and 2.15, respectively.

All these lemmas have been used in the sequential text of this book, and some of them have been used frequently. As a matter of fact, they have been also widely used in the literature by numerous researchers (see, e.g., VanAntwerp and Braatz 2000, Apkarian et al. 2001, Jin and Park 2001, Suplin et al. 2006).

## Exercises

**2.1** Let $c(x) \in \mathbb{R}^n$ and $P(x) = P^T(x) \in \mathbb{R}^{n \times n}$ depend affinely on $x$, and $P(x)$ is nonsingular for all $x$. Find the equivalent LMIs for the following constraints:

$$c^T(x)P(x)^{-1}c(x) < 1, \quad P(x) > 0.$$

**2.2** Let $P(x) \in \mathbb{S}^{n \times n}$ and $Q(x) \in \mathbb{R}^{n \times p}$ depend affinely on $x$. Convert the following constraints

$$\text{trace}\left(Q^T(x)P^{-1}(x)Q(x)\right) < 1, \quad P(x) > 0 \tag{2.63}$$

into a set of LMIs by introducing a new (slack) matrix variable $X \in \mathbb{S}^{p \times p}$.

**2.3** (Wang and Zhao 2007) Let

$$\Lambda = \left\{\alpha = [\alpha_1 \; \alpha_2]^T \in \mathbb{R}^2 \big| \alpha_1 + \alpha_2 = 1, \alpha_1, \alpha_2 \geq 0\right\}.$$

Show $P(\alpha) = \alpha_1 P_1 + \alpha_2 P_2 > 0$ for any $\alpha \in \Lambda$ if and only if $P_1 > 0$ and $P_2 > 0$.

**2.4** (Xu and Yang 2000) Let

$$M = \begin{bmatrix} M_1 & M_2 \\ M_3 & M_4 \end{bmatrix} \in \mathbb{R}^{n \times n}, \quad M_1 \in \mathbb{R}^{m \times m},$$

and $M_4$ be invertible. Show that $M + M^{\mathrm{T}} < 0$ implies

$$M_1 + M_1^{\mathrm{T}} - M_2 M_4^{-1} M_3 - M_3^{\mathrm{T}} M_4^{-\mathrm{T}} M_2^{\mathrm{T}} < 0.$$

**2.5** (Yu 2002, p. 128) Let $A$ be an arbitrary square matrix and $Q$ be some symmetric matrix. Show that there exists a $P > 0$ satisfying

$$A^{\mathrm{T}} P A - P + Q < 0$$

if and only if there exists a symmetric matrix $X$ such that

$$\begin{bmatrix} -X & AX \\ XA^{\mathrm{T}} & -X + XQX \end{bmatrix} < 0. \tag{2.64}$$

**2.6** Let

$$A = \begin{bmatrix} -2 & 1 & 0 \\ 1 & -3 & 1 \\ 0 & 1 & -1 \end{bmatrix}.$$

Work out by hand the following using the matrix inversion lemma and the Schur complement lemma:
1. Find out $\det(A)$ and $A^{-1}$ (if exists).
2. Judge the negative definiteness of $A$.

# Chapter 3

---

# Review of Optimization Theory

---

We mentioned in Chapter 1 that the two important advantages of LMI approaches are their global optimality and numerical reliability. In this chapter, we establish theoretical supports for these two important features of LMIs. To do this, in this chapter we will recall some facts and results about optimization theory and demonstrate theoretically the advantages of convex optimization. Gradually, we will lead to the important conclusion that LMIs are a special types of convex optimization problems.

## 3.1 Convex Sets

To achieve the aforementioned goal, we need first to present certain basic mathematical concepts involved in optimization theory.

### 3.1.1 Definitions and Properties

Let us start from the definition of convex sets.

**Definition 3.1**  A set $\mathbb{F}$ is called convex if for any $x_1, x_2 \in \mathbb{F}$ and $0 \le \theta \le 1$, there holds

$$\theta x_1 + (1 - \theta) x_2 \in \mathbb{F}.$$

Geometrically, a set is convex if and only if it contains the line segment between any two points in the set (see Figure 3.1).

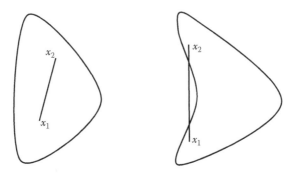

**Figure 3.1  Convex and nonconvex sets.**

**Definition 3.2**    Let $\mathbb{F} \subset \mathbb{R}^n$, then the intersection of all convex sets in $\mathbb{R}^n$ containing $\mathbb{F}$ is called the convex hull of $\mathbb{F}$ and is denoted by conv ($\mathbb{F}$).

Clearly, the convex hull of a set is the smallest convex set containing this set (see Figure 3.2).

**Definition 3.3**    A convex combination of points $z_1, z_2, \ldots, z_m$ from $\mathbb{R}^n$ is a linear combination

$$\alpha_1 z_1 + \alpha_2 z_2 + \cdots + \alpha_m z_m$$

with

$$\alpha_1 + \alpha_2 + \cdots + \alpha_m = 1,$$

and

$$\alpha_i \geq 0, \quad i = 1, 2, \ldots, m.$$

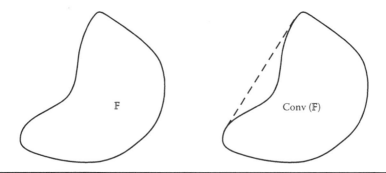

**Figure 3.2   Convex hull of set $\mathbb{F}$.**

**Definition 3.4**   Let $\mathbb{F} \subset \mathbb{R}^n$, and for some $x_0 \in \mathbb{R}^n$ and $\epsilon > 0$, denote

$$\mathbb{B}_\epsilon(x_0) = \{x \mid ||x - x_0|| < \epsilon\}.$$

Then,

- $x_0$ is called an interior point of $\mathbb{F}$ if there exists a scalar $\epsilon > 0$ such that $\mathbb{B}_\epsilon(x_0) \subset \mathbb{F}$.
- $\mathbb{F}$ is called open if every inner point of $\mathbb{F}$ is an interior point.
- $x_0$ is called a boundary point of $\mathbb{F}$ if for arbitrary small scalar $\epsilon > 0$ there holds

$$\mathbb{B}_\epsilon(x_0) \cap \mathbb{F} \neq \begin{cases} \emptyset \\ \mathbb{B}_\epsilon(x_0), \end{cases}$$

  and the set formed by all the boundary points of $\mathbb{F}$ is called the boundary set of $\mathbb{F}$, denoted by $\mathbb{F}_B$.
- $\mathbb{F}$ is called closed if it contains all its boundary points.
- $\mathbb{F}$ is called bounded if there exists a scalar $\gamma > 0$ such that $\mathbb{B}_\gamma(x_0) \supset \mathbb{F}$.
- $\mathbb{F}$ is called compact if it is both bounded and closed.
- When $\mathbb{F}$ is convex, $x_0 \in \mathbb{F}$ is said to be an extreme point of $\mathbb{F}$ if there are no two distinct points $x_1$ and $x_2$ in $\mathbb{F}$ such that $x_0 = \theta x_1 + (1 - \theta)x_2$ for some $\theta, 0 < \theta < 1$, and the set formed by all the extreme points of $\mathbb{F}$ is denoted by $\mathbb{F}_E$.

The following theorem gives the properties of convex sets and the convex hull of a set.

**Theorem 3.1**   Let $\mathbb{F} \subset \mathbb{R}^n$. Then,

1. $\mathbb{F}$ is convex if and only if any convex combination of points from $\mathbb{F}$ is again in $\mathbb{F}$.
2. conv($\mathbb{F}$) is the set of all convex combinations of points from $\mathbb{F}$, more concretely there holds

$$\text{conv}(\mathbb{F}) = \left\{ z \;\middle|\; z = \sum_{i=1}^{m} \alpha_i z_i, \; \alpha_i \geq 0, \right.$$

$$\left. \sum_{i=1}^{m} \alpha_i = 1, \quad \forall z_i \in \mathbb{F}, \quad i = 1, 2, \dots, m, \; m \in \mathbb{N} \right\}.$$

3. conv($\mathbb{F}$) is compact if $\mathbb{F}$ is compact.

*Proof*   Proof of conclusion 1.

Sufficiency directly follows from the definitions of convex sets and convex combinations of points in a set. Now we show the necessity.

Let $\mathbb{F}$ be convex, and $\forall z_i \in \mathbb{F}$, $i = 1, 2, \ldots, m$, it suffices to show

$$\alpha_1 z_1 + \alpha_2 z_2 + \cdots + \alpha_m z_m \in \mathbb{F},$$

where

$$\alpha_1 + \alpha_2 + \cdots + \alpha_m = 1, \quad \alpha_i \geq 0, \quad i = 1, 2, \ldots, m.$$

To do this, we adopt the method of mathematical induction.

It is obvious that the aforementioned fact holds when $m = 1$. Now suppose that this fact also holds when $m = k$, we need to show that it is also true when $m = k+1$. Choose arbitrarily $z_i \in \mathbb{F}$, $i = 1, 2, \ldots, k+1$, we now consider

$$z = \alpha_1 z_1 + \alpha_2 z_2 + \cdots + \alpha_{k+1} z_{k+1}, \tag{3.1}$$

with

$$\sum_{i=1}^{k+1} \alpha_i = 1, \quad \alpha_i \geq 0, \quad i = 1, 2, \ldots, k+1.$$

Without loss of generality, we can assume that $\alpha_{k+1} \neq 1$. In this case, we can define

$$\alpha'_i = \frac{\alpha_i}{1 - \alpha_{k+1}} \geq 0, \quad i = 1, 2, \ldots, k.$$

Let

$$z' = \alpha'_1 z_1 + \alpha'_2 z_2 + \cdots + \alpha'_k z_k,$$

then with these notations we can easily derive

$$z = (1 - \alpha_{k+1}) z' + \alpha_{k+1} z_{k+1}. \tag{3.2}$$

Since

$$\sum_{i=1}^{k} \alpha'_i = 1, \quad \alpha'_i \geq 0, \quad i = 1, 2, \ldots, k,$$

we know by the assumption for the case of $m = k$ that $z' \in \mathbb{F}$. Therefore, it follows from (3.2) that $z$ is a convex combination of two points in $\mathbb{F}$, and hence $z \in \mathbb{F}$. With this we complete the proof of conclusion 1.

Proof of conclusion 2.

Let $\bar{\mathbb{F}}$ be the set of all convex combinations of points from $\mathbb{F}$, that is,

$$\bar{\mathbb{F}} = \left\{ z \;\middle|\; z = \sum_{i=1}^{m} \alpha_i z_i, \; \sum_{i=1}^{m} \alpha_i = 1, \; \alpha_i \geq 0, \quad \forall z_i \in \mathbb{F}, \quad i = 1, 2, \ldots, m \right\}. \tag{3.3}$$

We need to prove $\bar{\mathbb{F}} = \text{conv}(\mathbb{F})$.

For arbitrary $z \in \bar{\mathbb{F}}$, there exist $z_i \in \mathbb{F}$, $i = 1, 2, \ldots, m$, satisfying

$$z = \sum_{i=1}^{m} \alpha_i z_i, \quad \sum_{i=1}^{m} \alpha_i = 1, \quad \alpha_i \geq 0, \quad i = 1, 2, \ldots, m.$$

Since $\text{conv}(\mathbb{F}) \supset \mathbb{F}$, and $z_i \in \mathbb{F}$, $i = 1, 2, \ldots, m$, we have $z_i \in \text{conv}(\mathbb{F})$, $i = 1, 2, \ldots, m$. Further, since $\text{conv}(\mathbb{F})$ is a convex set, it follows from the first conclusion that $z \in \text{conv}(\mathbb{F})$. This shows $\bar{\mathbb{F}} \subset \text{conv}(\mathbb{F})$.

Now let us further show $\text{conv}(\mathbb{F}) \subset \bar{\mathbb{F}}$. To achieve this, we first show that $\bar{\mathbb{F}}$ is convex. Arbitrarily choose $z_1, z_2 \in \bar{\mathbb{F}}$, then in view of (3.3) there exist $\mu_i \geq 0$, $x_i \in \mathbb{F}$, $i = 1, 2, \ldots, m_1$, and $v_j \geq 0$, $y_j \in \mathbb{F}$, $j = 1, 2, \ldots, m_2$, such that

$$z_1 = \mu_1 x_1 + \mu_2 x_2 + \cdots + \mu_{m_1} x_{m_1},$$
$$z_2 = v_1 y_1 + v_2 y_2 + \cdots + v_{m_2} y_{m_2},$$

with

$$\sum_{i=1}^{m_1} \mu_i = 1, \quad \sum_{j=1}^{m_2} v_j = 1.$$

Consider, for $0 \leq \lambda \leq 1$,

$$z = \lambda z_1 + (1 - \lambda) z_1$$
$$= \sum_{i=1}^{m_1} \lambda \mu_i x_i + \sum_{j=1}^{m_2} (1 - \lambda) v_j y_j.$$

Since

$$\lambda \mu_i \geq 0, \quad i = 1, 2, \ldots, m_1,$$

$$(1 - \lambda) v_j \geq 0, \quad j = 1, 2, \ldots, m_2,$$

$$\sum_{i=1}^{m_1} \lambda \mu_i + \sum_{j=1}^{m_2} (1 - \lambda) v_j = \lambda \sum_{i=1}^{m_1} \mu_i + (1 - \lambda) \sum_{j=1}^{m_2} v_j = 1,$$

we know that $z$ is a convex combination of points from $\mathbb{F}$. Thus, by the definition of $\bar{\mathbb{F}}$ we have $z \in \bar{\mathbb{F}}$. This shows that $\bar{\mathbb{F}}$ is convex.

Again by the definition of $\bar{\mathbb{F}}$ we have $\mathbb{F} \subset \bar{\mathbb{F}}$. Further recall the fact that the convex hull of a set is the smallest convex set containing this set, we clearly have $\text{conv}(\mathbb{F}) \subset \bar{\mathbb{F}}$.

Combining the two aspects gives $\bar{\mathbb{F}} = \text{conv}(\mathbb{F})$.

Proof of conclusion 3.

Let $\mathbb{F}$ be a compact set, then it is bounded and closed. The boundedness implies the existence of a scalar $\gamma > 0$ and $x_0 \in \mathbb{F}$, such that

$$\mathbb{B}_\gamma (x_0) = \{x | \|x - x_0\| < \gamma\} \supset \mathbb{F}.$$

Since $\text{conv}(\mathbb{F})$ is the smallest convex set containing $\mathbb{F}$, so

$$\text{conv}(\mathbb{F}) \subset \mathbb{B}_\gamma (x_0),$$

that is, $\text{conv}(\mathbb{F})$ is bounded. The closedness of $\text{conv}(\mathbb{F})$ directly follows from the second conclusion of the theorem. Thus, the third conclusion holds. ■

The following property of convex sets has particular importance.

**Theorem 3.2** A compact convex set in $\mathbb{R}^n$ is equal to the convex hull of its extreme points.

For a proof of the aforementioned result, please refer to Ruszczynski (2006).

## 3.1.2 Hyperplanes, Halfspaces, Polyhedrons, and Polytopes

Now in this section, let us clarify some important types of sets.

**Definition 3.5**  Let $0 \neq a \in \mathbb{R}^n$. Then,

- A set of the form $\mathbb{H}_p = \{x | a^\mathrm{T}x = b, \ x \in \mathbb{R}^n\}$, is called a hyperplane in $\mathbb{R}^n$.
- A set of the form $\mathbb{H}_s = \{x | a^\mathrm{T}x \leq b, \ x \in \mathbb{R}^n\}$, is called a halfspace in $\mathbb{R}^n$.
- In both cases, $a$ is called the normal vector.

**Definition 3.6**  The set $\mathbb{F} \subset \mathbb{R}^n$ is called affine if there exist an $x_0 \in \mathbb{R}^n$ and a linear space $\mathbb{F}_0 \subset \mathbb{R}^n$ such that

$$\mathbb{F} = \{x | \ x = x_0 + z, \ z \in \mathbb{F}_0\}.$$

It follows from the aforementioned definition, any affine set is the result of the shifting of some linear space.

About the properties of halfspaces and hyperplanes, we have the following theorem.

**Theorem 3.3**  Halfspaces are convex, and hyperplanes are affine and convex.

*Proof*  First let us show that the halfspace $\mathbb{H}_s = \left\{x \, \middle| \, a^\mathrm{T}x \leq b\right\}$ is convex. This requires to show that for any $x_1, x_2 \in \mathbb{H}_s$, the vector

$$y = \theta x_1 + (1 - \theta)x_2, \quad 0 \leq \theta \leq 1$$

is also in this halfspace. Since

$$a^\mathrm{T}x_1 \leq b, \quad a^\mathrm{T}x_2 \leq b,$$

then

$$a^\mathrm{T}y = a^\mathrm{T}(\theta x_1 + (1 - \theta)x_2)$$
$$= \theta a^\mathrm{T}x_1 + (1 - \theta)a^\mathrm{T}x_2$$
$$\leq \theta b + (1 - \theta)b$$
$$= b.$$

That is, $y \in \mathbb{H}_s$, then the halfspace is convex.

Similarly we can prove that the hyperplane $\mathbb{H}_p = \{x | a^\mathrm{T}x = b, \ x \in \mathbb{R}^n\}$ is convex. To complete the proof, we now only need to show $\mathbb{H}_p$ is affine. Let $x_0 \in \mathbb{R}^n$ be a specific point in $\mathbb{H}_p$. Define

$$\mathbb{H}_0 = \{x | \ a^\mathrm{T}x = 0, \ x \in \mathbb{R}^n\},$$

then we have

$$\mathbb{H}_p = \{x \mid x = x_0 + z, \ z \in \mathbb{H}_0\}.$$

Finally, notice that $\mathbb{H}_0$ is a linear space, we complete the proof. ■

**Definition 3.7**    A polyhedron in $\mathbb{R}^n$ is the intersection of finite number of halfspaces and hyperplanes in $\mathbb{R}^n$.

An illustration of the above definition is shown in Figure 3.3. According to this definition, a general polyhedron $\Pi$ can be expressed as

$$\Pi = \left\{ x \in \mathbb{R}^n \,\middle|\, a_i^T x = b_i, \ i = 1, 2, \ldots, m, \qquad g_j^T x < h_j, \quad j = 1, 2, \ldots, r \right\},$$

which can be also written in a more loose form as

$$\Pi = \left\{ x = [\ x_1 \quad x_2 \quad \cdots \quad x_n \ ]^T \,\middle|\, \sum_{k=1}^{n} a_{ik} x_k = b_i, \quad i = 1, 2, \ldots, m, \right.$$
$$\left. \sum_{k=1}^{n} g_{jk} x_k < h_j, \ j = 1, 2, \ldots, r \right\},$$

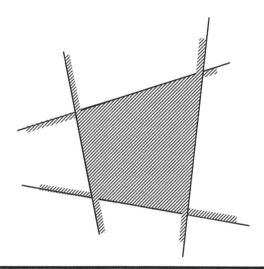

**Figure 3.3    A polyhedron, intersection of halfspaces.**

or in a more compact form as

$$\Pi = \{x \in \mathbb{R}^n \mid Ax = b, \ Gx \prec h\},$$

where $x \prec y$ represents the following relation between vectors $x$ and $y$ of the same size:

$$x_i < y_i, \quad i = 1, 2, \ldots, n.$$

**Definition 3.8**    A polytope $\mathbb{P} \subset \mathbb{R}^n$ is a set which is the convex hull of a nonempty finite set $\{z_1, z_2, \ldots, z_m\} \subset \mathbb{R}^n$, that is, $\mathbb{P} = \text{conv}\,\{z_1, z_2, \ldots, z_m\}$.

Obviously, for $\mathbb{P} = \text{conv}\,\{z_1, z_2, \ldots, z_m\}$, we have $\mathbb{P}_E \subset \{z_1, z_2, \ldots, z_m\}$, that is, any extreme point of $\mathbb{P}$ must be some $z_i, 1 \leq i \leq m$.

An illustration of the above definition is shown in Figure 3.4.

**Example 3.1**

The following set

$$\mathbb{P} = \left\{ x \,\middle|\, x = \begin{bmatrix} x_1 & x_2 & \cdots & x_n \end{bmatrix}^{\mathsf{T}}, \ x_i \in [c_i, d_i], \quad i = 1, 2, \ldots, n \right\}$$

is a polytope since it can be written as

$$\mathbb{P} = \text{conv} \left\{ x \,\middle|\, x = \begin{bmatrix} x_1 & x_2 & \cdots & x_n \end{bmatrix}^{\mathsf{T}}, \ x_i = c_i \text{ or } d_i, \quad i = 1, 2, \ldots, n \right\}.$$

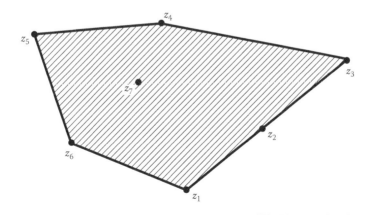

**Figure 3.4    A polytope, convex hull of a finite set.**

## 3.2 Convex Functions

Convex functions are a class of scalar functions of particular importance in optimization.

### 3.2.1 Definitions and Properties

**Definition 3.9**   Let $f : \Omega \subset \mathbb{R}^n \to \mathbb{R}$. Then,

1. $f$ is called convex if $\Omega$ is a convex set and

$$f\big(\theta x + (1 - \theta) y\big) \le \theta f(x) + (1 - \theta) f(y) \qquad (3.4)$$

  for all $x, y \in \Omega$ and $0 \le \theta \le 1$.
2. $f$ is called strictly convex if (3.4) holds strictly for all $x, y \in \Omega$, $x \ne y$, and $0 < \theta < 1$.
3. $f$ is called concave if $-f$ is convex.

A geometric illustration of a convex function is shown in Figure 3.5, with $z = \theta x + (1 - \theta) y,\ 0 \le \theta \le 1$.

**Example 3.2**

According to the aforementioned definition, the following types of functions can be easily shown to be convex:

- First-order polynomial functions: $ax + b$ on $\mathbb{R}$, for any $a, b \in \mathbb{R}$
- Exponential functions: $e^{ax}$, for any $a \in \mathbb{R}$
- Power functions of absolute value: $|x|^p$ on $\mathbb{R}$, for $p \ge 1$

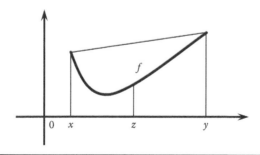

**Figure 3.5   A convex function.**

**Definition 3.10**    A function $f : \Omega \subset \mathbb{R}^n \to \mathbb{R}^m$ is called an affine function if there exist a vector $\zeta \in \mathbb{R}^m$ and a linear function $T : \Omega \subset \mathbb{R}^n \to \mathbb{R}^m$, such that

$$f(x) = \zeta + T(x).$$

It can be shown that scalar affine functions are in fact both convex and concave, and all norms are convex.

The following theorem gives a necessary and sufficient condition for convex functions, which can be viewed as an alternative definition.

**Theorem 3.4**    Let $f(x) : \Omega \subset \mathbb{R}^n \to \mathbb{R}$. Then, $f$ is a convex function if and only if for an arbitrary integer $l > 1$, and arbitrary $x_i \in \Omega, i = 1, 2, \ldots, l$, there holds

$$f(\alpha_1 x_1 + \alpha_2 x_2 + \cdots + \alpha_l x_l) \le \alpha_1 f(x_1) + \alpha_2 f(x_2) + \cdots + \alpha_l f(x_l), \quad (3.5)$$

where

$$\alpha_i \ge 0, \quad i = 1, 2, \ldots, l, \quad \text{and} \quad \sum_{i=1}^{l} \alpha_i = 1. \quad (3.6)$$

Furthermore, the inequality (3.5) holds strictly if and only if $f(x)$ is strictly convex.

*Proof*    Sufficiency is obvious. Now we show the necessity, that is, to prove, under the convex assumption of $f(x)$, that for arbitrary integer $l > 1$, and arbitrary $x_i \in \Omega$, $i = 1, 2, \ldots, l$, the relation (3.5) holds for an arbitrary set of $\alpha_i, i = 1, 2, \ldots, l$, satisfying (3.6). To do this, we adopt the method of mathematical induction.

Since $f(x)$ is convex, the aforementioned fact holds when $l = 2$ directly following the definition. Now suppose that this fact also holds when $l = k$, we proceed to show that it is also true when $l = k + 1$. Arbitrarily choose $x_i \in \Omega, i = 1, 2, \ldots, k + 1$, and let

$$x = \alpha_1 x_1 + \alpha_2 x_2 + \cdots + \alpha_{k+1} x_{k+1},$$

where

$$\alpha_i \ge 0, \quad i = 1, 2, \ldots, k + 1, \quad \text{and} \quad \sum_{i=1}^{k+1} \alpha_i = 1. \quad (3.7)$$

Without loss of generality, we can assume that $\alpha_{k+1} \ne 1$. In this case, we can define

$$\alpha_i' = \frac{\alpha_i}{1 - \alpha_{k+1}}, \quad i = 1, 2, \ldots, k,$$

then it is clear that

$$\alpha_i' \geq 0, \quad i = 1, 2, \ldots, k, \quad \sum_{i=1}^{k} \alpha_i' = 1.$$

Let

$$x' = \alpha_1' x_1 + \alpha_2' x_2 + \cdots + \alpha_k' x_k,$$

then by the assumption for the case of $l = k$ we have

$$f(x') \leq \frac{\alpha_1}{1 - \alpha_{k+1}} f(x_1) + \frac{\alpha_2}{1 - \alpha_{k+1}} f(x_2) + \cdots + \frac{\alpha_k}{1 - \alpha_{k+1}} f(x_k). \qquad (3.8)$$

Further, noticing

$$x = (1 - \alpha_{k+1})x' + \alpha_{k+1}x_{k+1}.$$

we have from the definition of convex function and relation (3.8) that

$$\begin{aligned} f(x) &= f((1 - \alpha_{k+1})x' + \alpha_{k+1}x_{k+1}) \\ &\leq (1 - \alpha_{k+1})f(x') + \alpha_{k+1}f(x_{k+1}) \\ &\leq \alpha_1 f(x_1) + \alpha_2 f(x_2) + \cdots + \alpha_{k+1}f(x_{k+1}). \end{aligned}$$

This shows the necessity of the condition.

Proof of the strictness directly follows from the definition of strictly convex functions and the earlier process. ■

For properties of convex functions, we first have the following obvious results.

**Proposition 3.1** Let $f_i$, $i = 1, 2, \ldots, m$, be a group of convex functions defined over the convex set $\Omega$, then for arbitrary scalars $\alpha_i \geq 0$, $i = 1, 2, \ldots, m$, the nonnegative combination $g = \sum_{i=1}^{m} \alpha_i f_i$ is also a convex function over $\Omega$.

**Proposition 3.2** Let $f$ be a convex function defined over the convex set $\Omega$, then $f$ is continuous over $\Omega$, and for an arbitrary scalar $\alpha$ the set $\{x| \ x \in \Omega, \ f(x) < \alpha\}$ is convex.

The following theorem further gives another important property of convex functions and convex sets.

**Theorem 3.5** Let $f$ be a convex function defined over the compact convex set $\Omega$. Then, $f(x) < 0$ holds in $\Omega$ if and only if it holds on all the extreme points of $\Omega$, that is,

$$f(x) < 0, \quad \forall x \in \Omega \Longleftrightarrow f(x) < 0, \quad \forall x \in \Omega_E.$$

*Proof* The necessity is obvious. To show the sufficiency, we arbitrarily choose $x \in \Omega$, $x \notin \Omega_E$. Since $\Omega$ is compact and convex, it follows from Theorem 3.2 that $\Omega = \text{conv}(\Omega_E)$. Therefore, there exist $e_i \in \Omega_E$, $i = 1, 2, \ldots, l$, such that

$$x = \alpha_1 e_1 + \alpha_2 e_2 + \cdots + \alpha_l e_l,$$

where

$$\alpha_i \geq 0, \quad i = 1, 2, \ldots, l, \quad \text{and} \quad \sum_{i=1}^{l} \alpha_i = 1.$$

Since, by assumption,

$$f(e_i) < 0, \quad i = 1, 2, \ldots, l,$$

we have, using Theorem 3.4,

$$f(x) = \alpha_1 f(e_1) + \alpha_2 f(e_2) + \cdots + \alpha_l f(e_l) < 0.$$

Thus, the sufficiency also holds. ∎

The aforementioned theorem clearly implies the following corollary.

**Corollary 3.1** Let $f$ be a convex function defined over $\Omega = \text{conv}\{z_i, i = 1, 2, \ldots, l\}$. Then,

$$f(x) < 0, \quad \forall x \in \Omega \Longleftrightarrow f(z_i) < 0, \quad i = 1, 2, \ldots, l.$$

## 3.2.2 Criteria

The following theorem gives a criterion for convexity of a function $f : \mathbb{R}^n \to \mathbb{R}$ in terms of its gradient

$$\nabla f(x) = \left[ \frac{\partial f}{\partial x_1} \quad \frac{\partial f}{\partial x_2} \quad \cdots \quad \frac{\partial f}{\partial x_n} \right]^{\mathrm{T}}.$$

**Theorem 3.6** Suppose $f : \Omega \subset \mathbb{R}^n \to \mathbb{R}$ is differentiable in $\Omega \neq \emptyset$, and $\Omega$ is convex. Then, $f(x)$ is convex if and only if

$$f(y) \geq f(x) + \nabla^T f(x)(y - x) \tag{3.9}$$

for all $x, y \in \Omega$.

*Proof* Sufficiency. Suppose that for all $x, y \in \Omega$ the relation (3.9) holds. Taking $y_1, y_2 \in \Omega$, $y_1 \neq y_2$, and define

$$z = \theta y_1 + (1 - \theta) y_2, \quad 0 \leq \theta \leq 1, \tag{3.10}$$

then we also have $z \in \Omega$ by the convexity of $\Omega$. Therefore, by assumption, we have

$$f(y_1) \geq f(z) + \nabla^T f(x)(y_1 - z), \tag{3.11}$$

and

$$f(y_2) \geq f(z) + \nabla^T f(x)(y_2 - z). \tag{3.12}$$

Multiplying (3.11) both sides by $\theta$, and (3.12) by $(1 - \theta)$, gives

$$\theta f(y_1) \geq \theta f(z) + \theta \nabla^T f(x)(y_1 - z),$$

and

$$(1 - \theta) f(y_2) \geq (1 - \theta) f(z) + (1 - \theta) \nabla^T f(x)(y_2 - z),$$

respectively. Adding the two inequalities side by side, and using (3.10), yields

$$\begin{aligned}
\theta f(y_1) + (1 - \theta) f(y_2) &\geq \theta f(z) + \theta \nabla^T f(x)(y_1 - z) \\
&\quad + (1 - \theta) f(z) + (1 - \theta) \nabla^T f(x)(y_2 - z) \\
&= f(z) + \nabla^T f(x) \left( \theta y_1 + (1 - \theta) y_2 - z \right) \\
&= f(z).
\end{aligned}$$

This states that $f(x)$ is convex.

Necessity. Suppose that $f(x)$ is convex, then for arbitrary $0 \leq \theta \leq 1$, and $x, y \in \Omega$, there holds

$$f\left(\theta y + (1 - \theta)x\right) \leq \theta f(y) + (1 - \theta) f(x).$$

Through shifting the term $f(x)$ on the right-hand side to the left, we obtain

$$f(\theta y + (1 - \theta)x) - f(x) \le \theta f(y) - \theta f(x).$$

This gives

$$\frac{f(\theta y + (1 - \theta)x) - f(x)}{\theta} \le f(y) - f(x),$$

that is

$$\frac{f(x + \theta(y - x)) - f(x)}{\theta} \le f(y) - f(x). \tag{3.13}$$

Note that

$$\lim_{\theta \to 0} \frac{f(x + \theta(y - x)) - f(x)}{\theta} = \nabla^T f(x)(y - x),$$

we immediately have from (3.13) the relation (3.9). ■

The aforementioned theorem provides us with a criterion using the gradient of the function. The geometric illustration is shown in Figure 3.6. The following theorem gives a criterion for convexity of a function $f : \mathbb{R}^n \to \mathbb{R}$ in terms of its Hessian matrix

$$\nabla^2 f(x) = \begin{bmatrix} \dfrac{\partial^2 f}{\partial x_1^2} & \dfrac{\partial^2 f}{\partial x_1 \partial x_2} & \cdots & \dfrac{\partial^2 f}{\partial x_1 \partial x_n} \\[2ex] \dfrac{\partial^2 f}{\partial x_2 \partial x_1} & \dfrac{\partial^2 f}{\partial x_2^2} & \cdots & \dfrac{\partial^2 f}{\partial x_2 \partial x_n} \\[2ex] \vdots & \vdots & \ddots & \vdots \\[2ex] \dfrac{\partial^2 f}{\partial x_n \partial x_1} & \dfrac{\partial^2 f}{\partial x_n \partial x_2} & \cdots & \dfrac{\partial^2 f}{\partial x_n^2} \end{bmatrix}.$$

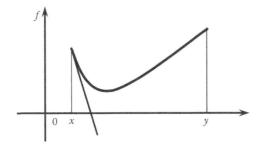

**Figure 3.6** Convex function and its derivative.

**Theorem 3.7** Assume that $f : \Omega \subset \mathbb{R}^n \rightarrow \mathbb{R}$ has second-order derivatives, and $\Omega \neq \emptyset$ is an open convex set. Then, $f$ is convex if and only if its Hessian matrix is symmetric semi-positive definite, that is,

$$\nabla^2 f(x) \geq 0, \ \forall x \in \Omega. \tag{3.14}$$

*Proof* Sufficiency. Assume that (3.14) is met. Arbitrarily take $x, y \in \Omega$, and for some $\theta \in (0,1)$ define

$$z = y + \theta(x - y),$$

then $z \in \Omega$. According to the mean-value theorem, we have the following relation:

$$f(x) = f(y) + (x - y)^{\mathrm{T}} \nabla f(y)$$
$$+ (x - y)^{\mathrm{T}} \nabla^2 f(z)(x - y) + o(\|x - y\|^2). \tag{3.15}$$

Since (3.14) holds, we obviously have

$$(x - y)^{\mathrm{T}} \nabla^2 f(z)(x - y) + o(\|x - y\|^2) \geq 0,$$

therefore, it follows from (3.15) that

$$f(x) \geq f(y) + (x - y)^{\mathrm{T}} \nabla f(y). \tag{3.16}$$

Thus, by Theorem 3.6, we know that $f$ is convex.

Necessity. Assume $f$ is convex in $\Omega$. Then, according to Theorem 3.6, one has for arbitrary $x, y \in \Omega$ the relation

$$f(y) \geq f(x) + (y - x)^{\mathrm{T}} \nabla f(x). \tag{3.17}$$

Since $\Omega$ is an open convex set, for any $z \in \mathbb{R}^n$, $z \neq 0$, there always exists an $\epsilon > 0$ such that

$$x + \epsilon z \in \Omega.$$

Substituting the $y$ in (3.17) by $x + \epsilon z$ gives

$$f(x + \epsilon z) \geq f(x) + \epsilon z^{\mathrm{T}} \nabla f(x). \tag{3.18}$$

On the other hand, applying Taylor expansion to the function $f$ yields

$$f(x + \epsilon z) = f(x) + \epsilon z^{\mathrm{T}} \nabla f(x) + \epsilon z^{\mathrm{T}} \nabla^2 f(x) \epsilon z + o(\|\epsilon z\|^2). \tag{3.19}$$

Combining (3.18) and (3.19) produces

$$\epsilon z^{\mathrm{T}} \nabla^2 f(x) \epsilon z + o(\|\epsilon z\|^2) \geq 0.$$

By the definition of $o(\|\epsilon z\|^2)$, we know that for sufficiently small $\epsilon$, there holds

$$\epsilon z^{\mathrm{T}} \nabla^2 f(x) \epsilon z \geq 0,$$

which gives

$$z^{\mathrm{T}} \nabla^2 f(x) z \geq 0.$$

Recalling the arbitrariness of $z$, we know from the aforementioned relation that $\nabla^2 f(x)$ must be semi-positive definite. ∎

**Example 3.3**

Let $P \in \mathbb{S}^n$, and $P \geq 0$, then the function

$$f(x) = x^{\mathrm{T}} P x$$

is convex in $\mathbb{R}^n$.

In fact, in view of $P = [p_{ij}]_{n \times n}$, $p_{ij} = p_{ji}$,

$$f(x) = x^{\mathrm{T}} P x$$

$$= \begin{bmatrix} x_1 & x_2 & \cdots & x_n \end{bmatrix} \begin{bmatrix} x_1 p_{11} + x_2 p_{12} + \cdots + x_n p_{1n} \\ x_1 p_{21} + x_2 p_{22} + \cdots + x_n p_{2n} \\ \vdots \\ x_1 p_{n1} + x_2 p_{n2} + \cdots + x_n p_{nn} \end{bmatrix}$$

$$= x_1 \left( x_1 p_{11} + x_2 p_{12} + \cdots + x_n p_{1n} \right)$$
$$+ x_2 \left( x_1 p_{21} + x_2 p_{22} + \cdots + x_n p_{2n} \right)$$
$$+ \cdots + x_n \left( x_1 p_{n1} + x_2 p_{n2} + \cdots + x_n p_{nn} \right),$$

thus

$$\frac{\partial f}{\partial x_j} = x_1 p_{1j} + x_2 p_{2j} + \cdots + x_{j-1} p_{j-1,j}$$

$$+ x_1 p_{j1} + x_2 p_{j2} + \cdots + x_{j-1} p_{j,j-1} + 2 x_j p_{jj}$$
$$+ x_{j+1} p_{j,j+1} + \cdots + x_n p_{jn}$$
$$+ x_{j+1} p_{j+1,j} + \cdots + x_n p_{nj}$$
$$= 2 \left( x_1 p_{1j} + x_2 p_{2j} + \cdots + x_n p_{nj} \right), \quad j = 1, 2, \ldots, n,$$

and

$$\frac{\partial^2 f}{\partial x_i \partial x_j} = \frac{\partial}{\partial x_i}\left(\frac{\partial f}{\partial x_j}\right)$$

$$= 2\frac{\partial}{\partial x_i}(x_1 p_{1j} + x_2 p_{2j} + \cdots + x_n p_{nj})$$

$$= 2p_{ij}, \quad i, j = 1, 2, \ldots, n.$$

Therefore,

$$\nabla^2 f(x) = 2P \geq 0.$$

The conclusion thus follows from Theorem 3.7.

## 3.3 Mathematical Optimization

Mathematical optimization is a very important subject in applied mathematics and has applications in almost every scientific field. Mathematical optimization is often called mathematical programming.

The standard formulation of a mathematical optimization problem is stated as follows.

**Problem 3.1**  Given functions $f_i(x) : \mathbb{R}^n \to \mathbb{R}$, $i = 0, 1, 2, \ldots, m$, and $h_i(x)$: $\mathbb{R}^n \to \mathbb{R}$, $i = 1, 2, \ldots, p$, find an $x_0 \in \mathbb{R}^n$ satisfying

$$\begin{cases} f_0(x_0) \leq f_0(x), & \forall x \in \mathbb{R}^n \\ f_i(x_0) \leq 0, & i = 1, 2, \ldots, m \\ h_i(x_0) = 0, & i = 1, 2, \ldots, p. \end{cases}$$

Such a problem is typically written as follows:

$$\begin{cases} \min & f_0(x) \\ \text{s.t.} & f_i(x) \leq 0, \quad i = 1, 2, \ldots, m \\ & h_i(x) = 0, \quad i = 1, 2, \ldots, p, \end{cases} \tag{3.20}$$

where $x \in \mathbb{R}^n$ is the optimization variable vector; $f_0 : \mathbb{R}^n \to \mathbb{R}$ is the objective (or cost) function; $f_i : \mathbb{R}^n \to \mathbb{R}$, $i = 1, 2, \ldots, m$, are the inequality constraint functions; and $h_i : \mathbb{R}^n \to \mathbb{R}$, $i = 1, 2, \ldots, p$, are the equality constraint functions.

Furthermore, $x_0$ is called a (globally) optimal parameter (variable) vector, or a globally optimal solution, for problem (3.20), while $f_0(x_0)$ is the corresponding optimal index, and is also called a global minimum of problem (3.20).

Denote

$$\mathbb{F} = \left\{ x \, \middle| \, f_i(x) \leq 0, \ i = 1, 2, \ldots, m; \ h_j(x) = 0, \ j = 1, 2, \ldots, p \right\}, \tag{3.21}$$

then aforementioned problem (3.20) can be rewritten as

$$\begin{cases} \min & f_0(x) \\ \text{s.t.} & x \in \mathbb{F}, \end{cases} \tag{3.22}$$

and the following definition in terms of the set $\mathbb{F}$ can be given.

**Definition 3.11**    The set $\mathbb{F}$ given in (3.21) is called the feasible set or the constraint set of Problem 3.1, and $x$ is called a feasible solution or a constraint solution if $x \in \mathbb{F}$. Furthermore, $x^*$ is called a (globally) optimal solution if $x^* \in \mathbb{F}$ and $f_0(x^*) < f_0(x)$, $\forall x \in \mathbb{F}$, $x \neq x^*$.

Corresponding to a global solution, for the optimization Problem 3.1 often a locally optimal solution is derived.

**Definition 3.12**    A vector $x^*$ is a locally optimal solution for Problem 3.1 if there exists a scalar $\varepsilon > 0$ such that $x^*$ is optimal for

$$\begin{cases} \min & f_0(z) \\ \text{s.t.} & f_i(z) \leq 0, \quad i = 1, 2, \ldots, m \\ & h_i(z) = 0, \quad i = 1, 2, \ldots, p \\ & ||z - x^*|| \leq \varepsilon. \end{cases} \tag{3.23}$$

In view of (3.21), the optimization problem can be more compactly written as

$$\begin{cases} \min & f_0(x) \\ \text{s.t.} & x \in \mathbb{F} \cap \mathbb{B}_\varepsilon (x^*). \end{cases} \tag{3.24}$$

Obviously, a local optimal point becomes a global optimal point when $\varepsilon = \infty$.

For solution to a general optimization problem, often a local minimum is found. Generally speaking, finding a global minimum of a general optimization problem is very difficult. In the following of this section, we mention some well-known solvable optimization problems.

### 3.3.1 Least-Squares Programming

A standard least-squares problem is formulated as follows.

**Problem 3.2**    Given $A \in \mathbb{R}^{m \times n}, b \in \mathbb{R}^m$, with rank $(A) = n$, find $x \in \mathbb{R}^n$ to minimize

$$J(x) = ||Ax - b||_2^2 = (Ax - b)^{\mathrm{T}} (Ax - b). \tag{3.25}$$

For solution to the least-squares problem, we have the following conclusion.

**Theorem 3.8**  The least-squares problem (3.25) has an analytical solution:

$$x^* = \left(A^T A\right)^{-1} A^T b. \tag{3.26}$$

*Proof*  It is well known that the optimal solution $x^*$ to the problem satisfies

$$\left.\frac{d}{dx}J(x)\right|_{x=x^*} = 2A^T (Ax - b)\big|_{x=x^*} = 0,$$

that is,

$$A^T Ax^* = A^T b.$$

This gives the analytical solution (3.26).  ■

Regarding the least-squares problem, we make the following remarks:

- The analytical solution is useful for certain small-size problems and in certain analysis problems. However, for solving large-size problems, this analytical solution is seldom used due to numerical problems. In this case, one may use the existing reliable and efficient algorithms and software.
- The computation time for solving a least-squares problem is proportional to $mn^2$; it becomes less if the problem is structured.
- Least-squares techniques have been a mature technique and have found successful applications in many fields.
- In practical applications, least-squares problems are easy to recognize and to formulate. Furthermore, there exist a few standard techniques that increase the problem flexibility (e.g., including weights and adding regularization terms).

### 3.3.2 Linear Programming

A general linear programming problem is stated as follows.

**Problem 3.3**  Let $c$, $a_i \in \mathbb{R}^n$, $i = 1, 2, \ldots, m$. Solve

$$\begin{cases} \min & c^T x \\ \text{s.t.} & a_i^T x \le b_i, \quad i = 1, 2, \ldots, m. \end{cases} \tag{3.27}$$

Regarding the solution to linear programming, we have the following comments:

- Unlike the case for least squares, there does not exist an analytical formula for solution to a general linear programming problem. However, there exist reliable and efficient algorithms and software for solving linear programming problems. The elementary algorithm is the simplex algorithm (Bazaraa et al. 2006).

■ Like least squares, linear programming has been a mature technique and has also found many successful applications in many fields.

■ The computation time for solving a linear programming problem is proportional to $mn^2$ if $m \geq n$; it is less if the problem possesses a special structure.

■ In practical applications, linear programming problems are not as easy to recognize as least-squares problems. There exist a few standard tricks used to convert problems into linear programming (e.g., problems involving $l_1$ or $l_\infty$ norms and piecewise-linear functions).

### 3.3.3 Quadratic Programming

A quadratic programming (QP) is an optimization problem with a quadratic cost function and can be formulated as follows.

**Problem 3.4**   Solve

$$
\begin{cases}
\min & \frac{1}{2}x^\mathrm{T}Px + q^\mathrm{T}x + r \\
\text{s.t.} & Gx \preceq h \\
& Ax = b,
\end{cases} \tag{3.28}
$$

where
    $q, r, h$ and $b$ are vectors of appropriate dimensions
    $G, A, P$ are matrices of appropriate dimensions
    $P$ is symmetric positive definite

In the aforementioned problem formulation, the symbol "$\preceq$" denotes the element-wise "not greater than" relation in the vector elements.

By examining the problem, we have the following observations:

■ Since $P$ is symmetric positive definite, the objective is convex quadratic.
■ The problem is to minimize a convex quadratic function over a polyhedron.
■ Least square is a special QP problem.

## 3.4 Convex Optimization

### 3.4.1 Problem

Based on the concept of convex functions, we can now define the well-known convex optimization. A standard problem of convex optimization appears as follows.

**Problem 3.5** Solve

$$
\begin{cases}
\min & f_0(x) \\
\text{s.t.} & f_i(x) \leq 0, \quad i = 1, 2, \ldots, m \\
& Ax = b,
\end{cases}
\tag{3.29}
$$

where

$f_i, \ i = 0, 1, 2, \ldots, m$, are a group of convex functions
$A$ is a given matrix
$b$ is a vector of appropriate dimensions

According to the aforementioned formulation, the equality constraint in a convex optimization problem must be affine. Further, it is easy to see that least squares, linear programming and quadratic programming, are all special convex optimization problems.

**Example 3.4**

Consider the optimization problem

$$
\begin{cases}
\min & x_1^2 + x_2^2 \\
\text{s.t.} & x_1 \leq 0 \\
& x_1 + x_2 = 0.
\end{cases}
$$

It is easy to see that with this problem we have $m = 1$, $n = 2$, $A = [1 \ 1]$, $b = 0$ and

$$
f_0(x) = x_1^2 + x_2^2,
$$
$$
f_1(x) = x_1
$$

are all convex functions. Therefore, the problem is a convex one.
From the equality constraint, we have $x_2 = -x_1$, thus $f_0(x)$ becomes

$$
f_0(x) = x_1^2 + x_2^2 = 2x_1^2,
$$

which clearly achieves its minimum on the halfspace $x_1 \leq 0$ at $x_1 = 0$. Therefore, the original problem indeed has a global solution $x^* = [0, \ 0]^T$ with $f_0(x^*) = 0$.

Denote

$$
\mathbb{F}_1 = \{ x \mid f_i(x) \leq 0, \ i = 1, 2, \ldots, m \}, \quad \mathbb{F}_2 = \{ x \mid Ax = b \},
\tag{3.30}
$$

then

$$
\mathbb{F} = \{ x \mid f_i(x) \leq 0, \quad i = 1, 2, \ldots, m, \ Ax = b \}
$$
$$
= \mathbb{F}_1 \cap \mathbb{F}_2
\tag{3.31}
$$

is the feasible set or constraint set of Problem 3.5. The following theorem reveals a very important feature of convex optimization.

**Theorem 3.9**   Let $f_i$, $i = 1, 2, \ldots, m$, be a group of convex functions, and $A$ and $b$ a matrix and a vector of appropriate dimensions, respectively. Then, the sets $\mathbb{F}_1$, $\mathbb{F}_2$, and $\mathbb{F}$ given in (3.30) and (3.31) are convex.

*Proof*   For $\forall x_1, x_2 \in \mathbb{F}_1$, we have

$$f_i(x_1) \leq 0, \quad f_i(x_2) \leq 0, \quad i = 1, 2, \ldots, m. \tag{3.32}$$

Define

$$z = \theta x_1 + (1 - \theta)x_2, \quad 0 \leq \theta \leq 1.$$

Since $f_i(x)$, $i = 1, 2, \ldots, m$, are convex functions, using (3.32) we have, for $0 \leq \theta \leq 1$,

$$f_i(z) = f_i(\theta x_1 + (1 - \theta)x_2)$$

$$\leq \theta f_i(x_1) + (1 - \theta)f_i(x_2)$$

$$\leq 0.$$

Thus,

$$z = \theta x_1 + (1 - \theta)x_2 \in \mathbb{F}_1, \quad \forall 0 \leq \theta \leq 1.$$

This states that the set $\mathbb{F}_1$ is convex.
   Next, for $\forall x_1, x_2 \in \mathbb{F}_2$, we have

$$Ax_1 = b, \quad Ax_2 = b. \tag{3.33}$$

Using (3.33) gives, for arbitrary $0 \leq \theta \leq 1$,

$$A(\theta x_1 + (1 - \theta)x_2)$$

$$= \theta A x_1 + (1 - \theta)A x_2$$

$$= \theta b + (1 - \theta)b$$

$$= b.$$

Thus,

$$\theta x_1 + (1 - \theta)x_2 \in \mathbb{F}_2, \quad \forall 0 \le \theta \le 1.$$

This states that the set $\mathbb{F}_2$ is convex.

As the intersection of two convex sets, the set $\mathbb{F}$ is obviously also convex. ■

In view of the aforementioned theorem that the feasible set of a convex optimization is actually a convex set, a general convex optimization may be described as follows.

**Problem 3.6** Solve

$$\begin{cases} \min & f_0(x) \\ \text{s.t.} & x \in \Omega, \end{cases} \tag{3.34}$$

where $\Omega \subset \mathbb{R}^n$ is a convex set, while $f_0$ is a convex function defined on $\Omega$.

## 3.4.2 Local and Global Optima

The following result reveals the most important feature of convex optimization.

**Theorem 3.10** Any locally optimal point of the convex optimization problem (3.34) is globally optimal. Furthermore, the problem has a unique globally optimal point when the objective function $f_0(x)$ is strictly convex.

*Proof* Suppose $x$ is locally optimal and $y$ is globally optimal, then

$$f_0(y) < f_0(x), \tag{3.35}$$

and there exists a scalar $\epsilon > 0$ such that

$$f_0(x) \le f_0(z), \quad \forall z \in \Omega_\epsilon(x),$$

where

$$\mathbb{B}_\epsilon(x) = \{z \mid ||z - x|| \le \epsilon\}. \tag{3.36}$$

Let

$$z^* = \theta y + (1 - \theta)x,$$

with

$$\theta = \frac{\epsilon}{2||y - x||_2}.$$

Note that $y \notin \Omega_\epsilon(x)$ (otherwise contradicts with (3.35)), we have

$$||y - x||_2 > \epsilon,$$

thus

$$0 < \theta < \frac{1}{2}.$$

Further, it is clear that

$$||z^* - x||_2 = \theta||y - x||_2 = \frac{1}{2}\epsilon,$$

thus $z^* \in \Omega_\epsilon(x)$ (See the illustration in Figure 3.7). Therefore, using the convexity of $f$ and the relation (3.36), we have

$$f_0(z^*) \leq \theta f_0(y) + (1 - \theta)f_0(x)$$
$$< \theta f_0(x) + (1 - \theta)f_0(x)$$
$$= f_0(x),$$

which contradicts our assumption that $x$ is locally optimal. Therefore, any locally optimal point of the convex optimization problem (3.34) is globally optimal.

Now, let $f_0$ be strictly convex, and we assume that there exist two points $x, y \in \Omega$, with $x \neq y$, which are both the globally optimal point of the convex optimization problem (3.34). Therefore,

$$f_0(x) = f_0(y) \leq f_0(z), \quad \forall z \in \Omega. \tag{3.37}$$

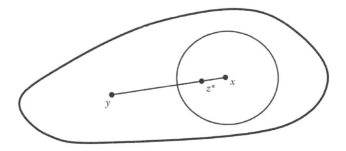

**Figure 3.7   Proof of Theorem 3.10.**

Since $f_0$ is strictly convex and $\Omega$ is convex, for $0 < \theta < 1$ we have

$$z_\theta = \theta y + (1 - \theta) x \in \Omega,$$

and

$$f_0(z_\theta) < \theta f_0(y) + (1 - \theta) f_0(x)$$
$$= \theta f_0(x) + (1 - \theta) f_0(x)$$
$$= f_0(x).$$

This contradicts with the assumption (3.37). Thus, the problem must have a unique globally optimal point when the objective function $f_0(x)$ is strictly convex. ■

The aforementioned theorem tells us that, for a convex optimization all the local and global minima are equal, or there is one local or global minimum only. Therefore, once we find a minimum for a convex optimization problem, it is a global minimum.

Regarding solution to convex optimization problems, we mention that there does not exist an analytical solution to a general convex optimization problem since it is generally nonlinear. However, due to the favorable property mentioned earlier, we have the following very positive facts:

- Reliable and efficient algorithms exist for solving convex optimization problems
- Convex optimization has been almost a technique

It is seen from the earlier text that the formulation of convex optimization is very simple. However, in practice convex optimization problems are often difficult to recognize. A main task of tackling an optimization problem is often to transform the problem into a convex one. Fortunately, by now there have been many tricks for transforming certain optimization problems into convex optimization problems, and with these tricks, surprisingly many practical problems can be formulated and solved via convex optimization.

**Remark 3.1** In the early 1970s, optimization problems were classified into two big categories: linear programming and nonlinear programming problems. At that time, linear programming problems were regarded as the type of solvable optimization problems, while nonlinear programming were generally regarded as unsolvable. Linearity was used as a dividing line. By now we have seen that all the convex optimizations, certainly including linear programming and many nonlinear programming problems, are such good problems that all their local and global minima are equal to each other. This is to say that such problems, as long as they are recognized, are easy to solve. From this sense, linearity in a problem is not of essence, while what really matters is the convexity.

## 3.5 LMI Problems

Having introduced the definition and the favorable features of convex optimization, we here look into the following optimization problem with an LMI constraint:

$$\begin{cases} \min & f_0(x) \\ \text{s.t.} & x \in \{x \mid A(x) < 0\} \cap \mathbb{F}, \end{cases} \tag{3.38}$$

with $A(x) < 0$ being an LMI, $f_0(x)$ a convex function, and $\mathbb{F}$ a convex set.

### 3.5.1 Convexity

In this section, we examine the covexity of the LMI constraint

$$A(x) = A_0 + x_1 A_1 + \cdots + x_n A_n < 0.$$

The following result confirms that an LMI constraint actually defines a convex set.

**Theorem 3.11**   Let $A_i \in \mathbb{S}^m$, $i = 0, 1, \ldots, n$, and

$$A(x) = A_0 + x_1 A_1 + \cdots + x_n A_n.$$

Then,

1. $\Phi_0 = \{x \mid A(x) < 0\}$ is an open convex set.
2. $\Phi_c = \{x \mid A(x) \leq 0\}$ is a closed convex set.

*Proof*   Choose

$$x = \begin{bmatrix} x_1 & x_2 & \cdots & x_n \end{bmatrix}^{\mathrm{T}} \in \Phi_0,$$

$$y = \begin{bmatrix} y_1 & y_2 & \cdots & y_n \end{bmatrix}^{\mathrm{T}} \in \Phi_0,$$

then

$$A(x) = A_0 + x_1 A_1 + \cdots + x_n A_n < 0,$$

and

$$A(y) = A_0 + y_1 A_1 + \cdots + y_n A_n < 0.$$

Therefore, for any $0 \leq \theta \leq 1$, we have

$$
\begin{aligned}
A\left(\theta x + (1-\theta)y\right) \\
&= A_0 + \left(\theta x_1 + (1-\theta)y_1\right)A_1 + \cdots + \left(\theta x_n + (1-\theta)y_n\right)A_n \\
&= \theta A_0 + \theta x_1 A_1 + \cdots + \theta x_n A_n \\
&\quad + (1-\theta)A_0 + (1-\theta)y_1 A_1 + \cdots + (1-\theta)y_n A_n \\
&= \theta A(x) + (1-\theta)A(y) \\
&< 0.
\end{aligned}
$$

This proves the convexity of $\Phi_0$. Similarly, we can show the convexity of $\Phi_c$.

Next, let us show the openness of $\Phi_0$. Arbitrarily choose $x^* \in \Phi_0$, then $A(x^*) < 0$.

Let

$$
\Delta x = \begin{bmatrix} \Delta x_1 & \Delta x_2 & \cdots & \Delta x_n \end{bmatrix}^{\mathrm{T}},
$$

then

$$
A\left(x^* + \Delta x\right) = A(x^*) + \Delta A\left(\Delta x\right),
$$

with

$$
\Delta A\left(\Delta x\right) = \Delta x_1 A_1 + \cdots + \Delta x_n A_n.
$$

Since $\Delta A\left(\Delta x\right)$ is linear with respect to $\Delta x$, we obviously have

$$
\lim_{\|\Delta x\| \to 0} \|\Delta A\left(\Delta x\right)\| = 0.
$$

Further, noticing $\lambda_{\max}(A(x^*)) < 0$ and the fact that the eigenvalues of a matrix are continuous with respect to the matrix elements, we can conclude that there exists a scalar $\epsilon > 0$, satisfying

$$
\lambda_{\max}(A(x^* + \Delta x)) < 0, \quad \|\Delta x\| < \epsilon.
$$

This is equivalent to

$$
A(x) < 0, \quad \|x - x^*\| < \epsilon.
$$

This states that

$$\{x \mid \|x - x^*\| < \epsilon\} \subset \Phi_0,$$

thus $\Phi_0$ is an open set.

Finally, we can similarly verify that $\bar{\Phi}_c = \{x \mid A(x) > 0\}$ is an open set. Therefore, $\Phi_c = \mathbb{R}^n \backslash \bar{\Phi}_c$ is closed. The proof is completed. ■

It follows from the earlier result that the optimization problem (3.38) with an LMI constraint is in fact a convex one; it thus eventually possesses the feature that any local minimum is a global one, or it has a unique globally optimal point when the objective function $f_0(x)$ is strictly convex.

## 3.5.2 Extreme Result

The following result tells us that an LMI holds over a compact and convex set if and only if it holds on the extreme points of the compact and convex set.

**Theorem 3.12**    Let $\mathbb{F}$ be a compact and convex set, $A_i \in \mathbb{S}^m$, $i = 0, 1, \ldots, n$, and $A(x) = A_0 + x_1 A_1 + \cdots + x_n A_n$. Then,

$$A(x) < 0, \quad x \in \mathbb{F} \Longleftrightarrow A(x) < 0, \quad x \in \mathbb{F}_E.$$

*Proof*    The necessity is obvious since $\mathbb{F}_E \subset \mathbb{F}$. To show the sufficiency, we arbitrarily choose $x \in \mathbb{F}$, $x \notin \mathbb{F}_E$. Since $\mathbb{F}$ is compact and convex, it follows from Theorem 3.2 that $\mathbb{F} = \text{conv}(\mathbb{F}_E)$. Therefore, there exist $e_i \in \mathbb{F}_E$, $i = 1, 2, \ldots, k$, such that

$$x = \alpha_1 e_1 + \alpha_2 e_2 + \cdots + \alpha_k e_k,$$

where

$$\alpha_i \geq 0, \quad i = 1, 2, \ldots, k, \quad \sum_{i=1}^{k} \alpha_i = 1.$$

Since $A(x)$ is linear, and by assumption $A(e_i) < 0$, $i = 1, 2, \ldots, k$, we have

$$A(x) = \sum_{i=1}^{k} \alpha_i A(e_i) < 0.$$

Thus, the condition is sufficient. The proof is completed. ■

With the aforementioned result, we know that, when $\mathbb{F}_E$ is further required to be bounded, the optimization problem (3.38) with an LMI constraint is equivalent to

$$\begin{cases} \min & f_0(x) \\ \text{s.t.} & x \in \{x|\ A(x) < 0\} \cap \mathbb{F}_E. \end{cases} \tag{3.39}$$

We point out that in certain cases the constraint set in the aforementioned problem is a finite one.

Remember that in Section 1.2 we have turned some algebraic problems into ones related to LMIs. Now we can recognize that these formulated problems (1.13), (1.15), and (1.17) are all convex optimization problems with LMI constraints.

## 3.5.3 *Standard Problems*

In this section, we introduce briefly the three standard problems involving LMIs. Many problems can be transformed into one of these standard forms. In the following, $A(x)$, $B(x)$, $C(x)$ and $L(x)$, $R(x)$ are assumed to be matrix functions of appropriate dimensions, which are all linear in vector $x$.

### 3.5.3.1 Feasibility Problem

**Problem 3.7** Find a solution $x \in \mathbb{R}^n$ satisfying the following LMI:

$$A(x) < B(x).$$

The corresponding function for solving this feasibility problem in the MATLAB® LMI Toolbox is `feasp`, which solves the following auxiliary convex optimization with LMI constraints:

$$\begin{cases} \min & t \\ \text{s.t.} & A(x) < B(x) + tI, \end{cases} \tag{3.40}$$

where $x$ and the scalar $t$ are the decision variables.

### 3.5.3.2 Convex Minimization Problem

**Problem 3.8** Given a convex function $f(x)$, find a solution $x \in \mathbb{R}^n$ to the following minimization problem with LMI constraints:

$$\begin{cases} \min & f(x) \\ \text{s.t.} & A(x) < B(x). \end{cases} \tag{3.41}$$

The corresponding function for solving this problem in the MATLAB LMI Toolbox is `mincx`, which solves the LMI problem:

$$\begin{cases} \min & c^{\mathrm{T}}x \\ \text{s.t.} & L(x) < R(x), \end{cases} \tag{3.42}$$

where $x$ is the vector of decision variables.

### 3.5.3.3 Generalized Eigenvalue Problem

**Problem 3.9**   Find a solution $x \in \mathbb{R}^n$ to the following minimization problem

$$\begin{cases} \min & \lambda \\ \text{s.t.} & A(x) < \lambda B(x) \\ & B(x) > 0 \\ & C(x) < 0, \end{cases} \tag{3.43}$$

where $x$ is the vector of (scalar) decision variables.

The corresponding function for solving this problem in the MATLAB LMI Toolbox is `gevp`.

In practical applications, many problems involving LMIs can be classified into these three types. The MATLAB LMI toolbox provides a very efficient and convenient package for solving these three types of problems involving LMIs. For details please refer to the appendix or the user's guide on LMI Control Toolbox for Use with MATLAB (Gahinet et al. 1995).

## 3.6 Notes and References

### 3.6.1 About This Chapter

This chapter introduces some basic facts and results about optimization theory. This chapter is necessary because, as we have seen from Section 3.5.3, LMI problems are generally optimization problems, and moreover, with this chapter we can fulfil the task to reveal theoretically the reasons behind the advantages of LMI problems.

The first two sections introduce convex sets and convex functions, respectively. An important fact we learn in these two sections is that a convex function defined over a compact convex set is less than zero if and only if it is less than zero on all the extreme points of the convex set. With sufficient preliminaries provided in these two sections, general mathematical optimization and convex optimization are presented in Sections 3.3 and 3.4, respectively. As we have seen in Section 3.4 that convex optimization problems possess an extremely good property, that is, any locally optimal point is globally optimal! It is such a property that really makes convex optimization so useful, so appealing and popular.

Finally, in Section 3.5, we turn to look into the LMI problems. It is shown that LMI constraints form a convex set, and an LMI holds within a compact convex set if and only if it holds on the extreme points of this set. Such facts tell us that an optimization problem with LMI constraints and a convex objective is a special convex optimization problem. So no wonder LMIs today have attracted so much attention in the control community.

For those readers who want to learn more about optimization theories and techniques, we recommend, among the many books on optimization, Torn and Zilinskas (1989) and Boyd and Vandenberghe (2004). Torn and Zilinskas (1989) stresses on global optimization and many of the methods that have been proposed for locating the global minimum of a function numerically are described. Boyd and Vandenberghe (2004) is about convex optimization, which includes least-squares and linear programming problems. It is well known that these problems can be solved numerically very efficiently, while this book has demonstrated to us that the same can be said for the larger class of convex optimization problems.

### 3.6.2 Open Source Software CVX

The open source software package, CVX, was designed especially for convex optimization by Michael Grant and Stephen Boyd and was implemented by Michael Grant (Grant et al. 2006). It incorporates ideas from earlier work by Crusius (2002), Wu and Boyd (2000), and some other researchers. The modeling language follows the spirit of AMPL (Fourer et al. 1999) or GAMS (Brooke et al. 1998). However, unlike these packages, CVX was designed from the beginning to fully exploit convexity. The specific method for implementing CVX in MATLAB draws heavily from YALMIP.

CVX is very easy to learn, and very easy to use, yet it is quick, reliable, and powerful. It works very efficiently and solves almost all of the convex optimization problems that one can formulate. Since most of the LMI problems are convex, instead of using the MATLAB LMI toolbox, one can also use CVX to solve LMI problems, but generally the reverse is not realizable. The latest version of CVX and the user's guide can be respectively retrieved from the following websites:

■ http://www.stanford.edu/~boyd/cvx
■ http://cvxr.com/cvx/cvx_usrguide.pdf

### 3.6.3 Counterexample for Numerical Reliability

Numerical reliability is a very important issue in both computational theory and applications. Regarding numerical reliability, we need to understand the following:

- Numerical reliability is a very important issue in numerical computation.
- Certain problems are good-conditioned (e.g., singular value decomposition), which may be solved efficiently using certain existing numerical methods with acceptable precision.
- Certain problems are ill-conditioned (e.g., some matrix inverses), solutions with acceptable precision to such problems are difficult to derive numerically.

While emphasizing the fact that convex optimization with LMI constraints is very numerically reliable, here we demonstrate by a pole assignment benchmark problem that ill-conditioned problems do exist. With this particular pole assignment problem, we will find that a solution with an acceptable precision to this problem is difficult to derive using the best algorithms at present.

The following problem is a well-known one and has been considered by several investigators (e.g., Mahmoud et al. 1998, Petres et al. 2007, Duan 2008).

**Problem 3.10** Given the following single input benchmark system

$$\dot{x} = Ax + Bu$$

with

$$A = \begin{bmatrix} 20 & & & \\ 20 & 19 & & \\ & \ddots & \ddots & \\ & & 20 & 1 \end{bmatrix}, \quad B = \begin{bmatrix} 1 \\ 0 \\ \vdots \\ 0 \end{bmatrix},$$

find a matrix

$$K = \begin{bmatrix} k_1 & k_2 & \cdots & k_{20} \end{bmatrix}$$

such that

$$\lambda(A + BK) = \{s_1, s_2, \ldots, s_{20}\},$$

where $s_i$, $i = 1, 2, \ldots, 20$, are a set of arbitrarily given self-conjugate complex numbers.

For this problem, we consider 14 different groups of prescribed closed-loop eigenvalues. They are selected in a most "natural" way and are listed in Table 3.1 (Duan 2008).

There are two well-known functions in MATLAB for solving pole assignment, `place` and `acker`. `place` is produced based on the pole assignment method

**Table 3.1  Prescribed Closed-Loop Eigenvalues**

| Case | Set of Eigenvalues, Γ |
|------|----------------------|
| 1 | $\{1,\ 2,\dots,\ 20\}$ |
| 2 | $\{-1,\ -2,\dots,\ -20\}$ |
| 3 | $\{1\pm i,\ 2\pm i,\dots,\ 10\pm i\}$ |
| 4 | $\{-1\pm i,\ -2\pm i,\dots,\ -10\pm i\}$ |
| 5 | $\{1\pm i,\ 2\pm 2i,\dots,\ 10\pm 10i\}$ |
| 6 | $\{-1\pm i,\ -2\pm 2i,\dots,\ -10\pm 10i\}$ |
| 7 | $\{1,\ 2,\dots,\ 10,\ 1\pm i,\ 2\pm i,\dots,\ 5\pm i\}$ |
| 8 | $\{-1,\ -2,\dots,\ -10,\ -1\pm i,\ -2\pm i,\dots,\ -5\pm i\}$ |
| 9 | $\{-1,\ -2,\dots,\ -10,\ 1\pm i,\ 2\pm i,\dots,\ 5\pm i\}$ |
| 10 | $\{1,\ 2,\dots,10,\ -1\pm i,\ -2\pm i,\dots,\ -5\pm i\}$ |
| 11 | $\{1,\ 2,\dots,10,\ 1\pm i,\ 2\pm 2i,\dots,\ 5\pm 5i\}$ |
| 12 | $\{-1,\ -2,\dots,-10,\ -1\pm i,\ -2\pm 2i,\dots,\ -5\pm 5i\}$ |
| 13 | $\{-1,\ -2,\dots,-10,\ 1\pm i,\ 2\pm 2i,\dots,5\pm 5i\}$ |
| 14 | $\{1,\ 2,\dots,10,\ -1\pm i,\ -2\pm 2i,\dots,-5\pm 5i\}$ |

proposed by Kautsky et al. (1985), which suits both single-input and multi-input systems, while `acker` is based on the well-known Ackermann's formula, which suits only single-input systems.

We have tried both functions for the 14 different cases, and the results are outlined as follows:

- For Case 1, both functions `place` and `acker` work well. In terms of precision, `place` gives slightly better results than `acker` when the derived gain matrix $K$ is rounded off to nine digits.
- For Cases 2 through 14, `place` works for Cases 5, 6, and 14. It refuses to produce a result for all the other cases. Furthermore, although `place` works for Cases 5, 6, and 14, it produces very poor results for these case. More precisely, for these three cases, even when the gain matrix $K$ is not rounded off, the closed-loop eigenvalues produced by this function have gone far away from the prescribed ones.
- Unlike function `place`, `acker` works for all Cases 2 through 14. However, like function `place`, it actually produces for each case a completely different set of closed-loop eigenvalues from the prescribed one.

To sum up, for the aforementioned simple and neat pole assignment problem, the two well-known functions in MATLAB for solving pole assignment, namely `place` and `acker`, in fact *gives correct results for only one case out of 14 ones*. These results can be very easily checked, and readers are strongly encouraged to try the solutions by themselves, the eigenvalues of the matrix $A + BK$ can be easily obtained using the MATLAB function `eig`. For a more detailed discussion on this example, please refer to Duan (2008).

These results tell us that the MATLAB functions `place` and `acker` for solving pole assignment, although have been quite complete, are not generally quite numerically reliable. On the other hand, this also indicates that this particular pole assignment problem is an ill-conditioned one. With this example, we wish to remind the readers to realize the fact that many numerical methods work theoretically but may not always practically!

Now come back to LMIs, unlike this problem, almost all convex optimization problems with LMI constraints can be solved efficiently and reliably.

### 3.6.4 Numerical Examples in the Book

Generally speaking, with complicated numerical computations, no matter how reliable an algorithm is used, there is always a computational error. The problem is really how large the computational error is.

From the next chapter on, we will encounter in this book some numerical examples related with the various proposed LMI problems. Although these problems are convex and can be solved numerically reliably, some of the produced results may still possess relatively large errors. Therefore, for some of the numerical examples in the book, the results maybe not repeatable by readers.

Taking Example 10.4 for example, it treats the problem of mixed robust $H_\infty/H_2$ designs with desired pole region. Using the function `mincx` in the MATLAB LMI toolbox, we produced a set of numerical results. Yet with the same MATLAB codes running on a different machine, we obtained a different set of results. It can be observed from Table 3.2 that these two sets of results really differ from each other with some degree. However, in spite of the considerable differences in these two sets of results, it is interesting to note that both sets of results produce the same feedback gain matrix

$$K = WX^{-1} = \begin{bmatrix} -1.0000 & 0.0000 & -1.0000 \\ -2.0000 & -1.0000 & 0.5000 \end{bmatrix}.$$

We point out that, besides Example 10.4, there are also some other ones, such as Examples 9.3 and 10.3, whose results maybe also not repeatable to a very high precision.

With the numerical examples in the book, there is also another problem, that is, the effective number of digits in display. Say, for example, with the values of the

**Table 3.2 Two Different Sets of Results for Example 10.4**

| | | Variables | | |
|---|---|---|---|---|
| Results in the book | $X = 10^8 \times$ | $\begin{bmatrix} 2.3553 & -2.6201 & 0.4250 \\ -2.6201 & 6.6853 & 1.0505 \\ 0.4250 & 1.0505 & 1.0143 \end{bmatrix}$ | | |
| | $W = 10^8 \times$ | $\begin{bmatrix} -2.7803 & 1.5696 & -1.4393 \\ -1.8779 & -0.9198 & -1.3935 \end{bmatrix}$ | | |
| | $Z = 1.1777 \times 10^{-4}$ | | | |
| | $\left( \gamma_\infty, \, \gamma_2^2 \right) = (0.0316, \, 0.0002)$ | | | |
| | $(\alpha, \, \beta) = \left( 2.6639 \times 10^8, \, 2.6618 \times 10^8 \right)$ | | | |
| The different results | $X = 10^8 \times$ | $\begin{bmatrix} 2.3724 & -2.7203 & 0.3856 \\ -2.7203 & 6.4923 & 0.8691 \\ 0.3856 & 0.8691 & 0.8908 \end{bmatrix}$ | | |
| | $W = 10^8 \times$ | $\begin{bmatrix} -2.7580 & 1.8512 & -1.2764 \\ -1.8317 & -0.6171 & -1.1948 \end{bmatrix}$ | | |
| | $Z = 9.5608 \times 10^{-6}$ | | | |
| | $\left( \gamma_\infty, \, \gamma_2^2 \right) = (0.0224, \, 0.00002)$ | | | |
| | $(\alpha, \, \beta) = \left( 2.8086 \times 10^8, \, 2.8030 \times 10^8 \right)$ | | | |

matrices $W$ and $X$ in Table 3.2 how many number of digits should be used? Maybe at least 10! From this point of view, please bear in mind when reading this book that some of the numerical results displayed in this book with a specific number of digits have only some format sense, while their real values with a proper satisfactory precision are really stored in the computer and are used in computation. This is why with the two sets of matrices $W$ and $X$ given in Table 3.2 the finally computed gain matrix $K$ could be the same.

Coming back to the numerical reliability problem, the aforementioned phenomenon appeared in Example 10.4 does not contradict with our theory that LMIs, as a type of convex optimization problems, can be solved numerically reliably. The two sets of results are still close to each other, and the gain matrix $K$ in which we are really interested is the same. We have seen with the aforementioned pole assignment benchmark problem in the aforementioned section that the solution to some bad-conditioned problems may have really gone meaningless.

After all, the phenomenon discussed in this section again highlights the importance of numerical reliability. It does exist in many numerical computations. It is partially the problem of numerical reliability that makes LMI techniques so important and attractive.

## Exercises

**3.1** Let $X$ be a linear space, $E_i \subset X$, $i = 1, 2, \ldots, n$, are convex sets. Find out whether $\bigcup_{i=1}^{n} E_i \left( \text{or } \bigcap_{i=1}^{n} E_i \right)$ is convex or not.

**3.2** Prove Proposition 3.1, that is, let $f_i$, $i = 1, 2, \ldots, m$ be a group of convex functions defined over the convex set $\Omega$, then for arbitrary scalars $\alpha_i \geq 0$, $i = 1, 2, \ldots, m$, the combination $g = \sum_{i=1}^{m} \alpha_i f_i$ is also a convex function over $\Omega$.

**3.3** Prove Proposition 3.2, that is, let $f$ be a convex function defined over the convex set $\Omega$, then for arbitrary scalar $\alpha$ the set $\Omega_\alpha = \{x \mid f(x) < \alpha\}$ is convex.

**3.4** (Feng 1995, p. 1) Given $b_i \in \mathbb{R}^n$, $\beta_i \in \mathbb{R}$, let

$$M_i = \left\{ x \in \mathbb{R}^n \mid b_i^{\mathrm{T}} x \leq \beta_i \right\}, \quad i = 1, 2.$$

Show that $M_1 \cap M_2$ is a convex set.

**3.5** (Feng 1995, p. 100) Let $f : \mathbb{R}^n \to \mathbb{R}$ be a positive homogeneous function, that is,

$$f(ax) = af(x), \quad \text{for } a \in \mathbb{R}, \ x \in \mathbb{R}^n.$$

Show that $f$ is a convex function if and only if

$$f(x + y) \leq f(x) + f(y) \tag{3.44}$$

for any $x, y \in \mathbb{R}^n$.

**3.6** Prove that the following functions are convex

$$f(x) = e^{ax}, \quad a \in R, \quad a \neq 0,$$
$$f(x) = |x|^p, \quad p \geq 1.$$

**3.7** Show that Problem 3.4 is a convex optimization problem.

# CONTROL SYSTEMS ANALYSIS

# Chapter 4

## Stability Analysis

Stability is the most important property of control systems. In this chapter, we examine the stability of control systems using LMI techniques. These problems include the well-known Hurwitz stability and Schur stability, and also the robust stability of uncertain systems as well as the stability of time-delay systems.

## 4.1 Hurwitz and Schur Stability

Stability is the most important property of dynamical systems. In this section, we will present LMI conditions for stability of both continuous- and discrete-time systems.

### 4.1.1 Hurwitz Stability

Consider a continuous-time linear system in the form of

$$\dot{x}(t) = Ax(t), \tag{4.1}$$

with $A \in \mathbb{R}^{n \times n}$. It is said to be stable if all the eigenvalues of the matrix $A$ have nonpositive real parts; in this case, we also say that the matrix $A$ is Hurwitz critically stable, and it is said to be asymptotically stable if all the eigenvalues of the matrix $A$ have negative real parts, and in this case we also say that the matrix $A$ is Hurwitz stable.

The problem of Hurwitz stability analysis can be stated as follows.

**Problem 4.1**  Given $A \in \mathbb{R}^{n \times n}$, check the Lyapunov stability of the continuous-time linear system (4.1), or equivalently, the Hurwitz stability of the matrix $A$.

According to the well-known Lyapunov stability theory, the solution to this problem can be immediately given.

**Proposition 4.1** The continuous-time system (4.1) is Hurwitz stable if and only if there exists a matrix $P \in \mathbb{S}^n$, such that

$$\begin{cases} P > 0 \\ A^T P + PA < 0. \end{cases} \tag{4.2}$$

*Proof* Let the continuous-time system (4.1) be stable, then by well-known results in linear systems theory, for arbitrarily given $Q \in \mathbb{S}^n$ and $Q > 0$ there exists a matrix $P \in \mathbb{S}^n$, $P > 0$ such that the following Lyapunov equation holds:

$$A^T P + PA = -Q. \tag{4.3}$$

This implies the relations in (4.2).

Conversely, let the relations in (4.2) hold. Define

$$Q = -\left(A^T P + PA\right),$$

then $Q > 0$ holds, and the Lyapunov matrix equation (4.3) holds. Again applying the Lyapunov stability theory we conclude that the continuous-time system (4.1) is stable. With this we complete the proof. ∎

By the aforementioned proposition, the problem of checking the stability of a continuous-time linear system is converted equivalently to a problem of finding a symmetric matrix satisfying a pair of LMIs.

**Example 4.1**

Consider the yaw/roll channel attitude system of an unrotating missile, whose linearized model at certain time is

$$\dot{x}(t) = \begin{bmatrix} -0.5 & 1 & 0 & 0 & 0 \\ -62 & -0.16 & 0 & 0 & 30 \\ 10 & 0 & -50 & 40 & 5 \\ 0 & 0 & 0 & -40 & 0 \\ 0 & 0 & 0 & 0 & -20 \end{bmatrix} x(t) + \begin{bmatrix} 0 & 0 \\ 0 & 0 \\ 0 & 0 \\ 4 & 0 \\ 0 & 20 \end{bmatrix} u(t).$$

This model has been considered by Duan and Wang (1992) and Jiang (1987). Applying the MATLAB® function feasp in MATLAB LMI toolbox to the corresponding LMIs problem (4.2) gives the following solution

$$P = \begin{bmatrix} 2.3840 & -0.0134 & 0.0014 & 0.0027 & -0.1635 \\ -0.0134 & 0.0386 & 0.0001 & 0.0002 & 0.0483 \\ 0.0014 & 0.0001 & 0.0124 & 0.0071 & 0.0013 \\ 0.0027 & 0.0002 & 0.0071 & 0.0219 & 0.0015 \\ -0.1635 & 0.0483 & 0.0013 & 0.0015 & 0.1021 \end{bmatrix} > 0.$$

Therefore, the system is stable. In fact, by the MATLAB function `eig`, it can be verified that the system poles are $-50$, $-0.3300 \pm 7.8722i$, $-40$ and $-20$, which are all stable ones.

## 4.1.2 Schur Stability

Parallely, the stability analysis problem for a discrete-time linear system can be stated as follows.

**Problem 4.2** Given $A \in \mathbb{R}^{n \times n}$, check the Lyapunov stability of the following discrete-time linear system

$$x(k+1) = Ax(k). \tag{4.4}$$

This is equivalent to check the Schur stability of the matrix $A$; in other words, this is equivalent to check if all the eigenvalues of the matrix $A$ are located within the unit circle of the complex plane.

Again, according to the well-known Lyapunov stability theory, we have, corresponding to Proposition 4.1, the following result about solution to the aforementioned problem.

**Proposition 4.2** The system (4.4) is Schur stable if and only if there exists a matrix $P \in \mathbb{S}^n$, such that one of the following three LMIs holds:

$$\begin{cases} P > 0 \\ APA^{\mathrm{T}} - P < 0, \end{cases} \tag{4.5}$$

$$M_1 = \begin{bmatrix} -P & PA^{\mathrm{T}} \\ AP & -P \end{bmatrix} < 0, \tag{4.6}$$

$$M_2 = \begin{bmatrix} -P & AP \\ PA^{\mathrm{T}} & -P \end{bmatrix} < 0. \tag{4.7}$$

*Proof* Similar to the proof of Proposition 4.1, the first condition (4.5) can be easily shown according to the Lyapunov stability theory for linear discrete-time systems.

Since the Schur complement of the element $-P$ (the (1,1) position) in the matrix appearing on the left-hand side of (4.6) is

$$-P - AP(-P)^{-1}PA^T = APA^T - P,$$

condition (4.6) can be easily shown to be equivalent to (4.5) with the help of the well-known Schur complement lemma. Using the Schur complement of the element $-P$ (the (2,2) position) in the matrix appearing on the left-hand side of (4.7) we can also show the equivalence between (4.7) and (4.5). ▪

**Example 4.2**

Consider the following discrete-time linear system:

$$x(k+1) = \begin{bmatrix} -0.5 & 2 & 0 \\ 0 & -0.25 & 0 \\ 1 & 0 & -0.5 \end{bmatrix} x(k).$$

Applying the MATLAB function `feasp` in the MATLAB LMI toolbox to the corresponding LMIs problem (4.6) gives the following solution

$$P = \begin{bmatrix} 9.6315 & 1.2549 & 0.1157 \\ 1.2549 & 1.0367 & -0.1235 \\ 0.1157 & -0.1235 & 22.7700 \end{bmatrix} > 0.$$

Therefore, the system is stable. In fact, by the MATLAB function `eig` it can be easily verified that the system poles are $-0.5000$, $-0.5000$, and $-0.2500$, which are all stable ones.

**Remark 4.1** It is obvious that $A^T$ is Schur stable if and only if $A$ is Schur stable. Substituting $A$ by $A^T$ in Proposition 4.2, we obtain another set of LMIs on the stability as follows:

$$\begin{cases} P > 0 \\ A^T PA - P < 0, \end{cases}$$

$$M_3 = \begin{bmatrix} -P & PA \\ A^T P & -P \end{bmatrix} < 0,$$

$$M_4 = \begin{bmatrix} -P & A^T P \\ PA & -P \end{bmatrix} < 0.$$

**Remark 4.2** After going over the results in this section, readers might ask the following question: While the eigenvalues of a matrix can be obtained as easily as typing `eig` in the MATLAB window and then pressing the enter key, then why LMIs?

The answers to this question are really of two aspects:

1. Solving the eigenvalues of a matrix is not generally a good-conditioned problem, and hence checking the stability by finding eigenvalues may have the risk of leading to wrong conclusions, while the LMIs can be always solved numerically efficiently and reliably, and hence can be fully trusted.
2. The LMI conditions proposed in this section can indeed be used to verify the Hurwitz or Schur stability of a given matrix as proven, while the more important point is that, as is demonstrated in Chapter 7, these LMI conditions are also useful in solving the problem of stabilization.

## 4.2 D-Stability

In the last section, we have presented LMI conditions for the well-known Hurwitz and Schur stability of matrices. In this section, we will introduce the concept of $\mathbb{D}$-stability, which is a generalization of Hurwitz stability and Schur stability. As in the sequential section, our main task is to establish the LMI conditions for the $\mathbb{D}$-stability of a matrix with different given $\mathbb{D}$-regions.

**Definition 4.1**   Let $\mathbb{D}$ be a domain on the complex plane, which is symmetric about the real axis. Then, a matrix $A \in \mathbb{R}^{n \times n}$ is said to be $\mathbb{D}$-stable if

$$\lambda_i(A) \in \mathbb{D}, \quad i = 1, 2, \ldots, n. \tag{4.8}$$

### 4.2.1 Special Cases

In the following, we consider a few typical cases.

### Case I. $\mathbb{D} = \mathbb{H}_{\alpha, \beta} = \{x + jy \mid -\beta < x < -\alpha\}$

In this case, the $\mathbb{D}$-stability condition (4.8) requires that all the eigenvalues of the matrix $A$ are located in the strip region shown in Figure 4.1.

**Proposition 4.3**   The matrix $A \in \mathbb{R}^{n \times n}$ is $\mathbb{H}_{\alpha, \beta}$-stable if and only if there exists a matrix $P \in \mathbb{S}^n$ satisfying

$$\begin{cases} P > 0 \\ A^{\mathrm{T}}P + PA + 2\alpha P < 0 \\ A^{\mathrm{T}}P + PA + 2\beta P > 0. \end{cases} \tag{4.9}$$

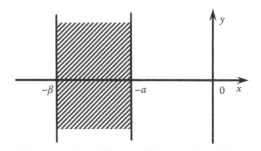

**Figure 4.1  Strip region** $\mathbb{H}_{\alpha,\beta}$**.**

*Proof*   The sufficiency holds obviously. Then, we only prove the necessity.

That the matrix $A \in \mathbb{R}^{n\times n}$ is $\mathbb{H}_{\alpha,\beta}$-stable is equivalent to the fact that both $(A+\alpha I)$ and $-(A+\beta I)$ are Hurwitz stable. According to Lyapunov stability theory, $(A+\alpha I)$ is Hurwitz stable if and only if for any $Q > 0$, there exists a $P_\alpha \in \mathbb{S}^n$, $P_\alpha > 0$, such that

$$(A+\alpha I)^{\mathrm{T}} P_\alpha + P_\alpha (A+\alpha I) = -Q,$$

which can be converted to

$$A^{\mathrm{T}} P_\alpha + P_\alpha A = -2\alpha P_\alpha - Q, \tag{4.10}$$

while $-(A+\beta I)$ is Hurwitz stable if and only if there exists a $P_\beta \in \mathbb{S}^n$, $P_\beta > 0$, such that

$$(-A - \beta I)^{\mathrm{T}} P_\beta + P_\beta (-A - \beta I) = -Q,$$

which can be converted to

$$A^{\mathrm{T}} P_\beta + P_\beta A = -2\beta P_\beta + Q. \tag{4.11}$$

Adding (4.10) and (4.11), we have

$$A^{\mathrm{T}} \left( P_\alpha + P_\beta \right) + \left( P_\alpha + P_\beta \right) A = -2\alpha P_\alpha - 2\beta P_\beta.$$

Letting

$$P = P_\alpha + P_\beta > 0,$$

yields

$$A^{\mathrm{T}} P + PA = -2\alpha P_\alpha - 2\beta P_\beta,$$

which means

$$A^{\mathrm{T}}P + PA + 2\alpha P = 2(\alpha - \beta)P_\beta,$$

and

$$A^{\mathrm{T}}P + PA + 2\beta P = 2(\beta - \alpha)P_\alpha.$$

Since $\alpha - \beta < 0$, then (4.9) holds obviously. With this we complete the proof of the result.  ■

When $\beta = \infty$, denote $\mathbb{H}_\alpha = \mathbb{H}_{\alpha,\infty} = \{x + jy \mid x < -\alpha\}$. In this case, the last inequality in (4.9) vanishes obviously.

**Case II.** $\mathbb{D} = \mathbb{D}_{(q,r)} = \{x + jy \mid (x + q)^2 + y^2 < r^2\}$

In this case, the $\mathbb{D}$-stability condition (4.8) requires that all the eigenvalues of the matrix $A$ are located in the circle shown in Figure 4.2.

**Proposition 4.4**  The matrix $A \in \mathbb{R}^{n \times n}$ is $\mathbb{D}_{(q,r)}$-stable if and only if there exists a matrix $P \in \mathbb{S}^n$, such that

$$\begin{bmatrix} -rP & qP + AP \\ qP + PA^{\mathrm{T}} & -rP \end{bmatrix} < 0. \tag{4.12}$$

*Proof*  The matrix $A \in \mathbb{R}^{n \times n}$ is $\mathbb{D}_{(q,r)}$-stable is equivalent to the fact that the matrix

$$\tilde{A} = \frac{1}{r}(A + qI)$$

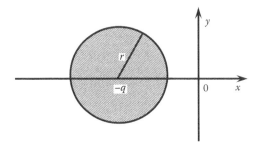

**Figure 4.2  Disk region $\mathbb{D}_{(q,r)}$.**

is Schur stable. Therefore, the problem has a solution if and only if there exists a matrix $P \in \mathbb{S}^n$, such that

$$\begin{cases} P > 0 \\ (A + qI)^{\mathrm{T}} P (A + qI) - r^2 P < 0. \end{cases}$$

By the well-known Schur complement lemma, this is easily shown to be further equivalent to the LMI in (4.12).  ■

**Remark 4.3**   In the case of $\mathbb{D} = \mathbb{H}_\alpha \cap \mathbb{D}_{(q,r)}$, both conditions (4.9) and (4.12) should be satisfied. Therefore, for this case, the matrix $A$ is $\mathbb{D}$-stable if and only if there exists a matrix $P \in \mathbb{S}^n$, such that

$$\begin{cases} (A + \alpha I)^{\mathrm{T}} P + P(A + \alpha I) < 0 \\ \begin{bmatrix} -rP & qP + AP \\ qP + PA^{\mathrm{T}} & -rP \end{bmatrix} < 0. \end{cases} \tag{4.13}$$

**Example 4.3**

Consider the $\mathbb{D}$-stability of the following linear system:

$$\dot{x}(t) = \begin{bmatrix} -4.2386 & -0.2026 & 0.7193 \\ 2.6649 & -2.8342 & 0.0175 \\ 0.0344 & 0.0005 & -3.1772 \end{bmatrix} x(t),$$

where $\mathbb{D} = \mathbb{H}_\alpha \cap \mathbb{D}_{(q,r)}$ with $\alpha = 2$, $q = 3$, and $r = 3$. Applying the MATLAB function feasp in the MATLAB LMI toolbox to the corresponding LMI problem (4.13) gives the following solution

$$P = \begin{bmatrix} 0.2046 & 0.0270 & 0.0211 \\ 0.0270 & 0.2376 & -0.0042 \\ 0.0211 & -0.0042 & 0.3462 \end{bmatrix} > 0.$$

Thus, we can conclude that the system is $\mathbb{D}$-stable. In fact, by the MATLAB function eig it can be easily verified that the system poles are $-3.5184$, $-3.4806$, and $-3.2510$, which are indeed located within the region $\mathbb{D} = \mathbb{H}_\alpha \cap \mathbb{D}_{(q,r)}$.

## 4.2.2 General LMI Regions

Besides the two types of regions, we have a more general type of regions, which are called LMI regions.

**Definition 4.2**    Let $\mathbb{D}$ be a region on the complex plane. If there exist matrices $L \in \mathbb{S}^m$, and $M \in \mathbb{R}^{m \times m}$ such that

$$\mathbb{D} = \{s \mid s \in \mathbb{C}, \ L + sM + \bar{s}M^\mathsf{T} < 0\}, \tag{4.14}$$

then $\mathbb{D}$ is called an LMI region and is usually denoted by $\mathbb{D}_{(L,M)}$,

$$F_{\mathbb{D}}(s) = L + sM + \bar{s}M^\mathsf{T}$$

is called the characteristic function of the LMI region $\mathbb{D}_{(L,M)}$.

The following proposition reveals a fundamental property of LMI regions.

**Proposition 4.5**    An LMI region is convex and symmetric about the real axis.

*Proof*    The symmetry property is obvious. To show the convexity, we choose arbitrarily two points $s_1, s_2 \in \mathbb{D}$, and define

$$s^* = \theta s_1 + (1 - \theta) s_2, \quad 0 \le \theta \le 1.$$

By definition, we have

$$L + s_1 M + \bar{s}_1 M^\mathsf{T} < 0,$$

and

$$L + s_2 M + \bar{s}_2 M^\mathsf{T} < 0.$$

Using these relations we can derive

$$
\begin{aligned}
L &+ s^* M + \bar{s}^* M^\mathsf{T} \\
&= L + (\theta s_1 + (1 - \theta) s_2) M + (\theta \bar{s}_1 + (1 - \theta) \bar{s}_2) M^\mathsf{T} \\
&= L + \theta s_1 M + \theta \bar{s}_1 M^\mathsf{T} + (1 - \theta) s_2 M + (1 - \theta) \bar{s}_2 M^\mathsf{T} \\
&= \theta \left( L + s_1 M + \bar{s}_1 M^\mathsf{T} \right) + (1 - \theta) \left( L + s_2 M + \bar{s}_2 M^\mathsf{T} \right) \\
&< 0, \quad \forall 0 \le \theta \le 1.
\end{aligned}
$$

This shows the convexity of the set $\mathbb{D}$ in (4.14). ∎

**Example 4.4**

The halfplane

$$\mathbb{H}_\alpha = \{x + jy \mid x < -\alpha < 0\}$$

is an LMI region because

$$\mathbb{H}_\alpha = \{x + jy \mid x < -\alpha\}$$
$$= \{s \mid \operatorname{Re}(s) < -\alpha\}$$
$$= \left\{ s \mid \frac{1}{2}(s + \bar{s}) < -\alpha \right\},$$

that is, $\mathbb{H}_\alpha$ is an LMI region with $L = 2\alpha$, $M = 1$. Thus, the characteristic function is

$$F_{\mathbb{H}_\alpha}(s) = 2\alpha + s + \bar{s}.$$

### Example 4.5

The strip

$$\mathbb{H}_{\alpha,\beta} = \{x + jy \mid -\beta < x < -\alpha < 0\}$$

is also an LMI region because

$$\mathbb{H}_{\alpha,\beta} = \{x + jy \mid -\beta < x < -\alpha\}$$
$$= \{s \mid -\beta < \operatorname{Re}(s) < -\alpha\}$$
$$= \left\{ s \mid -\beta < \frac{1}{2}(s + \bar{s}) < -\alpha \right\}$$
$$= \left\{ s \mid \frac{1}{2}\operatorname{diag}(s, -s) + \frac{1}{2}\operatorname{diag}(\bar{s}, -\bar{s}) + \operatorname{diag}(\alpha, -\beta) < 0 \right\},$$

that is, $\mathbb{H}_{\alpha,\beta}$ is an LMI region with

$$L = 2 \begin{bmatrix} \alpha & 0 \\ 0 & -\beta \end{bmatrix}, \quad M = \begin{bmatrix} 1 & 0 \\ 0 & -1 \end{bmatrix}.$$

Thus, the characteristic function is

$$F_{\mathbb{H}_{\alpha,\beta}}(s) = 2 \begin{bmatrix} \alpha & 0 \\ 0 & -\beta \end{bmatrix} + \begin{bmatrix} 1 & 0 \\ 0 & -1 \end{bmatrix} s + \begin{bmatrix} 1 & 0 \\ 0 & -1 \end{bmatrix} \bar{s}.$$

### Example 4.6

The disk

$$\mathbb{D}_{(q,r)} = \{x + jy \mid (x + q)^2 + y^2 < r^2\}$$

is an LMI region. In fact,

$$\mathbb{D}_{(q,r)} = \{x + jy \mid (x + q)^2 + y^2 < r^2\}$$
$$= \{s \mid (s + q)(\bar{s} + q) < r^2\}$$
$$= \left\{ s \mid \begin{bmatrix} -r & s + q \\ \bar{s} + q & -r \end{bmatrix} < 0 \right\}$$
$$= \left\{ s \mid \begin{bmatrix} -r & q \\ q & -r \end{bmatrix} + s \begin{bmatrix} 0 & 1 \\ 0 & 0 \end{bmatrix} + \bar{s} \begin{bmatrix} 0 & 1 \\ 0 & 0 \end{bmatrix}^{\mathsf{T}} < 0 \right\},$$

that is, $\mathbb{D}_{(q,r)}$ is an LMI region with

$$L = \begin{bmatrix} -r & q \\ q & -r \end{bmatrix}, \quad M = \begin{bmatrix} 0 & 1 \\ 0 & 0 \end{bmatrix}.$$

**Example 4.7**

The sector, shown in Figure 4.3,

$$\mathbb{D}_\theta = \{x + jy \mid |y| < -x \tan \theta\},$$

with $0 < \theta < \frac{\pi}{2}$, is an LMI region. In fact,

$$\mathbb{D}_\theta = \{x + jy \mid |y| < -x \tan \theta\}$$
$$= \{x + jy \mid y^2 < x^2 \tan^2 \theta, \quad x \tan \theta < 0\}$$
$$= \{x + jy \mid y^2 \cos^2 \theta < x^2 \sin^2 \theta, \quad x \sin \theta < 0\}$$

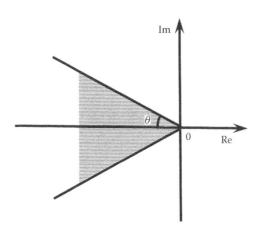

**Figure 4.3   Region $\mathbb{D}_\theta$.**

$$= \left\{ x + jy \left| \begin{bmatrix} x \sin \theta & iy \cos \theta \\ -iy \cos \theta & x \sin \theta \end{bmatrix} < 0 \right. \right\}$$

$$= \left\{ s \left| \begin{bmatrix} (s + \bar{s}) \sin \theta & (s - \bar{s}) \cos \theta \\ (-s + \bar{s}) \cos \theta & (s + \bar{s}) \sin \theta \end{bmatrix} < 0 \right. \right\},$$

that is, $\mathbb{D}_\theta$ is an LMI region with

$$L = 0, \quad M = \begin{bmatrix} \sin \theta & \cos \theta \\ -\cos \theta & \sin \theta \end{bmatrix}.$$

The following proposition further shows an important property of an LMI region.

**Proposition 4.6**   Let $\mathbb{D}_1$ and $\mathbb{D}_2$ be two LMI regions with characteristic functions $F_{\mathbb{D}_1}$ and $F_{\mathbb{D}_2}$, respectively. Then, $\mathbb{D} = \mathbb{D}_1 \cap \mathbb{D}_2$ is also an LMI region whose characteristic function is

$$F_{\mathbb{D}} = \text{diag}(F_{\mathbb{D}_1}, F_{\mathbb{D}_2}).$$

*Proof*   It follows from the given conditions that

$$\mathbb{D}_1 = \{s \mid F_{\mathbb{D}_1}(s) < 0\},$$
$$\mathbb{D}_2 = \{s \mid F_{\mathbb{D}_2}(s) < 0\},$$

and

$$\mathbb{D} = \mathbb{D}_1 \cap \mathbb{D}_2$$
$$= \{s \mid F_{\mathbb{D}_1}(s) < 0, \ F_{\mathbb{D}_2}(s) < 0\}$$
$$= \{s \mid \text{diag}(F_{\mathbb{D}_1}(s), F_{\mathbb{D}_2}(s)) < 0\}.$$

Thus, the conclusion clearly follows.                                          ■

### 4.2.3  Generalized Lyapunov Theorem

Recall that the Kronecker product of a pair of matrices $A \in \mathbb{R}^{m \times n}$ and $B \in \mathbb{R}^{p \times q}$ is defined as follows:

$$A \otimes B = \begin{bmatrix} a_{11}B & a_{12}B & \cdots & a_{1n}B \\ a_{21}B & a_{22}B & \cdots & a_{2n}B \\ \vdots & \vdots & \ddots & \vdots \\ a_{m1}B & a_{m2}B & \cdots & a_{mn}B \end{bmatrix} \in \mathbb{R}^{mp \times nq}.$$

The following lemma lists some manipulation rules related to Kronecker products.

**Lemma 4.1**   Let $A, B, C$ be matrices with appropriate dimensions. Then, the Kronecker product has the following properties:

- $1 \otimes A = A$;
- $(A + B) \otimes C = A \otimes C + B \otimes C$
- $(A \otimes B)(C \otimes D) = (AC) \otimes (BD)$
- $(A \otimes B)^{\mathrm{T}} = A^{\mathrm{T}} \otimes B^{\mathrm{T}}$
- $(A \otimes B)^{-1} = A^{-1} \otimes B^{-1}$
- $\lambda(A \otimes B) = \{\lambda_i(A)\lambda_j(B)\}$

In terms of Kronecker products, the following theorem gives the $\mathbb{D}$-stability condition for the general LMI region case.

**Theorem 4.1**   Let $\mathbb{D} = \mathbb{D}_{(L,M)}$ be an LMI region, whose characteristic function is

$$F_{\mathbb{D}} = L + sM + \bar{s}M^{\mathrm{T}}.$$

Then, a matrix $A \in \mathbb{R}^{n \times n}$ is $\mathbb{D}_{(L,M)}$-stable if and only if there exists a symmetric positive definite matrix $P$ such that

$$R_{\mathbb{D}}(A, P) = L \otimes P + M \otimes (AP) + M^{\mathrm{T}} \otimes (AP)^{\mathrm{T}} < 0, \qquad (4.15)$$

where $\otimes$ represents the Kronecker product.

The aforementioned theorem can be viewed as a generalization of the well-known continuous- and discrete-time Lyapunov theorems. We remark that this result is really amazing since it not only covers both the well-known continuous- and discrete-time Lyapunov theorems, but also gives truly essential generalization to both of them. For a proof of this theorem, please refer to Appendix A. Although the proof is a bit lengthy, we strongly encourage the reader to go it over since this result is so important.

According to the definition of the Kronecker product, the generalized Lyapunov inequality (4.15) can also be written in the following form:

$$R_{\mathbb{D}}(A, P) = \left[ l_{ij}P + m_{ij}(AP) + m_{ji}(AP)^{\mathrm{T}} \right]_{(\upsilon n) \times (\upsilon n)} < 0, \qquad (4.16)$$

where

- $l_{ij}$ and $m_{ij}$ are the $i$th row and $j$th column element in the matrices $L$ and $M$, respectively
- $\upsilon$ is the number of rows in $L$

**Lemma 4.2** Given two LMI regions $\mathbb{D}_1$ and $\mathbb{D}_2$, a matrix $A$ is both $\mathbb{D}_1$-stable and $\mathbb{D}_2$-stable if and only if there exists a positive definite matrix $P$, such that $R_{\mathbb{D}_1}(A, P) < 0$ and $R_{\mathbb{D}_2}(A, P) < 0$.

*Proof* Denote the associated characteristic functions of the LMI regions $\mathbb{D}_1$ and $\mathbb{D}_2$ respectively as

$$F_{\mathbb{D}_1} = L_1 + sM_1 + \bar{s}M_1^{\mathrm{T}},$$

and

$$F_{\mathbb{D}_2} = L_2 + sM_2 + \bar{s}M_2^{\mathrm{T}}.$$

According to Proposition 4.6, the characteristic function of the LMI region $\mathbb{D} = \mathbb{D}_1 \cap \mathbb{D}_2$ can be represented by

$$F_{\mathbb{D}} = \mathrm{diag}(F_{\mathbb{D}_1}, F_{\mathbb{D}_2}) = L + sM + \bar{s}M^{\mathrm{T}},$$

where

$$L = \mathrm{diag}(L_1, L_2), \quad M = \mathrm{diag}(M_1, M_2).$$

Then,

$$
\begin{aligned}
R_{\mathbb{D}}(A, P) &= L \otimes P + M \otimes (AP) + M^{\mathrm{T}} \otimes (AP)^{\mathrm{T}} \\
&= \mathrm{diag}(L_1, L_2) \otimes P + \mathrm{diag}(M_1, M_2) \otimes (AP) + \mathrm{diag}(M_1, M_2)^{\mathrm{T}} \otimes (AP)^{\mathrm{T}} \\
&= \mathrm{diag}\big(L_1 \otimes P + M_1 \otimes (AP) + M_1^{\mathrm{T}} \otimes (AP)^{\mathrm{T}}, \\
&\qquad\quad L_2 \otimes P + M_2 \otimes (AP) + M_2^{\mathrm{T}} \otimes (AP)^{\mathrm{T}}\big) \\
&= \mathrm{diag}\big(R_{\mathbb{D}_1}(A, P), R_{\mathbb{D}_2}(A, P)\big),
\end{aligned}
$$

which means

$$R_{\mathbb{D}}(A, P) < 0 \Leftrightarrow R_{\mathbb{D}_1}(A, P) < 0 \text{ and } R_{\mathbb{D}_2}(A, P) < 0.$$

Thus, the conclusion holds. ∎

### Example 4.8

Consider the following LMI region

$$\mathbb{S}(\alpha, r, \theta) = \big\{(x, y) \,\big|\, x < -\alpha < 0, \ |x + jy| < r, \ |y| < -x\tan\theta \big\} \qquad (4.17)$$

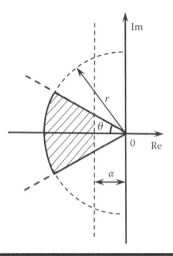

**Figure 4.4    LMI region $\mathbb{S}(\alpha, r, \theta)$.**

shown in Figure 4.4. We will show that a matrix $A$ is $\mathbb{S}(\alpha, r, \theta)$-stable if and only if there exists a symmetric matrix $P$, such that the following three LMIs are satisfied:

$$2\alpha P + AP + PA^\mathsf{T} < 0, \tag{4.18}$$

$$\begin{bmatrix} -rP & AP \\ PA^\mathsf{T} & -rP \end{bmatrix} < 0, \tag{4.19}$$

$$\begin{bmatrix} \left(AP + PA^\mathsf{T}\right)\sin\theta & \left(AP - PA^\mathsf{T}\right)\cos\theta \\ \left(PA^\mathsf{T} - AP\right)\cos\theta & \left(AP + PA^\mathsf{T}\right)\sin\theta \end{bmatrix} < 0. \tag{4.20}$$

This example has clearly a practical meaning. It is known that the transient response of a linear system is related to the location of its poles. By constraining poles to lie in a prescribed region, specific bounds can be put on these quantities to ensure a satisfactory transient response. Confining the closed-loop poles to $\mathbb{S}(\alpha, r, \theta)$ region ensures a minimum decay rate $\alpha$, a minimum damping ratio $\zeta = \cos\theta$, and a maximum undamped natural frequency $\omega_d = r\sin\theta$. This in turn bounds the maximum overshoot, the frequency of oscillatory modes, the delay time, the rise time, and the settling time.

Note that the region $\mathbb{S}(\alpha, r, \theta)$ in (4.17) can be represented in the intersection of three LMI regions, that is,

$$\mathbb{S}(\alpha, r, \theta) = \mathbb{H}_\alpha \cap \mathbb{D}_r \cap \mathbb{D}_\theta, \tag{4.21}$$

where their associated characteristic functions are

$$F_{\mathbb{H}_\alpha} = L_1 + sM_1 + \bar{s}M_1^\mathsf{T},$$

$$F_{\mathbb{D}_r} = L_2 + sM_2 + \bar{s}M_2^\mathsf{T},$$

and

$$F_{\mathbb{D}_\theta} = L_3 + sM_3 + \bar{s}M_3^\mathsf{T},$$

with

$$L_1 = 2\alpha, \quad M_1 = 1,$$

$$L_2 = \begin{bmatrix} -r & 0 \\ 0 & -r \end{bmatrix}, \quad M_2 = \begin{bmatrix} 0 & 1 \\ 0 & 0 \end{bmatrix},$$

$$L_3 = 0, \quad M_3 = \begin{bmatrix} \sin\theta & \cos\theta \\ -\cos\theta & \sin\theta \end{bmatrix}.$$

According to Proposition 4.6, the characteristic functions of LMI region $\mathbb{S}(\alpha, r, \theta)$ can be represented by

$$F_{\mathbb{S}(\alpha,r,\theta)} = \mathrm{diag}\left(F_{\mathbb{H}_\alpha}, F_{\mathbb{D}_r}, F_{\mathbb{D}_\theta}\right) = L + sM + \bar{s}M^\mathsf{T},$$

where

$$L = \mathrm{diag}(L_1, L_2, L_3), \quad M = \mathrm{diag}(M_1, M_2, M_3),$$

and it is easily shown that

$$R_{\mathbb{H}_\alpha}(A, P) = 2\alpha P + AP + PA^\mathsf{T},$$

$$R_{\mathbb{D}_r}(A, P) = \begin{bmatrix} -rP & AP \\ PA^\mathsf{T} & -rP \end{bmatrix},$$

and

$$R_{\mathbb{D}_\theta}(A, P) = \begin{bmatrix} \left(AP + PA^\mathsf{T}\right)\sin\theta & \left(AP - PA^\mathsf{T}\right)\cos\theta \\ \left(PA^\mathsf{T} - AP\right)\cos\theta & \left(AP + PA^\mathsf{T}\right)\sin\theta \end{bmatrix}.$$

Therefore, a matrix $A$ is $\mathbb{S}(\alpha, r, \theta)$-stable, according to Lemma 4.2, if and only if there exists a symmetric matrix $P$ satisfying the three LMIs (4.18) through (4.20).

## 4.3 Quadratic Stability

In this section, we treat the quadratic Hurwitz and Schur stability analysis problem. Different from the previous sections, here the stability of families of systems is considered instead of a single one.

### 4.3.1 Family of Systems

Consider the following uncertain system

$$\rho x(t) = A(\delta(t))x(t), \tag{4.22}$$

with $\rho$ being chosen as in one of the following two cases:

**Case 1**: $\rho$ represents the differential operator. In this case, the system is a continuous-time one, and possesses the following form:

$$\dot{x}(t) = A(\delta(t))x(t),$$

and here $t$ is a continuous variable.

**Case 2**: $\rho$ represents the one step forward shift operator. In this case, the system is a discrete-time one, and possesses the following form:

$$x(t+1) = A(\delta(t))x(t),$$

and here $t$ is a discrete variable.

The system coefficient matrix takes the form of

$$A(\delta(t)) = A_0 + \Delta A(\delta(t)), \tag{4.23}$$

where $A_0 \in \mathbb{R}^{n \times n}$ is a known matrix, which represents the nominal system matrix, while

$$\Delta A(\delta(t)) = \delta_1(t)A_1 + \delta_2(t)A_2 + \cdots + \delta_k(t)A_k, \tag{4.24}$$

is the system matrix perturbation, where

- $A_i \in \mathbb{R}^{n \times n}$, $i = 1, 2, \ldots, k$, are known matrices, which represent the perturbation directions.
- $\delta_i(t)$, $i = 1, 2, \ldots, k$, are arbitrary time functions, which represent the uncertain parameters in the system.
- $\delta(t) = \begin{bmatrix} \delta_1(t) & \delta_2(t) & \cdots & \delta_k(t) \end{bmatrix}^{\mathrm{T}}$ is the uncertain parameter vector, which is often assumed to be within a certain compact and convex set $\Delta$, that is

$$\delta(t) = \begin{bmatrix} \delta_1(t) & \delta_2(t) & \cdots & \delta_k(t) \end{bmatrix} \in \Delta, \tag{4.25}$$

and this set $\Delta$ is called the set of perturbation parameters.

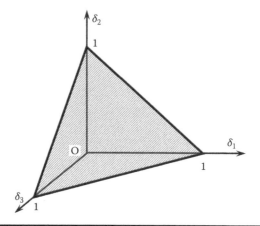

**Figure 4.5** Set Δ, case of $k = 3$.

In practical applications, often two types of perturbation parameters sets are used. One is the following regular polyhedron:

$$\Delta_I = \left\{ \delta(t) \, \big| \, \delta_i(t) \in [\delta_i^-, \, \delta_i^+], \quad i = 1, 2, \ldots, k \right\}. \tag{4.26}$$

In this case, the system (4.22) through (4.24) is called a type of interval systems. The other type of perturbation parameter set is the following polytope

$$\Delta_P = \left\{ \delta(t) \, \Bigg| \, \sum_{i=1}^{k} \delta_i(t) = 1, \quad \delta_i(t) \geq 0, \quad i = 1, 2, \ldots, k \right\}, \tag{4.27}$$

which is clearly a regular polyhedron in $\mathbb{R}^k$ (see Figure 4.5). In this case, the system (4.22) through (4.24) is called a type of polytopic systems.

### 4.3.2 Quadratic Hurwitz Stability

For a family of systems, the so-called quadratic Hurwitz stability is defined as follows.

**Definition 4.3**    The system (4.22) through (4.24), with $\forall \delta(t) \in \Delta$, is called quadratically Hurwitz stable if there exists a symmetric positive definite matrix $P$, such that

$$A^{\mathrm{T}}(\delta(t))P + PA(\delta(t)) < 0, \quad \forall \delta(t) \in \Delta. \tag{4.28}$$

The following theorem reveals the relation between quadratic Hurwitz stability and Lyapunov stability.

**Theorem 4.2** Let the system (4.22) through (4.24), with $\forall \delta(t) \in \Delta$, be quadratically Hurwitz stable. Then the system (4.22) through (4.24) is uniformly and asymptotically stable in the Lyapunov sense for every $\delta(t) \in \Delta$.

*Proof* If the system (4.22) through (4.24), with $\forall \delta(t) \in \Delta$, is quadratically Hurwitz stable, then (4.28) holds for some $P > 0$. It thus follows from (4.28) that $V(x) = x^T P x$ is a quadratic form Lyapunov function of system (4.22) through (4.24) and satisfies

$$\dot{V} = \frac{dV(x(t))}{dt}$$
$$= x^T \left[ A^T(\delta(t))P + PA(\delta(t)) \right] x$$
$$< 0.$$

Therefore, $V(x)$ is a monotone decreasing function and satisfies

$$V(x) \leq x^T(t_0) P x(t_0).$$

So $V(x)$ is bounded. On the other hand, since $P$ is a symmetric positive definite matrix, we have $V(x) = x^T P x > 0$. Finally, in view of the fact that $V(x)$ is independent of time $t$, it follows from the well-known Lyapunov stability theories that the system (4.22) through (4.24) is uniformly and asymptotically stable in the Lyapunov sense for every $\delta(t) \in \Delta$. ■

The following result reveals the fact that a family of systems is quadratically Hurwitz stable over a compact and convex set of parameters if and only if it is quadratically Hurwitz stable on all of the extremes of the set of parameters.

**Theorem 4.3** The system (4.22) through (4.24) is quadratically Hurwitz stable if and only if there exists a symmetric positive definite matrix $P$, such that

$$A^T(\delta)P + PA(\delta) < 0, \quad \forall \delta \in \Delta_E. \tag{4.29}$$

where $\Delta_E$ represents the set of all the extreme points of $\Delta$.

*Proof* Choose

$$V(x) = x^T P x,$$

then

$$\dot{V}(\delta(t)) = x^T \left[ A^T(\delta(t))P + PA(\delta(t)) \right] x.$$

It is easy to see that $\dot{V}(\delta(t))$ is linear in $\delta(t)$, and thus convex with respect to $\delta(t)$. Thus, it follows from Theorem 3.5 that for $\forall x \neq 0$,

$$\dot{V}(\delta(t)) < 0, \quad \forall \delta \in \Delta \Longleftrightarrow \dot{V}(\delta(t)) < 0, \quad \forall \delta \in \Delta_E.$$

This equivalently implies that (4.28) holds if and only if (4.29) holds. The proof is then completed. ■

In view of the fact that, when $\Delta = \Delta_I$, we have

$$\Delta_E = \left\{ \delta = \begin{bmatrix} \delta_1 & \delta_2 & \cdots & \delta_k \end{bmatrix} \middle| \delta_i = \delta_i^- \text{ or } \delta_i^+, \ i = 1, 2, \ldots, k \right\}.$$

Thus, the following result for the quadratic Hurwitz stability of interval systems immediately follows from the Theorem 4.3.

**Corollary 4.1** The system (4.22) through (4.24), with $\Delta = \Delta_I$ given by (4.26), is quadratically Hurwitz stable if and only if there exists a symmetric positive definite matrix $P$, such that

$$A^T(\delta)P + PA(\delta) < 0, \quad \forall \delta_i = \delta_i^- \text{ or } \delta_i^+, \quad i = 1, 2, \ldots, k. \tag{4.30}$$

When $\Delta = \Delta_P$, it is easy to see, referring to Figure 4.5, that

$$\Delta_E = \left\{ e_1^T, e_2^T, \ldots, e_k^T \right\} \subset \mathbb{R}^k, \tag{4.31}$$

where $e_i \in \mathbb{R}^k$ is the vector with the $i$th element being 1 while all the others being zero. Further, recalling

$$A(\delta(t)) = A_0 + \delta_1(t)A_1 + \cdots + \delta_k(t)A_k,$$

we clearly have

$$A(e_i) = A_0 + A_i, \quad i = 1, 2, \ldots, k.$$

Again following from Theorem 4.3, we now have the following result for the stability of polytopic systems.

**Corollary 4.2** The system (4.22) through (4.24), with $\Delta = \Delta_P$ given by (4.27), is quadratically Hurwitz stable if and only if there exists a symmetric positive definite matrix $P$, such that

$$(A_0 + A_i)^T P + P(A_0 + A_i) < 0, \quad i = 1, \ldots, k. \tag{4.32}$$

The importance of the two corollaries lies in the fact that an infinite problem is now converted into a finite problem.

**Example 4.9**

Consider the following continuous-time linear system:

$$\dot{x}(t) = A(\delta(t))x(t),$$

with

$$A(\delta(t)) = A_0 + \delta_1(t)A_1,$$

where

$$A_0 = \begin{bmatrix} -3 & 0 & 1 \\ 0 & -3 & 0 \\ 1 & 0 & -4 \end{bmatrix}, \quad A_1 = \begin{bmatrix} 1 & 0 & 1 \\ 0 & -1 & 0 \\ 0 & 1 & 1 \end{bmatrix},$$

and

$$\delta_1(t) \in [-0.1, 0.1].$$

Applying the MATLAB function `feasp` in the MATLAB LMI toolbox to the corresponding LMI problem (4.30) gives the solution

$$P = \begin{bmatrix} 0.2623 & -0.0000 & 0.0677 \\ -0.0000 & 0.2393 & 0.0001 \\ 0.0677 & 0.0001 & 0.1944 \end{bmatrix} > 0.$$

Thus, the system is quadratically Hurwitz stable.

**Example 4.10**

Consider a mass-spring-damper system with unit mass. The state–space representation of the system is (Gahinet et al. 1996, Yu 2002)

$$\dot{x}(t) = A\left(d(t), k(t)\right)x(t), \quad A(d, k) = \begin{bmatrix} 0 & 1 \\ -k(t) & -d(t) \end{bmatrix}, \tag{4.33}$$

where the viscosity $d(t)$ and stiffness $k(t)$ are uncertain and possibly time-varying, and their values are assumed to be within certain intervals:

$$k(t) \in [k_1, k_2], \quad d(t) \in [d_1, d_2].$$

According to Corollary 4.1, the system is quadratically stable if and only if there exists a symmetric positive definite matrix $P$ satisfying

$$A^{\mathsf{T}}(d, k)P + PA(d, k) < 0, \tag{4.34}$$

for

$$\begin{cases} (d, k) = (d_1, k_1) \\ (d, k) = (d_2, k_1) \\ (d, k) = (d_1, k_2) \\ (d, k) = (d_2, k_2). \end{cases} \tag{4.35}$$

Thus, to verify the quadratic stability of the system, we need only to solve the feasibility problem (4.34) and (4.35). And to do this, the function `feasp` in the MATLAB LMI toolbox can be readily used.

For the case of

$$k(t) \in [9.51, 12.49], \quad d(t) \in [0.451, 0.749],$$

using `feasp` in the MATLAB LMI toolbox, we obtain a feasible solution

$$P = \begin{bmatrix} 26.7632 & 0.5439 \\ 0.5439 & 2.4332 \end{bmatrix} > 0.$$

Thus the system is quadratically stable according to Corollary 4.1.

Unfortunately, for the case of

$$k(t) \in [4, 8], \quad d(t) \in [0.1, 0.3],$$

using `feasp` in the MATLAB LMI toolbox, we failed to find a feasible solution to (4.34) and (4.35). Therefore, in this case, the system is not quadratically stable.

**Remark 4.4** When the coefficients $k$ and $d$ are constants, it is well known that the system (4.33) is asymptotically stable for arbitrary $k > 0$ and $d > 0$, and hence for $k \in [4, 8]$, $d \in [0.1, 0.3]$. However, it can be seen that the system with $k \in [4, 8]$ and $d \in [0.1, 0.3]$ is not quadratically stable. Such a fact demonstrates that the converse of Theorem 4.2 is not true.

### 4.3.3 Quadratic Schur Stability

In this subsection, we treat the quadratic Schur stability analysis problem, which is concerned with the stability of the uncertain discrete-time linear system (4.22) through (4.24). For this family of systems, the so-called quadratic Schur stability is defined as follows.

**Definition 4.4**  The system (4.22) through (4.24), with $\forall \delta(t) \in \Delta$, is called quadratically Schur stable if there exists a symmetric positive definite matrix $P$, such that,

$$A(\delta(t))PA^{\mathrm{T}}(\delta(t)) - P < 0, \quad \forall \delta(t) \in \Delta. \tag{4.36}$$

The following theorem reveals the relation between quadratic Schur stability and Schur stability.

**Theorem 4.4**  Let the system (4.22) through (4.24), with $\forall \delta(t) \in \Delta$, be quadratically Schur stable. Then, the system (4.22) through (4.24) is Schur stable for every $\delta(t) \in \Delta$.

*Proof*  If the system (4.22) through (4.24), with $\forall \delta(t) \in \Delta$, is quadratically Schur stable, then (4.36) holds for some $P > 0$, which means, according to Proposition 4.2, that the system (4.22) through (4.24) is Schur stable for every $\delta(t) \in \Delta$.  ∎

The following result reveals the fact that a family of systems is quadratically Schur stable over a compact and convex set of parameters if and only if it is quadratically Schur stable on all of the extremes of the set of parameters.

**Theorem 4.5**  The system (4.22) through (4.24) is quadratically Schur stable if and only if there exists a symmetric matrix $P$, such that one of the following conditions holds:

$$\begin{cases} P > 0 \\ A(\delta(t))PA^{\mathrm{T}}(\delta(t)) - P < 0, \quad \forall \delta(t) \in \Delta_E, \end{cases} \tag{4.37}$$

$$\begin{bmatrix} -P & A(\delta(t))P \\ PA^{\mathrm{T}}(\delta(t)) & -P \end{bmatrix} < 0, \quad \forall \delta(t) \in \Delta_E, \tag{4.38}$$

$$\begin{bmatrix} -P & PA^{\mathrm{T}}(\delta(t)) \\ A(\delta(t))P & -P \end{bmatrix} < 0, \quad \forall \delta(t) \in \Delta_E. \tag{4.39}$$

*Proof*  It is easy to see that the LMI (4.36) is equivalent to, according to the Schur complement lemma,

$$\begin{bmatrix} -P & A(\delta(t))P \\ PA^{\mathrm{T}}(\delta(t)) & -P \end{bmatrix} < 0, \quad \forall \delta(t) \in \Delta, \tag{4.40}$$

which is linear in $\delta(t)$, and thus convex with respect to $\delta(t)$. Thus, it follows from Theorem 3.5 that (4.40) is equivalent to (4.38).

Using the Schur complement of the element $-P$ (the (1,1) position) in the matrix appearing on the left-hand side of (4.38) we can show the equivalence between (4.37) and (4.38). Similarly, using the Schur complement of the element $-P$ (the (2,2) position) in the matrix appearing on the left-hand side of (4.39) we can show the equivalence between (4.37) and (4.39). Then the conclusion holds.   ▪

The following result for the quadratic Schur stability of interval systems immediately follows from Theorem 4.5.

**Corollary 4.3**  The system (4.22) through (4.24), with $\Delta = \Delta_I$ given by (4.26), is quadratically Schur stable if and only if there exists a symmetric matrix $P$, such that one of the following conditions holds for $\delta_i = \delta_i^-$ or $\delta_i^+$, $i = 1, 2, \ldots, k$:

$$\begin{cases} P > 0 \\ A(\delta)PA^{\mathrm{T}}(\delta) - P < 0, \end{cases}$$

$$\begin{bmatrix} -P & A(\delta)P \\ PA^{\mathrm{T}}(\delta) & -P \end{bmatrix} < 0, \tag{4.41}$$

$$\begin{bmatrix} -P & PA^{\mathrm{T}}(\delta) \\ A(\delta)P & -P \end{bmatrix} < 0.$$

Again following from Theorem 4.5 we now have the following result for the quadratic Schur stability of polytopic systems.

**Corollary 4.4**  The system (4.22) through (4.24), with $\Delta = \Delta_P$ given by (4.27), is quadratically Schur stable if and only if there exists a symmetric matrix $P$, such that one of the following conditions holds

$$\begin{cases} P > 0 \\ (A_0 + A_i)P(A_0 + A_i)^{\mathrm{T}} - P < 0, & i = 1, \ldots, k, \end{cases}$$

$$\begin{bmatrix} -P & (A_0 + A_i)P \\ P(A_0 + A_i)^{\mathrm{T}} & -P \end{bmatrix} < 0, \quad i = 1, \ldots, k, \tag{4.42}$$

$$\begin{bmatrix} -P & P(A_0 + A_i)^{\mathrm{T}} \\ (A_0 + A_i)P & -P \end{bmatrix} < 0, \quad i = 1, \ldots, k.$$

The importance of the two corollaries lies in the fact that an infinite problem is now converted into a finite problem.

**Example 4.11**

Consider the following discrete-time linear system

$$x(k + 1) = A(\delta(k))x(k),$$

with

$$A(\delta(k)) = A_0 + \delta_1(k)A_1 + \delta_2(k)A_2,$$

where

$$A_0 = \begin{bmatrix} 9 & 6 & -6 \\ -5 & -2 & 2.5 \\ 7 & 7 & 6.5 \end{bmatrix}, \quad A_1 = \begin{bmatrix} 0 & -1 & 1 \\ -2 & -1 & 1 \\ -4 & -4 & 4 \end{bmatrix}, \quad A_2 = \begin{bmatrix} -2 & -1 & 1 \\ 2 & 1 & -1 \\ 0 & 0 & 0 \end{bmatrix}$$

and

$$\delta_1(k) \in [1.2, 1.8], \quad \delta_2(k) \in [4.2, 4.8].$$

Applying the MATLAB function `feasp` in the MATLAB LMI toolbox to the corresponding LMI problem (4.41) gives the solution

$$P = \begin{bmatrix} 2.3466 & -1.5238 & 1.1474 \\ -1.5238 & 3.3631 & 1.8344 \\ 1.1474 & 1.8344 & 4.4675 \end{bmatrix} > 0.$$

Thus, the system is quadratically Schur stable.

**Example 4.12**

Consider the following discrete-time linear system

$$x(k + 1) = A(\delta(k))x(k),$$

with

$$A(\delta(k)) = A_0 + \delta_1(k)A_1 + \delta_2(k)A_2,$$

where

$$A_0 = \begin{bmatrix} -3.5 & 1.2 \\ -9 & 3.1 \end{bmatrix}, \quad A_1 = \begin{bmatrix} 5 & -2 \\ 15 & -6 \end{bmatrix}, \quad A_2 = \begin{bmatrix} 6 & -2 \\ 15 & -5 \end{bmatrix}$$

and

$$\delta_1(k) \geq 0, \quad \delta_2(k) \geq 0, \quad \delta_1(k) + \delta_2(k) = 1.$$

Applying the MATLAB function `feasp` in the MATLAB LMI toolbox to the corresponding LMI problem (4.42) gives the solution

$$P = \begin{bmatrix} 35.4036 & -12.5823 \\ -12.5823 & 4.6416 \end{bmatrix} > 0.$$

Thus, the system is quadratically Schur stable.

## 4.4 Quadratic $\mathbb{D}$-Stability

In this section, we continue to investigate the stability of the family of systems in (4.22) through (4.24). For families of systems, the so-called quadratic $\mathbb{D}$-stability is introduced, which is indeed a combination of $\mathbb{D}$-stability and quadratic stability. As in the previous sections, we aim to establish LMI conditions for quadratic $\mathbb{D}$-stability of families of systems with different $\mathbb{D}$ regions.

### 4.4.1 Definition and Main Results

In view of Theorem 4.1, we can introduce for families of systems the so-called quadratic $\mathbb{D}$-stability as follows.

**Definition 4.5**    Let $\mathbb{D} = \mathbb{D}_{(L,M)}$ be an LMI region, whose characteristic function is

$$F_{\mathbb{D}} = L + sM + \bar{s}M^{\mathrm{T}}.$$

The system (4.22) through (4.24), with $\forall \delta(t) \in \Delta$, is called quadratically $\mathbb{D}_{(L,M)}$-stable if there exists a symmetric positive definite matrix $P$, such that

$$R_{\mathbb{D}}(A(\delta(t)), P) = L \otimes P + M \otimes (A(\delta(t))P)$$

$$+ M^{\mathrm{T}} \otimes (A(\delta(t))P)^{\mathrm{T}} < 0, \quad \forall \delta(t) \in \Delta. \qquad (4.43)$$

The following theorem reveals the relation between quadratic $\mathbb{D}$-stability and $\mathbb{D}$-stability.

**Theorem 4.6**    Let the system (4.22) through (4.24), with $\forall \delta(t) \in \Delta$, be quadratically $\mathbb{D}$-stable. Then, the system (4.22) through (4.24) is $\mathbb{D}$-stable for every $\delta(t) \in \Delta$.

*Proof*    If the system (4.22) through (4.24), with $\forall \delta(t) \in \Delta$, is quadratically $\mathbb{D}$-stable, then (4.43) holds for some $P > 0$. It thus follows from Theorem 4.1 that the system (4.22) through (4.24) is $\mathbb{D}$-stable for every $\delta(t) \in \Delta$. ∎

The following result reveals the fact that a family of systems is quadratically $\mathbb{D}$-stable over a compact and convex set of parameters if and only if it is $\mathbb{D}$-stable on all of the extremes of the set of parameters.

**Theorem 4.7**   The system (4.22) through (4.24) is quadratically $\mathbb{D}_{(L,M)}$-stable if and only if there exists a symmetric positive definite matrix $P$, such that

$$R_{\mathbb{D}}(A(\delta(t)), P) = L \otimes P + M \otimes (A(\delta(t))P)$$

$$+ M^{\mathrm{T}} \otimes (A(\delta(t))P)^{\mathrm{T}} < 0, \quad \delta(t) \in \Delta_E. \tag{4.44}$$

*Proof*   It is easy to see that $R_{\mathbb{D}}(A(\delta(t)), P)$ is linear in $\delta(t)$, and thus convex with respect to $\delta(t)$. Further recalling that $\Delta$ is a compact and convex set, thus it follows from Theorem 3.5 that for matrix $P$ (4.43) holds if and only if (4.44) holds. Then, the proof is then completed.  ∎

**Lemma 4.3**   Given the system (4.22) through (4.24), with $\forall \delta(t) \in \Delta$, and two LMI regions $\mathbb{D}_1$ and $\mathbb{D}_2$. Then, the system (4.22) through (4.24) is both quadratically $\mathbb{D}_1$-stable and quadratically $\mathbb{D}_2$-stable if and only if there exists a positive definite matrix $P$, such that

$$R_{\mathbb{D}_1}(A(\delta(t)), P) < 0 \quad \text{and} \quad R_{\mathbb{D}_2}(A(\delta(t)), P) < 0, \quad \delta(t) \in \Delta.$$

*Proof*   Denote the associated characteristic functions of the LMI regions $\mathbb{D}_1$ and $\mathbb{D}_2$ as

$$F_{\mathbb{D}_1} = L_1 + sM_1 + \bar{s}M_1^{\mathrm{T}},$$

and

$$F_{\mathbb{D}_2} = L_2 + sM_2 + \bar{s}M_2^{\mathrm{T}},$$

respectively. According to Proposition 4.6, the characteristic function of the LMI region $\mathbb{D} = \mathbb{D}_1 \cap \mathbb{D}_2$ can be represented by

$$F_{\mathbb{D}} = \mathrm{diag}(F_{\mathbb{D}_1}, F_{\mathbb{D}_2}) = L + sM + \bar{s}M^{\mathrm{T}},$$

where

$$L = \mathrm{diag}(L_1, L_2), \quad M = \mathrm{diag}(M_1, M_2).$$

Then,

$$R_{\mathbb{D}}(A(\delta(t)), P) = L \otimes P + M \otimes (A(\delta(t))P) + M^{\mathrm{T}} \otimes (A(\delta(t))P)^{\mathrm{T}}$$

$$= \mathrm{diag}(L_1 \otimes P + M_1 \otimes (A(\delta(t))P) + M_1^{\mathrm{T}} \otimes (A(\delta(t))P)^{\mathrm{T}},$$

$$L_2 \otimes P + M_2 \otimes (A(\delta(t))P) + M_2^{\mathrm{T}} \otimes (A(\delta(t))P)^{\mathrm{T}})$$

$$= \mathrm{diag}(R_{\mathbb{D}_1}(A(\delta(t)), P), R_{\mathbb{D}_2}(A(\delta(t)), P)),$$

which means

$$R_{\mathbb{D}}(A(\delta(t)), P) < 0 \Leftrightarrow R_{\mathbb{D}_1}(A(\delta(t)), P) < 0 \quad \text{and} \quad R_{\mathbb{D}_2}(A(\delta(t)), P) < 0.$$

Thus, the conclusion holds. ∎

The following result for quadratic $\mathbb{D}$-stability of interval systems immediately follows from Theorem 4.7.

**Corollary 4.5** The system (4.22) through (4.24), with $\Delta = \Delta_I$ given by (4.26), is quadratically $\mathbb{D}_{(L,M)}$-stable if and only if there exists a symmetric positive definite matrix $P$, such that

$$L \otimes P + M \otimes (A(\delta)P) + M^{\mathrm{T}} \otimes (A(\delta)P)^{\mathrm{T}} < 0,$$

$$\delta_i = \delta_i^- \text{ or } \delta_i^+, \quad i = 1, 2, \ldots, k. \tag{4.45}$$

When $\Delta = \Delta_P$, following from Theorem 4.7, we now have the following result for the stability of polytopic systems.

**Corollary 4.6** The system (4.22) through (4.24), with $\Delta = \Delta_P$ given by (4.27), is quadratically $\mathbb{D}_{(L,M)}$-stable if and only if there exists a symmetric positive definite matrix $P$, such that

$$L \otimes P + M \otimes ((A_0 + A_i) P) + M^{\mathrm{T}} \otimes ((A_0 + A_i) P)^{\mathrm{T}} < 0, \quad i = 1, \ldots, k. \tag{4.46}$$

**Example 4.13**

Consider again the system in Example 4.9 and the following LMI region

$$\mathbb{H}_{1,5} = \{x + jy \mid -5 < x < -1 < 0\}.$$

It is easy to find that the characteristic function of $\mathbb{H}_{1,5}$ is

$$F_{\mathbb{H}_{1,5}}(s) = L + sM + \bar{s}M^{\mathrm{T}}$$

with

$$L = 2 \begin{bmatrix} 1 & 0 \\ 0 & -5 \end{bmatrix}, \quad M = \begin{bmatrix} 1 & 0 \\ 0 & -1 \end{bmatrix}.$$

Applying the MATLAB function `feasp` in the MATLAB LMI toolbox to the corresponding LMI problem (4.45) gives the solution

$$P = \begin{bmatrix} 0.4183 & -0.0003 & 0.0680 \\ -0.0003 & 0.5037 & -0.0004 \\ 0.0680 & -0.0004 & 0.3489 \end{bmatrix} > 0.$$

Thus, the system is quadratically $\mathbb{H}_{1,5}$-stable.

**Example 4.14**

Consider the following LMI region

$$\mathbb{D}_{(4,8)} = \{x + jy \mid (x+4)^2 + y^2 < 8^2\},$$

and a system in the form of (4.22) through (4.24), with $\Delta = \Delta_P$ given by (4.27), where

$$A_0 = \begin{bmatrix} -2.1333 & -1.3333 \\ -0.6667 & -1.4667 \end{bmatrix}, \quad A_1 = \begin{bmatrix} 0.2 & 0.1 \\ 0.1 & 0.2 \end{bmatrix}, \quad A_2 = \begin{bmatrix} 0.1 & 0.2 \\ 0.2 & 0.1 \end{bmatrix}.$$

Further, let us assume that the set of perturbation parameters form a polytope in the form of (4.27), with $k = 2$.

It is easy to find that the characteristic function of $\mathbb{D}_{(4,8)}$ is

$$F_{\mathbb{D}_{4,8}}(s) = L + sM + \bar{s}M^\mathsf{T}$$

with

$$L = \begin{bmatrix} -8 & 4 \\ 4 & -8 \end{bmatrix}, \quad M = \begin{bmatrix} 0 & 1 \\ 0 & 0 \end{bmatrix}.$$

Applying the MATLAB function `feasp` in the MATLAB LMI toolbox to the corresponding LMI problem (4.46) gives the solution

$$P = \begin{bmatrix} 0.0945 & 0.0091 \\ 0.0091 & 0.0862 \end{bmatrix} > 0.$$

Thus, the system is quadratically $\mathbb{D}_{(4,8)}$-stable.

## *4.4.2 Some Special Cases*

In this subsection, we further examine the quadratic $\mathbb{D}$-stability of system (4.22) through (4.24), with $\forall \delta(t) \in \Delta$, and $\mathbb{D}$ being some special LMI regions.

**Corollary 4.7**   Given the system (4.22) through (4.24), with $\forall \delta(t) \in \Delta$, and the LMI region

$$\mathbb{H}_\alpha = \{x + jy \mid x < -\alpha\},$$

the system (4.22) through (4.24) is quadratically $\mathbb{H}_\alpha$-stable if and only if there exists a positive definite matrix $P$, such that

$$2\alpha P + A\left(\delta(t)\right) P + PA^{\mathrm{T}}\left(\delta(t)\right) < 0, \quad \forall \delta(t) \in \Delta_E.$$

*Proof*   For the $\mathbb{H}_\alpha$ region, it has been shown in the earlier equations that the associated matrices in its characteristic function are

$$L = 2\alpha, \quad M = 1,$$

thus

$$R_{\mathbb{H}_\alpha}\left(A(\delta(t)), P\right) = 2\alpha P + A(\delta(t))P + PA^{\mathrm{T}}(\delta(t)).$$

Then, the conclusion holds according to Theorem 4.7.   ▪

**Corollary 4.8**   Given the system (4.22) through (4.24), with $\forall \delta(t) \in \Delta$, and the LMI region

$$\mathbb{D}_{(q,r)} = \{x + jy \mid (x+q)^2 + y^2 < r^2\},$$

the system (4.22) through (4.24) is quadratically $\mathbb{D}_{(q,r)}$-stable if and only if there exists a symmetric matrix $P$, such that

$$\begin{bmatrix} -rP & qP + A(\delta(t))P \\ qP + PA^{\mathrm{T}}(\delta(t)) & -rP \end{bmatrix} < 0, \quad \forall \delta(t) \in \Delta_E.$$

*Proof*   For the $\mathbb{D}_{(q,r)}$ region, it has been shown earlier that the associated matrices in its characteristic function are

$$L = \begin{bmatrix} -r & q \\ q & -r \end{bmatrix}, \quad M = \begin{bmatrix} 0 & 1 \\ 0 & 0 \end{bmatrix},$$

thus

$$R_{\mathbb{D}_{(q,r)}}\left(A(\delta(t)), P\right) = \begin{bmatrix} -rP & qP + A(\delta(t))P \\ qP + PA^{\mathrm{T}}(\delta(t)) & -rP \end{bmatrix}.$$

Then, the conclusion holds according to Theorem 4.7. ■

**Corollary 4.9** Given the system (4.22) through (4.24), with $\forall \delta(t) \in \Delta$, and the LMI region

$$\mathbb{D}_\theta = \{x + jy \mid |y| < -x \tan \theta\},$$

the system (4.22) through (4.24) is quadratically $\mathbb{D}_\theta$-stable if and only if there exists a positive definite matrix $P$, such that

$$\begin{bmatrix} \left(A(\delta(t))P + PA^{\mathrm{T}}(\delta(t))\right)\sin\theta & \left(A(\delta(t))P - PA^{\mathrm{T}}(\delta(t))\right)\cos\theta \\ \left(PA^{\mathrm{T}}(\delta(t)) - A(\delta(t))P\right)\cos\theta & \left(A(\delta(t))P + PA^{\mathrm{T}}(\delta(t))\right)\sin\theta \end{bmatrix} < 0,$$

$$\forall \delta(t) \in \Delta_E.$$

*Proof* For the $\mathbb{D}_\theta$ region, it has been shown earlier that the associated matrices in its characteristic function are

$$L = 0, \quad M = \begin{bmatrix} \sin\theta & \cos\theta \\ -\cos\theta & \sin\theta \end{bmatrix},$$

thus

$$
\begin{aligned}
& R_{\mathbb{D}_\theta}\left(A(\delta(t)), P\right) \\
&= M \otimes (A(\delta(t))P) + M^{\mathrm{T}} \otimes (A(\delta(t))P)^{\mathrm{T}} \\
&= \begin{bmatrix} \sin\theta & \cos\theta \\ -\cos\theta & \sin\theta \end{bmatrix} \otimes (A(\delta(t))P) + \begin{bmatrix} \sin\theta & -\cos\theta \\ \cos\theta & \sin\theta \end{bmatrix} \otimes (A(\delta(t))P)^{\mathrm{T}} \\
&= \begin{bmatrix} \left(A(\delta(t))P + PA^{\mathrm{T}}(\delta(t))\right)\sin\theta & \left(A(\delta(t))P - PA^{\mathrm{T}}(\delta(t))\right)\cos\theta \\ \left(PA^{\mathrm{T}}(\delta(t)) - A(\delta(t))P\right)\cos\theta & \left(A(\delta(t))P + PA^{\mathrm{T}}(\delta(t))\right)\sin\theta \end{bmatrix}.
\end{aligned}
$$

Then, the conclusion holds according to Theorem 4.7. ■

**Corollary 4.10** Given the system (4.22) through (4.24), with $\forall \delta(t) \in \Delta$, and the LMI region $\mathbb{S}(\alpha, r, \theta)$ defined in (4.21), the system (4.22) through (4.24) is quadratic $\mathbb{S}(\alpha, r, \theta)$-stable if and only if there exists a symmetric matrix $P$, such that

$$2\alpha P + A(\delta(t))P + PA^{\mathrm{T}}(\delta(t)) < 0, \quad \forall \delta(t) \in \Delta_E,$$

$$\begin{bmatrix} -rP & A(\delta(t))P \\ PA^{\mathrm{T}}(\delta(t)) & -rP \end{bmatrix} < 0, \quad \forall \delta(t) \in \Delta_E,$$

$$\begin{bmatrix} \big(A(\delta(t))P + PA^{\mathrm{T}}(\delta(t))\big) \sin\theta & \big(A(\delta(t))P - PA^{\mathrm{T}}(\delta(t))\big) \cos\theta \\ \big(PA^{\mathrm{T}}(\delta(t)) - A(\delta(t))P\big) \cos\theta & \big(A(\delta(t))P + PA^{\mathrm{T}}(\delta(t))\big) \sin\theta \end{bmatrix} < 0,$$

$$\forall \delta(t) \in \Delta_E.$$

*Proof*   The conclusion directly follows from Theorem 4.3 and Corollaries (4.7) through (4.9).   ■

## 4.5  Time-Delay Systems

The problem of stability analysis for a time-delay system can be stated as follows.

**Problem 4.3**   Given matrices $A, A_d \in \mathbb{R}^{n \times n}$, check the stability of the following linear time-delay system

$$\begin{cases} \dot{x}(t) = Ax(t) + A_d x(t-d) \\ x(t) = \phi(t), \ t \in [-d, 0], \ 0 < d \leq \bar{d}, \end{cases} \tag{4.47}$$

where
   $\phi(t)$ is the initial condition
   $d$ represents the time-delay
   $\bar{d}$ is a known upper bound of $d$

### 4.5.1  Delay-Independent Condition

The following theorem gives a sufficient condition for the stability problem in terms of an LMI.

**Theorem 4.8**   The system (4.47) is asymptotically stable if there exist two symmetric matrices $P, S \in \mathbb{S}^n$, such that

$$\begin{cases} P > 0 \\ \begin{bmatrix} A^{\mathrm{T}}P + PA + S & PA_d \\ A_d^{\mathrm{T}}P & -S \end{bmatrix} < 0. \end{cases} \tag{4.48}$$

*Proof* Let (4.48) hold, then both $P$ and $S$ are positive definite. Therefore, we can define the following Lyapunov function

$$V(x_t) = x^{\mathrm{T}}(t)\,Px(t) + \int_{t-d}^{t} x^{\mathrm{T}}(\tau)\,Sx(\tau)\,d\tau,$$

with

$$x_t = x(t+\theta), \quad \theta \in [-d \ \ 0].$$

Again using (4.48), we have

$$
\begin{aligned}
\dot{V}(x_t) &= x^{\mathrm{T}}(t)\,A^{\mathrm{T}}Px(t) + x^{\mathrm{T}}(t-d)\,A_d^{\mathrm{T}}Px(t) \\
&\quad + x^{\mathrm{T}}(t)\,PAx(t) + x^{\mathrm{T}}(t)\,PA_d x(t-d) \\
&\quad + x^{\mathrm{T}}(t)\,Sx(t) - x^{\mathrm{T}}(t-d)\,Sx(t-d) \\
&= x^{\mathrm{T}}(t)\left(A^{\mathrm{T}}P + PA + S\right)x(t) + x^{\mathrm{T}}(t)\,PA_d x(t-d) \\
&\quad + x^{\mathrm{T}}(t-d)\,A_d^{\mathrm{T}}Px(t) - x^{\mathrm{T}}(t-d)\,Sx(t-d) \\
&= \begin{bmatrix} x(t) \\ x(t-d) \end{bmatrix}^{\mathrm{T}} \begin{bmatrix} A^{\mathrm{T}}P + PA + S & PA_d \\ A_d^{\mathrm{T}}P & -S \end{bmatrix} \begin{bmatrix} x(t) \\ x(t-d) \end{bmatrix} \\
&< 0, \quad \forall x(t) \neq 0, \ t \geq 0.
\end{aligned}
$$

The conclusion clearly follows from Lyapunov stability theory. ▪

**Remark 4.5** By the Schur complement lemma, the second inequality in (4.48) is clearly seen to be equivalent to the following Riccati inequality

$$A^{\mathrm{T}}P + PA + PA_d S^{-1} A_d^{\mathrm{T}}P + S < 0. \tag{4.49}$$

**Example 4.15**

The following time-delay linear system has been considered in Fu et al. (2004):

$$\dot{x}(t) = \begin{bmatrix} -1 & 0 \\ 0 & -2 \end{bmatrix} x(t) + \begin{bmatrix} 1 & 0 \\ 0 & 1 \end{bmatrix} x(t - d(t)) + \begin{bmatrix} 1 \\ 0 \end{bmatrix} u(t) + \begin{bmatrix} 1 \\ 2 \end{bmatrix} w(t).$$

For this system, applying the function `feasp` in the MATLAB LMI toolbox to the corresponding LMI problem (4.48) yields the following solutions:

$$P = \begin{bmatrix} 6176.6157 & 0 \\ 0 & 4325.0202 \end{bmatrix} > 0,$$

and

$$S = \begin{bmatrix} 6176.6157 & 0 \\ 0 & 8650.0405 \end{bmatrix} > 0.$$

Thus, this time-delay system is asymptotically stable.

**Example 4.16**

Consider the following time-delay linear system:

$$\dot{x}(t) = \begin{bmatrix} -2 & 0 & 1 \\ 0 & -3 & 0 \\ 1 & 0 & -2 \end{bmatrix} x(t) + \begin{bmatrix} -1 & 1 & 1 \\ 2 & -1 & 1 \\ 0 & 0 & -1 \end{bmatrix} x(t - d). \tag{4.50}$$

Applying the function `feasp` in the MATLAB LMI toolbox verifies that the corresponding LMIs problem (4.48) is infeasible with respect to the two symmetric matrices $P, S \in \mathbb{S}^3$. Thus, for this system condition, (4.48) fails to give a definite conclusion about the stability of the system.

## 4.5.2 Delay-Dependent Condition

In the last section, we have considered the delay-independent condition for the stability of the time delay system (4.47). In this section, we further present a delay-dependent condition. For this purpose, we first rewrite the system as

$$\dot{x}(t) = Ax(t) + A_d x(t - d)$$
$$= (A + A_d)x(t) - A_d \left( x(t) - x(t - d) \right)$$
$$= (a + A_d)x(t) - A_d \int_{-d}^{0} \dot{x}(t + s)\, ds$$
$$= (A + A_d)x(t) - A_d \int_{-d}^{0} [Ax(t + s) + A_d x(t - d + s)]\, ds. \tag{4.51}$$

The main result in this section is given as follows.

**Theorem 4.9**   The time-delay system (4.47) is uniformly asymptotically stable if there exist a symmetric positive definite matrix $X$ and a scalar $0 < \beta < 1$, such that

$$\begin{bmatrix} \Phi(X) & \bar{d}XA^T & \bar{d}XA_d^T \\ \bar{d}AX & -\bar{d}\beta I & 0 \\ \bar{d}A_d X & 0 & -\bar{d}(1 - \beta)I \end{bmatrix} < 0, \tag{4.52}$$

where

$$\Phi(X) = X(A + A_d)^{\mathrm{T}} + (A + A_d)X + \bar{d}A_d A_d^{\mathrm{T}}.$$

*Proof* Introduce the following positive definite functional

$$V(x_t) = x^{\mathrm{T}}(t)Px(t)$$

$$+ \frac{1}{\beta} \int_{-d}^{0} \int_{t+s}^{t} x^{\mathrm{T}}(\theta)A^{\mathrm{T}}Ax(\theta)d\theta \, ds$$

$$+ \frac{1}{1 - \beta} \int_{-d}^{0} \int_{t-d+s}^{t} x^{\mathrm{T}}(\theta)A_d^{\mathrm{T}}A_d x(\theta)d\theta \, ds,$$

then the time-derivative of $V(x_t)$ along the solution of (4.51) is given by

$$\dot{V}(x_t) = x^{\mathrm{T}}(t)\left(P(A + A_d) + (A + A_d)^{\mathrm{T}}P\right)x(t)$$

$$- 2x^{\mathrm{T}}(t)PA_d \int_{t-d}^{t} Ax(t+s)ds$$

$$- 2x^{\mathrm{T}}(t)PA_d \int_{t-d}^{t} A_d x(t-d+s)ds + \frac{d}{\beta}x^{\mathrm{T}}(t)A^{\mathrm{T}}Ax(t)$$

$$- \frac{1}{\beta}\int_{-d}^{0} x^{\mathrm{T}}(t+s)A^{\mathrm{T}}Ax(t+s)ds + \frac{d}{1-\beta}x^{\mathrm{T}}(t)A_d^{\mathrm{T}}A_d x(t)$$

$$- \frac{1}{1-\beta}x^{\mathrm{T}}(t-d+s)A_d^{\mathrm{T}}A_d x(t-d+s). \tag{4.53}$$

Applying Lemma 2.1, we have, for any $0 < \beta < 1$,

$$-2x^{\mathrm{T}}(t)PA_d \int_{-d}^{0} Ax(t+s)ds$$

$$\leq d\beta x^{\mathrm{T}}(t)PA_d A_d^{\mathrm{T}}Px(t) + \frac{1}{\beta}\int_{-d}^{0} x^{\mathrm{T}}(t+s)A^{\mathrm{T}}Ax(t+s)ds,$$

and

$$
- 2x^{\mathrm{T}}(t) P A_d \int_{-d}^{0} A_d x(t + s - d) ds
$$

$$
\leq d(1 - \beta) x^{\mathrm{T}}(t) P A_d A_d^{\mathrm{T}} P x(t)
$$

$$
+ \frac{1}{1 - \beta} \int_{-d}^{0} x^{\mathrm{T}} (t + s - d) A_d^{\mathrm{T}} A_d x(t + s - d) ds.
$$

Substituting the two relations into (4.53), and using $0 < d \leq \bar{d}$, yields

$$
\dot{V}(x_t) \leq x^{\mathrm{T}}(t) \left( P(A + A_d) + (A + A_d)^{\mathrm{T}} P \right) x(t)
$$

$$
+ x^{\mathrm{T}}(t) \left( d P A_d A_d^{\mathrm{T}} P + \frac{d}{\beta} A^{\mathrm{T}} A + \frac{d}{1 - \beta} A_d^{\mathrm{T}} A_d \right) x(t)
$$

$$
\leq x^{\mathrm{T}}(t) \left( P(A + A_d) + (A + A_d)^{\mathrm{T}} P \right) x(t)
$$

$$
+ x^{\mathrm{T}}(t) \left( \bar{d} P A_d A_d^{\mathrm{T}} P + \frac{\bar{d}}{\beta} A^{\mathrm{T}} A + \frac{\bar{d}}{1 - \beta} A_d^{\mathrm{T}} A_d \right) x(t)
$$

Thus, it follows from the Lyapunov's method that the system (4.51), or equivalently, the system (4.47), is uniformly asymptotically stable if

$$
P(A + A_d) + (A + A_d)^{\mathrm{T}} P + \bar{d} P A_d A_d^{\mathrm{T}} P + \frac{\bar{d}}{\beta} A^{\mathrm{T}} A + \frac{\bar{d}}{1 - \beta} A_d^{\mathrm{T}} A_d < 0.
$$

Pre- and postmultiplying the aforementioned inequality by $P^{-1}$ and letting $X = P^{-1}$ yields

$$
\Phi(X) + \frac{\bar{d}}{\beta} X A^{\mathrm{T}} A X + \frac{\bar{d}}{1 - \beta} X A_d^{\mathrm{T}} A_d X < 0,
$$

that is

$$
\Phi(X) + \begin{bmatrix} \bar{d} X A^{\mathrm{T}} & \bar{d} X A_d^{\mathrm{T}} \end{bmatrix} \begin{bmatrix} -\bar{d} \beta I & 0 \\ 0 & -\bar{d}(1 - \beta) I \end{bmatrix}^{-1} \begin{bmatrix} \bar{d} A X \\ \bar{d} A_d X \end{bmatrix} < 0.
$$

Finally, applying Schur complement lemma to the aforementioned inequality, gives the equivalent condition (4.52). The proof is then completed. ■

It is important to note that the condition (4.52) implies

$$
\Phi(X) = X(A + A_d)^{\mathrm{T}} + (A + A_d) X + \bar{d} A_d A_d^{\mathrm{T}} < 0.
$$

This is to say that the Hurwitz stability of the matrix $A + A_d$ is necessary for condition (4.52) to hold true.

**Example 4.17**

Consider the time-delay linear system (4.50) again with $0 \leq d \leq 0.3$. Applying the function `feasp` in the MATLAB LMI toolbox to the corresponding LMIs problem (4.52) yields the following solution:

$$X = \begin{bmatrix} 0.3500 & 0.0574 & -0.0199 \\ 0.0574 & 0.4290 & -0.0024 \\ -0.0199 & -0.0024 & 0.2615 \end{bmatrix} > 0,$$

$$\beta = 0.5261.$$

Thus, it follows from Theorem 4.9, that this time-delay system is uniformly asymptotically stable with $0 \leq d \leq 0.3$.

**Remark 4.6**    In Example 4.16, we have verified that there do not exist two symmetric matrices $P, S \in \mathbb{S}^n$, such that LMI (4.48) is satisfied. Thus, with (4.48), we cannot judge the delay-independent stability of the time-delay system. But in Example 4.17 we know it is uniformly asymptotically stable when $0 \leq d \leq 0.3$.

# 4.6  Notes and References

## 4.6.1  Summary and References

This chapter provides LMI conditions for certain types of stability. Relations among these several types of stability are shown in Figure 4.6, and the criteria in terms of LMIs are given in Tables 4.1 through 4.3.

Stability is a huge topic in control theory. It has such a diverse content since stability itself is composed of different types, and meanwhile there are so many different types of systems. In the last two decades, a huge amount of results on stability analysis based on LMIs have been reported. Here, we can only mention a few.

LMI techniques have been applied to prove the stability of various types of systems, such as time-delay systems (Gu 1997, Jing et al. 2004, Han 2005, Xu and Lam 2006, Kwon et al. 2008, Zhu et al. 2009), time-varying polytopic systems (He et al. 2004c, Zhong and Duan 2007, 2009), pendulum-like systems (Lu and Yang 2009), 2-D systems (Dumitrescu 2008, Kaczorek 2009), and stochastic systems with time-varying delay, nonlinearity, and Markovian switching (Yue and Han 2005).

For robust stability analysis using LMIs, a great amount of work is also concentrated on time-delay systems (Li and deSouza 1997, Geromel et al. 1998, Park 1999, Kim 2001, Wu et al. 2004a,b, He et al. 2004b). Meanwhile, we also have LMI

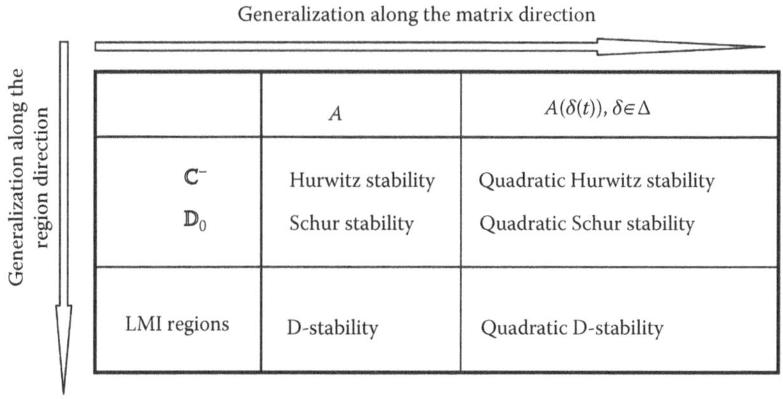

Generalization along the matrix direction

|  | $A$ | $A(\delta(t)), \delta \in \Delta$ |
|---|---|---|
| $\mathbf{C}^-$ <br> $\mathbf{D}_0$ | Hurwitz stability <br> Schur stability | Quadratic Hurwitz stability <br> Quadratic Schur stability |
| LMI regions | D-stability | Quadratic D-stability |

Generalization along the region direction

**Figure 4.6  Relations among different types of stability.**

criteria for robust stability of discrete-time systems (de Oliveira et al. 1999), interval systems (Gahinet et al. 1996), linear time-varying systems (Xie et al. 1997), stochastic systems with time delay and nonlinear uncertainties (Yue and Won 2001), uncertain continuous-time linear systems (Ramos and Peres 2002), uncertain continuous-time delayed fuzzy systems (He et al. 2006b), and discrete-time descriptor polytopic systems (Gao et al. 2011b). In Particular, Peaucelle et al. (2000) investigate the robust $\mathbb{D}$-stability condition for systems with real convex polytopic uncertainty.

### 4.6.2 Affine Quadratic Stability

As we have seen from Sections 4.3 and 4.4, quadratic stability of a set of systems requires to seek an associated common constant positive definite matrix $P$, or a common Lyapunov function for the set of systems, while affine quadratic stability of a set of systems seeks a Lyapunov function that is dependent on the perturbations. A consequence of such a Lyapunov function is certainly the less conservatism.

The following is a brief statement of the main result about affine quadratic stability, which is due to Gahinet et al. (1996).

Consider the following uncertain system

$$\dot{x} = A(\delta(t))x, \tag{4.54}$$

with

$$A(\delta(t)) = A_0 + \delta_1(t)A_1 + \delta_2(t)A_2 + \cdots + \delta_k(t)A_k, \tag{4.55}$$

where $A_i \in \mathbb{R}^{n \times n}$, $i = 0, 1, 2, \ldots, k$, are known matrices, and

$$\delta(t) = \begin{bmatrix} \delta_1(t) & \delta_2(t) & \cdots & \delta_k(t) \end{bmatrix},$$

**Table 4.1  Criteria for $\mathbb{D}$-Stability of Matrix $A$**

| Regions | Criteria | Variables |
|---|---|---|
| $\mathbb{H}_0$ (Hurwitz) | $P > 0$ <br> $A^\mathsf{T}P + PA < 0$ | $P$ |
| $\mathbb{H}_\alpha$ | $P > 0$ <br> $A^\mathsf{T}P + PA + 2\alpha P < 0$ | $P$ |
| $\mathbb{H}_{\alpha,\beta}$ | $P > 0$ <br> $A^\mathsf{T}P + PA + 2\alpha P < 0$ <br> $A^\mathsf{T}P + PA + 2\beta P > 0$ | $P$ |
| $\mathbb{D}_{(0,1)}$ (Schur) | $P > 0$ <br> $A^\mathsf{T}PA - P < 0$ | $P$ |
|  | $\begin{bmatrix} -P & AP \\ PA^\mathsf{T} & -P \end{bmatrix} < 0$ | $P$ |
| $\mathbb{D}_{(q,r)}$ | $P > 0$ <br> $(A + qI)^\mathsf{T}P(A + qI) - r^2 P < 0$ | $P$ |
|  | $\begin{bmatrix} -rP & qP + AP \\ qP + PA^\mathsf{T} & -rP \end{bmatrix} < 0$ | $P$ |
| $\mathbb{D}_{(L,M)}$ | $P > 0$ <br> $L \otimes P + M \otimes (AP) + M^\mathsf{T} \otimes (AP)^\mathsf{T} < 0$ | $P$ |

with $\delta_i(t)$, $i = 1, 2, \ldots, k$, being some time functions, represent the uncertain parameter vector in the system. The uncertain parameter vector $\delta$ and its variation rate $\dot{\delta}$ are restricted to be within two rectangles $\Omega$ and $\Omega_0$, that is,

$$\delta(t) \in \Omega = \left\{ \delta(t) \,\middle|\, \delta_i(t) \in [\delta_i^-, \, \delta_i^+], \, i = 1, 2, \ldots, k \right\}, \qquad (4.56)$$

and

$$\dot{\delta}(t) \in \Omega_0 = \left\{ \dot{\delta}(t) \,\middle|\, \dot{\delta}_i(t) \in [\kappa_i^-, \, \kappa_i^+], \, i = 1, 2, \ldots, k \right\}, \qquad (4.57)$$

where $\delta_i^-$, $\delta_i^+$, $\kappa_i^-$, $\kappa_i^+$, $i = 1, 2, \ldots, k$, are groups of given scalars.

**Table 4.2  Criteria for Quadratic Stability of Matrix *A***

| Stabilities | Criteria | Variables |
|---|---|---|
| Hurwitz | $P > 0$ <br><br> $A^{\mathsf{T}}(\delta)P + PA(\delta) < 0, \ \forall \delta \in \Delta_E$ | $P$ |
| Schur | $P > 0$ <br><br> $A(\delta(t))PA^{\mathsf{T}}(\delta(t)) - P < 0, \ \forall \delta(t) \in \Delta_E$ | $P$ |
| | $\begin{bmatrix} -P & A(\delta(t))P \\ PA^{\mathsf{T}}(\delta(t)) & -P \end{bmatrix} < 0, \ \forall \delta \in \Delta_E$ | $P$ |
| | $\begin{bmatrix} -P & PA^{\mathsf{T}}(\delta(t)) \\ A(\delta(t))P & -P \end{bmatrix} < 0, \ \forall \delta \in \Delta_E$ | $P$ |
| $\mathbb{D}$-stability | $P > 0$ <br><br> $L \otimes P + \langle M \otimes (A(\delta(t))P) \rangle_s < 0, \ \delta(t) \in \Delta_E$ | $P$ |

**Table 4.3  Stability Criteria for Time-Delay Systems**

| Conditions | Criteria | Variables |
|---|---|---|
| Delay independent | $P > 0$ <br><br> $\begin{bmatrix} A^{\mathsf{T}}P + PA + S & PA_d \\ A_d^{\mathsf{T}}P & -S \end{bmatrix} < 0$ | $P, S$ |
| Delay dependent | $0 < \beta < 1$ <br><br> $\begin{bmatrix} \Phi(X) & \bar{d}XA^{\mathsf{T}} & \bar{d}XA_d^{\mathsf{T}} \\ \bar{d}AX & -\bar{d}\beta I & 0 \\ \bar{d}A_dX & 0 & -\bar{d}(1-\beta)I \end{bmatrix} < 0$ <br><br> $\Phi(X) = X(A + A_d)^{\mathsf{T}} + (A + A_d)X + \bar{d}A_dA_d^{\mathsf{T}}$ | $X, \beta$ |

A parameter-dependent Lyapunov function for the system (4.54) and (4.55) can be chosen as

$$V(x) = x^{\mathsf{T}}P(\delta(t))x, \tag{4.58}$$

where

$$P(\delta(t)) = P_0 + \delta_1(t)P_1 + \cdots + \delta_k(t)P_k > 0, \quad \forall \delta(t) \in \Omega, \tag{4.59}$$

with $P_i$, $i = 0, 1, \ldots, k$, being a group of symmetric matrices, and

$$\frac{\mathrm{d}}{\mathrm{d}t} V(x) = x^{\mathrm{T}} Q(\delta(t)) x < 0, \tag{4.60}$$

where

$$
\begin{aligned}
Q(\delta, \dot{\delta}) &= A^{\mathrm{T}}(\delta(t)) P(\delta(t)) + P(\delta(t)) A(\delta(t)) + \frac{\mathrm{d} P(\delta(t))}{\mathrm{d}t} \\
&= A^{\mathrm{T}}(\delta(t)) P(\delta(t)) + P(\delta(t)) A(\delta(t)) + P(\dot{\delta}(t)) - P_0.
\end{aligned}
$$

In view of the equivalence of (4.60) with $Q(\delta(t)) < 0$, we can introduce the following concept of affine quadratic Hurwitz stability.

**Definition 4.6** The system (4.54) and (4.55), with $\forall \delta(t) \in \Omega$, $\dot{\delta}(t) \in \Omega_0$, is called affine quadratically stable if there exist a group of symmetric matrices $P_i$, $i = 0, 1, \ldots, k$, satisfying (4.59) and

$$Q(\delta, \dot{\delta}) = A^{\mathrm{T}}(\delta) P(\delta) + P(\delta) A(\delta) + P(\dot{\delta}) - P_0 < 0, \quad \forall \delta \in \Omega, \quad \dot{\delta} \in \Omega_0. \tag{4.61}$$

It is clearly seen from the aforementioned equation that affine quadratic stability implies Lyapunov asymptotical stability. While the following reveals its relation to quadratic stability.

Let the system (4.54) and (4.55), with $\delta \in \Omega$, be quadratically stable. Then, by definition, there exists a symmetric positive definite matrix $P$, satisfying (4.28). Now, for system (4.54) and (4.55) with $\delta \in \Omega$, taking

$$P_0 = P, \; P_i = 0, \quad i = 1, 2, \ldots, k,$$

gives

$$P(\delta(t)) = P > 0.$$

Further, using (4.28) we also have

$$
\begin{aligned}
Q(\delta) &= A^{\mathrm{T}}(\delta(t)) P(\delta(t)) + P(\delta(t)) A(\delta(t)) + P_0 - P_0 \\
&= A^{\mathrm{T}}(\delta(t)) P + P A(\delta(t)) \\
&< 0, \quad \forall \delta(t) \in \Delta.
\end{aligned}
$$

This shows that quadratic stability implies affine quadratic stability.

The following theorem gives the LMI conditions for the affine quadratic stability of the uncertain linear system (4.54) and (4.55).

**Theorem 4.10** Consider the linear system (4.54) and (4.55), with the uncertain parameter vector $\delta$ satisfying (4.56) and (4.57). Denote the set of extreme points of $\Omega$ and $\Omega_0$ by $\Omega^E$ and $\Omega_0^E$, respectively, and define

$$\delta_{\text{mean}} = \begin{bmatrix} \dfrac{\delta_1^- + \delta_1^+}{2} & \dfrac{\delta_2^- + \delta_2^+}{2} & \cdots & \dfrac{\delta_k^- + \delta_k^+}{2} \end{bmatrix}. \tag{4.62}$$

Then, the system is affinely quadratically stable if $A(\delta_{\text{mean}})$ is stable, and there exist a group of symmetric matrices $P_i$, $i = 0, 1, \ldots, k$, satisfying

$$A_i^{\text{T}} P_i + P_i A_i \geq 0, \quad i = 1, 2, \ldots, k, \tag{4.63}$$

and

$$Q(\delta, \tau) = A^{\text{T}}(\delta)P(\delta) + P(\delta)A(\delta) + P(\tau) - P_0 < 0, \quad \forall \delta \in \Omega^E, \quad \tau \in \Omega_0^E, \tag{4.64}$$

where $P(\delta)$ is defined as in (4.59).

For a proof of the result and many other related helpful results and comments, one can refer to the original paper by Gahinet et al. (1996).

## Exercises

**4.1** What are the relations among the concepts of robust stability, quadratic stability, and affine quadratic stability?

**4.2** Let $A \in \mathbb{R}^{n \times n}$ and $B = \dfrac{1}{2}(A + A^{\text{T}})$. Show that
   1. If $B$ is Hurwitz stable, then $A$ is also Hurwitz stable.
   2. If $B$ is Schur stable, and $A$ is symmetric positive definite, then $A$ is also Schur stable.

**4.3** Verify using the LMI technique the stability of the following continuous-time linear system

$$\dot{x}(t) = \begin{bmatrix} -0.0180 & -0.2077 & -0.7150 \\ -0.5814 & -4.2900 & 0 \\ 1.0670 & 4.2730 & -6.6540 \end{bmatrix} x(t).$$

**4.4** Show that the following regions are LMI ones:

$$\mathbb{D}_1 = \{x + yi \mid -r < y < r,\ r > 0\},$$

$$\mathbb{D}_2 = \{x + yi \mid x > ay^2 + c,\ a > 0\},$$

$$\mathbb{D}_3 = \{x + yi \mid (x + q)^2 + p^2 y^2 < r^2\}.$$

**4.5** Consider the following advanced (CCV-type) fighter aircraft system (Syrmos and Lewis 1993, Duan 2003):

$$\dot{x}(t) = Ax(t) + Bu(t)$$

with

$$A = \begin{bmatrix} -1.3410 & 0.9933 & 0 & -0.1689 & -0.2518 \\ 43.2230 & -0.8693 & 0 & -17.2510 & -1.5766 \\ 1.3410 & 0.0067 & 0 & 0.1689 & 0.2518 \\ 0 & 0 & 0 & -20.0000 & 0 \\ 0 & 0 & 0 & 0 & -20.0000 \end{bmatrix},$$

$$B = \begin{bmatrix} 0 & 0 \\ 0 & 0 \\ 0 & 0 \\ 20 & 0 \\ 0 & 20 \end{bmatrix}.$$

Judge the open-loop stability of the aforementioned system.

**4.6** Consider a system in the form of (4.22) through (4.24), with $\Delta = \Delta_P$ given by (4.27), where

$$A_0 = \begin{bmatrix} -2.1333 & -1.3333 \\ -0.6667 & -1.4667 \end{bmatrix}, \quad A_1 = \begin{bmatrix} 0.2 & 0.1 \\ 0.1 & 0.2 \end{bmatrix}, \quad A_2 = \begin{bmatrix} 0.1 & 0.2 \\ 0.2 & 0.1 \end{bmatrix}.$$

Further, let us assume that the set of perturbation parameters form a polytope in the form of (4.27), with $k = 2$. Judge the quadratic stability of the system.

**4.7** Verify using the LMI technique the stability of the following time-delay linear system

$$\dot{x}(t) = \begin{bmatrix} -3 & 0 & 1 \\ 0 & -3 & 0 \\ 1 & 0 & -4 \end{bmatrix} x(t) + \begin{bmatrix} 1 & 0 & 1 \\ 2 & 1 & 1 \\ 0 & 0 & -1 \end{bmatrix} x(t - d).$$

**4.8** Judge the stability of the following time-delay linear system

$$\dot{x}(t) = \begin{bmatrix} -5 & 1 \\ 0 & -10 \end{bmatrix} x(t) + \begin{bmatrix} 4 & 0 \\ 2 & 6 \end{bmatrix} x(t - d(t)) + \begin{bmatrix} 2 \\ 0 \end{bmatrix} u(t).$$

# Chapter 5

# H∞/H₂ Performance

In this chapter, we introduce the $H_\infty$ norm and $H_2$ norm of rational matrices. Particularly, for the transfer matrix function of a linear system, we give LMI conditions for these norms to be bounded by a certain level. The results in this chapter form a base for $H_\infty$ and $H_2$ control systems design problems addressed in the next part.

## 5.1 H∞ and H₂ Indices

We are aware of norms of constant matrices, while $H_\infty$ and $H_2$ norms are defined for rational matrices. A rational matrix $G(s)$ is a matrix function in the form of

$$G(s) = N(s)D^{-1}(s),$$

or

$$G(s) = L^{-1}(s)H(s),$$

where

- $s$ is a complex variable
- $N(s)$, $D(s)$, $L(s)$ and $H(s)$ are polynomial matrices of appropriate dimensions

Furthermore, these polynomial matrices satisfy

$$\deg(N(s)) < \deg(D(s)),$$

$$\deg(H(s)) < \deg(L(s)).$$

**139**

### 5.1.1 $H_\infty$ Index

For a constant matrix $A$, we know that its spectral norm is defined as

$$||A||_2 = \left(\lambda_{\max}\left(AA^\mathrm{T}\right)\right)^{1/2} = \sigma_{\max}(A).$$

As an extension, the $H_\infty$ norm of a rational matrix is defined as follows.

**Definition 5.1**    Let $G(s)$ be a rational matrix. Then, its $H_\infty$ norm, denoted by $||G(s)||_\infty$, is given by

$$||G(s)||_\infty = \sup_\omega \left\{\sigma_{\max}\left(G(j\omega)\right)\right\}.$$

The following proposition gives an equivalent condition for the $H_\infty$ norm of a rational matrix.

**Proposition 5.1**    Let $G(s)$ be a rational matrix, which does not have poles on the closed right-half complex plane. Then,

$$||G(s)||_\infty = \sup_s\{\sigma_{\max}(G(s)) \,|\, \mathrm{Re}(s) > 0\}.$$

*Proof*    Let

$$f(s) = \sigma_{\max}(G(s)).$$

Since $G(s)$ does not have poles on the closed right-half complex plane, it is analytic in the closed right half complex plane. Therefore, by the maximum modulus principle, that is, Lemma 2.14, we have

$$\sup_s f(s) = \sup_s\{\sigma_{\max}(G(s))\}$$

over $\mathbb{C}^+ = \{s \,|\, \mathrm{Re}(s) > 0\}$ must be reached on the boundary of $\mathbb{C}^+$, that is,

$$\sup\{\sigma_{\max}(G(s)) \,|\, \mathrm{Re}(s) > 0\}$$
$$= \sup_\omega \sigma_{\max}(G(j\omega))$$
$$= ||G(s)||_\infty.$$

The proof is then completed.    ■

**Remark 5.1**    The condition in the aforementioned proposition, that is, $G(s)$ does not have poles on the closed right-half complex plane, is not a constraint since we will soon find out that $||G(s)||_\infty$ does not exist if $G(s)$ has poles on the closed right-half complex plane.

## 5.1.2 $H_2$ *Index*

Like $H_\infty$ norm, $H_2$ norm is also defined for rational matrices. The strict definition is as follows.

**Definition 5.2**    Suppose $G(s)$ is a matrix function such that

$$\int_{-\infty}^{\infty} G(j\omega)G^H(j\omega)d\omega < \infty,$$

then the $H_2$ norm of $G(s)$ is defined as

$$||G(s)||_2 = \left( \text{trace} \left( \frac{1}{2\pi} \int_{-\infty}^{\infty} G(j\omega)G^H(j\omega)d\omega \right) \right)^{1/2}$$

$$= \left( \text{trace} \left( \frac{1}{2\pi} \int_{-\infty}^{\infty} G^H(j\omega)G(j\omega)d\omega \right) \right)^{1/2}. \qquad (5.1)$$

For a constant matrix $A \in \mathbb{R}^{n \times n}$, we know that its Frobenius norm is defined as

$$||A||_F = \left( \text{trace} \left( AA^T \right) \right)^{1/2} = \left( \text{trace} \left( A^T A \right) \right)^{1/2}.$$

Thus it is clear that the $H_2$ norm for a matrix function is a generalization of the Frobenius norm for a constant matrix.

**Remark 5.2**    If $u(s)$ is a vector function such that the integral $\int_{-\infty}^{\infty} u^H(j\omega)u(j\omega)d\omega$ exists, then, obviously the $H_2$ norm of $u(s)$ reduces to

$$||u(s)||_2 = \left( \text{trace} \left( \frac{1}{2\pi} \int_{-\infty}^{\infty} u(j\omega)u^H(j\omega)d\omega \right) \right)^{1/2}$$

$$= \left( \frac{1}{2\pi} \int_{-\infty}^{\infty} u^H(j\omega)u(j\omega)d\omega \right)^{1/2}.$$

In control systems context, we need to consider the following transfer function

$$G(s) = C(sI - A)^{-1}B + D, \tag{5.2}$$

where $A \in \mathbb{R}^{n \times n}$, $B \in \mathbb{R}^{n \times r}$, $C \in \mathbb{R}^{m \times n}$, $D \in \mathbb{R}^{m \times r}$.

Before presenting the expressions for the $H_2$ norm of this function, we first give a preliminary result which can be found in many textbooks for control systems theory, for example, Duan (2004a).

**Proposition 5.2**   Let $A$, $Q \in \mathbb{R}^{n \times n}$, and $A$ be stable, $Q$ semi-positive definite, then the following Lyapunov algebraic equation

$$A^{\mathrm{T}}P + PA = -Q$$

has a unique positive definite solution, which can be expressed as

$$P = \int_0^\infty e^{A^{\mathrm{T}}t} Q e^{At} \, dt.$$

Based on this proposition, we can prove the following result.

**Lemma 5.1**   Let $G(s)$ be given by (5.2).

1. $||G(s)||_2$ exists if and only if $A$ is stable and $D = 0$.
2. When $A$ is stable and $D = 0$, there holds

$$||G(s)||_2^2 = \mathrm{trace}\left(CXC^{\mathrm{T}}\right)$$

with $X > 0$ being the unique solution to the following Lyapunov matrix equation:

$$AX + XA^{\mathrm{T}} + BB^{\mathrm{T}} = 0, \tag{5.3}$$

or

$$||G(s)||_2^2 = \mathrm{trace}\left(B^{\mathrm{T}}YB\right)$$

with $Y > 0$ being the unique solution to the following Lyapunov matrix equation:

$$A^{\mathrm{T}}Y + YA + C^{\mathrm{T}}C = 0. \tag{5.4}$$

*Proof* By the well-known Parseval lemma,

$$||G(s)||_2^2 = \text{trace} \left( \frac{1}{2\pi} \int_{-\infty}^{\infty} G(j\omega)G^H(j\omega)d\omega \right)$$

$$= \text{trace} \int_0^{\infty} \left(Ce^{At}B + D\right) \left(Ce^{At}B + D\right)^T dt. \tag{5.5}$$

From this we see that $||G(s)||_2$ exists only if $A$ is stable and $D = 0$. Thus, the first conclusion is proven.

Let $A$ be stable and $D = 0$, we have

$$||G(s)||_2^2 = \text{trace} \left( \frac{1}{2\pi} \int_{-\infty}^{\infty} G(j\omega)G^H(j\omega)d\omega \right)$$

$$= \text{trace} \int_0^{\infty} Ce^{At}BB^T e^{A^T t}C^T dt$$

$$= \text{trace} \left(CXC^T\right), \tag{5.6}$$

where

$$X = \int_0^{\infty} e^{At}BB^T e^{A^T t}dt. \tag{5.7}$$

$X$ happens to be the solution of the Lyapunov matrix equation (5.3) in view of Proposition 5.2.

Parallely, when $A$ is stable and $D = 0$, we have

$$||G(s)||_2^2 = \text{trace} \left( \frac{1}{2\pi} \int_{-\infty}^{\infty} G^H(j\omega)G(j\omega)d\omega \right)$$

$$= \text{trace} \int_0^{\infty} B^T e^{A^T t}C^T Ce^{At}Bdt$$

$$= \text{trace} \left(B^T YB\right), \tag{5.8}$$

where

$$Y = \int_0^{\infty} e^{A^T t}C^T Ce^{At} dt. \tag{5.9}$$

This happens to be the solution of the Lyapunov matrix equation (5.4) in view of Proposition 5.2. Thus, the second conclusion also holds.  ■

### 5.1.3 Equivalent Definitions

#### 5.1.3.1 $H_\infty$ Index

The following gives an equivalent definition for the $H_\infty$ norm of a rational matrix $G(s)$ based on the $H_2$ norm of vector signals.

**Theorem 5.1**   Let $G(s)$ be a stable $m \times r$ transfer function matrix. Then,

$$\|G(s)\|_\infty = \sup_{\|u(s)\|_2 = 1, \, u(s) \in \mathcal{H}_2} \{\|G(s)\,u(s)\|_2\}$$

$$= \sup_{\|u(s)\|_2 \neq 0, \, u(s) \in \mathcal{H}_2} \left\{ \frac{\|G(s)\,u(s)\|_2}{\|u(s)\|_2} \right\},$$

where

$$\mathcal{H}_2 = \{u(s) \,|\, \|u(s)\|_2 < \infty\}.$$

For a proof of this theorem, please refer to Section A.2 in Appendix A. In view of this fact, we also say that the $H_\infty$ norm of a rational matrix $G(s)$ is induced from the $H_2$ norm of vector signals. This is parallel to the fact that the spectral norm of a matrix is induced from the Euclidean norm of vectors.

This theorem obviously gives the following corollary.

**Corollary 5.1**   Let

$$y(s) = G(s)u(s), \tag{5.10}$$

then

$$\|y(s)\|_2 \leq \|G(s)\|_\infty \|u(s)\|_2. \tag{5.11}$$

**Remark 5.3**   The aforementioned corollary clearly tells us that $\|G(s)\|_\infty$ can be viewed as an amplifier between the $H_2$ norms (energy) of the input $u(t)$ and the output $y(t)$.

#### 5.1.3.2 $H_2$ Index

Let $u(t)$ be a function of time. The power of the signal $u(t)$ is defined as

$$\|u(t)\|_P = \left( \lim_{T \to \infty} \frac{1}{2T} \int_{-T}^{T} \|u(t)\|^2 \, dt \right)^{1/2}.$$

The following gives an equivalent definition for the $H_2$ norm of a rational matrix $G(s)$ based on the power of vector signals.

**Theorem 5.2**   Let $G(s)$ be a stable $m \times r$ transfer function matrix. Then,

$$\|G(s)\|_2 = \sup_{\|u(t)\|_P = 1, u(t) \in \mathcal{P}} \{\|G(s)u(s)\|_P\}$$

$$= \sup_{\|u(t)\|_P \neq 0,\ u(t) \in \mathcal{P}} \left\{ \frac{\|G(s)u(s)\|_P}{\|u(t)\|_P} \right\},$$

where $\mathcal{P} = \left\{ f(t) \,\middle|\, \|f(t)\|_P < \infty \right\}$.

For a proof of this theorem, please refer to Section A.3 in Appendix A. In view of this fact, we also say that the $H_2$ norm of a rational matrix $G(s)$ is induced from the power of vector signals.

This theorem obviously gives the following corollary.

**Corollary 5.2**   Given (5.10), then

$$\|y(s)\|_P \leq \|G(s)\|_2 \|u(s)\|_P. \tag{5.12}$$

**Remark 5.4**   The aforementioned corollary clearly tells us that $\|G(s)\|_2$ can be viewed as an amplifier between the power of the input $u(t)$ and the output $y(t)$.

**Remark 5.5**   Theorem 5.1 shows that $\|G(s)\|_\infty$ is an induced norm, but Theorem 5.2 shows that $\|G(s)\|_2$ is not because $\|u(t)\|_P$, unlike $\|u(t)\|_2$, is not a norm.

Due to Remarks 5.3 and 5.4, in the next two sections we will examine the conditions for the $H_\infty$ norm and the $H_2$ norm of the rational matrix $G(s)$ in the form of (5.2) to be bounded by a given level.

## 5.2 LMI Conditions for $H_\infty$ Index

In this section, we consider the condition for the $H_\infty$ norm of the following transfer function matrix

$$G(s) = C(sI - A)^{-1}B + D \tag{5.13}$$

to be bounded by a given level $\gamma$, that is,

$$\|G(s)\|_\infty < \gamma. \tag{5.14}$$

This problem has a clear practical meaning. Let a realization of the transfer function be

$$\begin{cases} \dot{x} = Ax + Bw \\ y = Cx + Dw, \end{cases} \tag{5.15}$$

where $x$, $y$, and $w$ are the system state vector, output vector, and the disturbance vector, respectively. Thus, in the frequency domain, we have

$$y(s) = G(s)w(s).$$

Therefore, by Remark 5.3, the scalar $\gamma$ represents a level for the attenuation of the disturbance in the system.

### 5.2.1 Basic Conditions

**Theorem 5.3**    Let $G(s)$ be given by (5.13). Then, $||G(s)||_\infty < \gamma$ if and only if there exists a matrix $P > 0$, such that one of the following three inequalities holds:

$$\begin{bmatrix} A^{\mathrm{T}}P + PA & PB & C^{\mathrm{T}} \\ B^{\mathrm{T}}P & -\gamma I & D^{\mathrm{T}} \\ C & D & -\gamma I \end{bmatrix} < 0, \tag{5.16}$$

$$\begin{bmatrix} PA + A^{\mathrm{T}}P & PB & C^{\mathrm{T}} \\ B^{\mathrm{T}}P & -\gamma^2 I & D^{\mathrm{T}} \\ C & D & -I \end{bmatrix} < 0, \tag{5.17}$$

and

$$\begin{bmatrix} PA + A^{\mathrm{T}}P + C^{\mathrm{T}}C & PB + C^{\mathrm{T}}D \\ B^{\mathrm{T}}P + D^{\mathrm{T}}C & D^{\mathrm{T}}D - \gamma^2 I \end{bmatrix} < 0. \tag{5.18}$$

*Proof*    First, let us show the equivalence between the conditions (5.17) and (5.18). Rewrite condition (5.18) as

$$\begin{bmatrix} PA + A^{\mathrm{T}}P & PB \\ B^{\mathrm{T}}P & -\gamma^2 I \end{bmatrix} + \begin{bmatrix} C^{\mathrm{T}} \\ D^{\mathrm{T}} \end{bmatrix} [C \ D] < 0.$$

Applying Schur complement lemma, the inequality can be equivalently converted into (5.17).

Second, let us show the equivalence between the conditions (5.17) and (5.16). Let

$$T = \mathrm{diag}\left(\gamma^{-1/2}I, \ \gamma^{-1/2}I, \ \gamma^{1/2}I\right), \qquad P' = \gamma^{-1}P,$$

then it is easy to verify

$$T \begin{bmatrix} PA + A^{\mathrm{T}}P & PB & C^{\mathrm{T}} \\ B^{\mathrm{T}}P & -\gamma^2 I & D^{\mathrm{T}} \\ C & D & -I \end{bmatrix} T^{\mathrm{T}} = \begin{bmatrix} P'A + A^{\mathrm{T}}P' & P'B & C^{\mathrm{T}} \\ B^{\mathrm{T}}P' & -\gamma I & D^{\mathrm{T}} \\ C & D & -\gamma I \end{bmatrix}. \quad (5.19)$$

Thus, the equivalence between conditions (5.16) and (5.17) is obvious.

In the rest of the proof, let us show that $||G(s)||_\infty < \gamma$ if and only if there exists a matrix $P > 0$ satisfying (5.16).

Sufficiency: Suppose that there exists a matrix $P > 0$ satisfying (5.16). From (5.16), we clearly have

$$Q \triangleq -(PA + A^{\mathrm{T}}P) > 0,$$

which indicates, in view of the Lyapunov stability theorem, that $A$ is stable. Further, it again follows from (5.16) that

$$\begin{bmatrix} -\gamma I & D^{\mathrm{T}} \\ D & -\gamma I \end{bmatrix} < 0,$$

which implies

$$\sigma_{\max}(D) < \gamma.$$

Since (5.16) is equivalent to (5.18), and by Schur complement lemma again, it can be shown that these two conditions are further equivalent to

$$PA + A^{\mathrm{T}}P + C^{\mathrm{T}}C + (PB + C^{\mathrm{T}}D)R^{-1}(B^{\mathrm{T}}P + D^{\mathrm{T}}C) < 0,$$

with

$$R = \gamma^2 I - D^{\mathrm{T}}D > 0.$$

Adding

$$-j\omega P - \overline{j\omega}P = \left( -j\omega + j\omega \right) P = 0$$

to the left-hand side of the aforementioned inequality, gives

$$-P(j\omega I - A) - (j\omega I - A)^{\mathrm{H}}P + C^{\mathrm{T}}C + (PB + C^{\mathrm{T}}D)R^{-1}(B^{\mathrm{T}}P + D^{\mathrm{T}}C) < 0.$$

For any nonzero vector $u$, pre- and postmultiplying the aforementioned formula with $u^{\mathrm{T}}B^{\mathrm{T}}(j\omega I - A)^{-\mathrm{H}}$ and $(j\omega I - A)^{-1}Bu$, respectively, implies that

$$||C(j\omega I - A)^{-1}Bu||^2$$
$$\leq u^{\mathrm{T}}B^{\mathrm{T}}(j\omega I - A)^{-\mathrm{H}}PBu + u^{\mathrm{T}}B^{\mathrm{T}}P(j\omega I - A)^{-1}Bu$$
$$- ||R^{-1/2}(B^{\mathrm{T}}P + D^{\mathrm{T}}C)(j\omega I - A)^{-1}Bu||^2. \quad (5.20)$$

Since

$$\det\left(j\omega I - A\right) \neq 0, \quad \forall \omega \in \mathbb{R},$$

then the equality in the aforementioned formula holds only if $Bu = 0$. Further, adding

$$u^{\mathrm{T}}[D^{\mathrm{T}}D + D^{\mathrm{T}}C(j\omega I - A)^{-1}B + B^{\mathrm{T}}(j\omega I - A)^{-\mathrm{H}}C^{\mathrm{T}}D]u$$

to both sides of the aforementioned formula (5.20), and using

$$D^{\mathrm{T}}D = \gamma^2 I - R,$$

yields

$$||G(j\omega)u||^2 \leq \gamma^2 ||u||^2 - ||R^{-1/2}W(j\omega)u||^2, \tag{5.21}$$

with

$$W(s) = -\left(B^{\mathrm{T}}P + D^{\mathrm{T}}C\right)(sI - A)^{-1}B + R.$$

Similar to the case of inequality (5.20), the equality in (5.21) holds only if $Bu = 0$, that is,

$$||G(j\omega)u||^2 \begin{cases} = \gamma^2 ||u||^2 - ||R^{-1/2}W(j\omega)u||^2, & Bu = 0 \\ < \gamma^2 ||u||^2 - ||R^{-1/2}W(j\omega)u||^2, & Bu \neq 0 \end{cases} . \tag{5.22}$$

Therefore, when $Bu \neq 0$, we clearly have

$$||G(j\omega)u||^2 < \gamma^2 ||u||^2. \tag{5.23}$$

Further, noting that

$$||R^{-1/2}W(j\omega)u||^2 > 0, \quad \text{when } Bu = 0,$$

we know from (5.22) that (5.23) still holds when $Bu = 0$. Therefore, (5.23) holds for arbitrary $\omega$ and $u \neq 0$. From this and the definition of $H_\infty$ norm, we get $||G(s)||_\infty < \gamma$.

Necessity: Now suppose $||G(s)||_\infty < \gamma$. Then, obviously, the matrix $A$ is stable. Note that $\lim_{\omega \to \infty} G(j\omega) = D$, by the definition of $H_\infty$ norm, we have

$$\sigma_{\max}(D) < \gamma,$$

which implies that

$$R = \gamma^2 I - D^{\mathrm{T}}D > 0.$$

Define the following Hamiltonian matrix

$$H = \begin{bmatrix} A + BR^{-1}D^\mathsf{T}C & BR^{-1}B^\mathsf{T} \\ -C^\mathsf{T}(I + DR^{-1}D^\mathsf{T})C & -(A + BR^{-1}D^\mathsf{T}C)^\mathsf{T} \end{bmatrix}.$$

Then, we can obtain, after some deduction, that

$$\Phi(s) \triangleq \gamma^2 I - G(-s)^\mathsf{H}G(s)$$
$$= \begin{bmatrix} D^\mathsf{T}C & B^\mathsf{T} \end{bmatrix} \Gamma^{-1}(s) \begin{bmatrix} -B \\ C^\mathsf{T}D \end{bmatrix} + R, \tag{5.24}$$

with

$$\Gamma(s) = \begin{bmatrix} sI - A & 0 \\ C^\mathsf{T}C & sI + A^\mathsf{T} \end{bmatrix},$$

and the zeros and uncontrollable or unobservable poles of $\Phi(s)$ are just the eigenvalues of $A$. Since $A$ is stable, then $\Phi(s)$ does not have uncontrollable or unobservable poles in the imaginary axis. Further, it follows from (5.24) that $||G(s)||_\infty < \gamma$ is equivalent to $\Phi(j\omega) > 0$. Thus, under the given assumption, $\Phi(s)$ does not possess zeros in the imaginary axis, and hence the matrix $H$ does not have eigenvalues in the imaginary axis. In addition, in view of the stability of $A$, it is easy to show that the matrix pair $(A + BR^{-1}D^\mathsf{T}C, \ BR^{-1/2})$ is stabilizable in the imaginary axis. Thus, according to the theory about Riccati algebraic equations, the following Riccati equation

$$\langle P(A + BR^{-1}D^\mathsf{T}C) \rangle_s + PBR^{-1}BP + C^\mathsf{T}(I + DR^{-1}D^\mathsf{T})C = 0$$

has a real symmetric solution $P \geq 0$, where $\langle A \rangle_s$ represents the sum of matrix $A$ and its transpose, that is, $\langle A \rangle_s = A + A^\mathsf{T}$. It is important to note that, through some simplifications, the Riccati equation can be turned equivalently into

$$PA + A^\mathsf{T}P + C^\mathsf{T}C + (B^\mathsf{T}P + D^\mathsf{T}C)^\mathsf{T}R^{-1}(B^\mathsf{T}P + D^\mathsf{T}C) = 0. \tag{5.25}$$

Now consider

$$\hat{G}(s) = \begin{bmatrix} C \\ \varepsilon I \end{bmatrix} (sI - A)^{-1} \begin{bmatrix} B & \varepsilon I \end{bmatrix} + \begin{bmatrix} D & 0 \\ 0 & 0 \end{bmatrix}.$$

It is clear that, if $\varepsilon > 0$ is small enough, $||\hat{G}(s)||_\infty < \gamma$ when $||G(s)||_\infty < \gamma$. Thus, by the same reasoning, we know that the Riccati equation in the form of (5.25) corresponding to $\hat{G}(s)$ appearing as

$$P_1 A + A^\mathsf{T}P_1 + C^\mathsf{T}C + (B^\mathsf{T}P_1 + D^\mathsf{T}C)^\mathsf{T}R^{-1}(B^\mathsf{T}P_1 + D^\mathsf{T}C)$$
$$= -\varepsilon\gamma^2 P_1^2 - \varepsilon^2 I < 0, \tag{5.26}$$

has a symmetric solution $P_1 \geq 0$. Further, we show that this solution $P_1$ is in fact a positive definite one. Otherwise, there exists a nonzero vector $z$ such that $P_1 z = 0$, then, pre- and postmultiplying both sides of the earlier formula by $z^T$ and $z$, respectively, yields

$$z^T C^T (I + DR^{-1}D^T)Cz < 0,$$

which turns out to be a contradiction. Thus $P_1 > 0$ must hold. This is to say that the inequality (5.26) has a symmetric positive definite solution $P_1 > 0$.

By Schur complement lemma, it is easily shown that the inequality (5.26) is equivalent to

$$\begin{bmatrix} P_1 A + A^T P_1 + C^T C & P_1 B + C^T D \\ B^T P_1 + D^T C & D^T D - \gamma^2 I \end{bmatrix} < 0,$$

that is,

$$\begin{bmatrix} P_1 A + A^T P_1 & P_1 B \\ B^T P_1 & -\gamma^2 I \end{bmatrix} + \begin{bmatrix} C^T \\ D^T \end{bmatrix} \begin{bmatrix} C & D \end{bmatrix} < 0,$$

which becomes, when multiplied by the scalar $\gamma^{-1}$, the following with $P = P_1/\gamma$ :

$$\begin{bmatrix} PA + A^T P & PB \\ B^T P & -\gamma I \end{bmatrix} + \begin{bmatrix} C^T \\ D^T \end{bmatrix} (\gamma I)^{-1} \begin{bmatrix} C & D \end{bmatrix} < 0.$$

Finally, using Schur complement lemma again, the aforementioned inequality can be shown to be equivalent to (5.21). Therefore, the inequality (5.21) also possesses a symmetric positive definite solution. The necessity is proven. The whole proof is done. ∎

Parallel to the first two LMI conditions in the aforementioned theorem, we also have the following result.

**Theorem 5.4**  Let $G(s)$ be given by (5.13). Then, $||G(s)||_\infty < \gamma$ if and only if there exists matrix $X > 0$, such that one of the following three inequalities holds:

$$\begin{bmatrix} XA^T + AX & B & XC^T \\ B^T & -\gamma I & D^T \\ CX & D & -\gamma I \end{bmatrix} < 0, \tag{5.27}$$

$$\begin{bmatrix} AX + XA^T & B & XC^T \\ B^T & -\gamma^2 I & D^T \\ CX & D & -I \end{bmatrix} < 0, \tag{5.28}$$

and

$$\begin{bmatrix} AX + XA^\mathrm{T} + BB^\mathrm{T} & XC^\mathrm{T} + BD^\mathrm{T} \\ CX + DB^\mathrm{T} & -\gamma^2 I + DD^\mathrm{T} \end{bmatrix} < 0. \tag{5.29}$$

*Proof*  We first prove (5.27).

Define

$$T = \begin{bmatrix} P^{-1} & 0 & 0 \\ 0 & I & 0 \\ 0 & 0 & I \end{bmatrix},$$

then, noticing

$$T \begin{bmatrix} A^\mathrm{T}P + PA & PB & C^\mathrm{T} \\ B^\mathrm{T}P & -\gamma I & D^\mathrm{T} \\ C & D & -\gamma I \end{bmatrix} T^\mathrm{T} = \begin{bmatrix} P^{-1}A^\mathrm{T} + AP^{-1} & B & P^{-1}C^\mathrm{T} \\ B^\mathrm{T} & -\gamma I & D^\mathrm{T} \\ CP^{-1} & D & -\gamma I \end{bmatrix},$$

we know that (5.16) is equivalent to

$$\begin{bmatrix} P^{-1}A^\mathrm{T} + AP^{-1} & B & P^{-1}C^\mathrm{T} \\ B^\mathrm{T} & -\gamma I & D^\mathrm{T} \\ CP^{-1} & D & -\gamma I \end{bmatrix} < 0.$$

Further substituting in the aforementioned inequality $P^{-1}$ by $X$ gives (5.27).

(5.28) can be proven similarly. Now let us show the equivalence between (5.28) and (5.29).

Let

$$S = \begin{bmatrix} \gamma I & 0 & 0 \\ 0 & 0 & \gamma I \\ 0 & \gamma^{-1}I & 0 \end{bmatrix},$$

then, in view of

$$S \begin{bmatrix} AX + XA^\mathrm{T} & B & XC^\mathrm{T} \\ B^\mathrm{T} & -\gamma^2 I & D^\mathrm{T} \\ CX & D & -I \end{bmatrix} S^\mathrm{T} = \begin{bmatrix} \gamma^2 \left( AX + XA^\mathrm{T} \right) & \gamma^2 XC^\mathrm{T} & B \\ \gamma^2 CX & -\gamma^2 I & D \\ B^\mathrm{T} & D^\mathrm{T} & -I \end{bmatrix},$$

we know that (5.28) is equivalent to

$$\begin{bmatrix} \gamma^2 \left( AX + XA^{\mathrm{T}} \right) & \gamma^2 XC^{\mathrm{T}} & B \\ \gamma^2 CX & -\gamma^2 I & D \\ B^{\mathrm{T}} & D^{\mathrm{T}} & -I \end{bmatrix} < 0.$$

Applying Schur complement lemma, the aforementioned inequality can be equivalently converted into

$$\begin{bmatrix} \gamma^2 \left( AX + XA^{\mathrm{T}} \right) & \gamma^2 XC^{\mathrm{T}} \\ \gamma^2 CX & -\gamma^2 I \end{bmatrix} + \begin{bmatrix} B \\ D \end{bmatrix} \begin{bmatrix} B^{\mathrm{T}} & D^{\mathrm{T}} \end{bmatrix} < 0.$$

Letting $X' = \gamma^2 X$ in the aforementioned relation, yields, equivalently,

$$\begin{bmatrix} AX' + X'A^{\mathrm{T}} + BB^{\mathrm{T}} & X'C^{\mathrm{T}} + BD^{\mathrm{T}} \\ CX' + DB^{\mathrm{T}} & -\gamma^2 I + DD^{\mathrm{T}} \end{bmatrix} < 0.$$

This shows the equivalence between (5.28) and (5.29). ■

The following corollary immediately follows from the aforementioned theorem.

**Corollary 5.3**  Let $G(s)$ be given by (5.13). Then, $||G(s)||_\infty < \gamma$ for some $\gamma > 0$ only if the matrix $A$ is asymptotically stable.

In practical applications, solving the minimal $\gamma$ satisfying $||G(s)||_\infty < \gamma$ is often of interest. Following Theorem 5.3, such a problem can be equivalently converted into the following minimization problem with LMI constraints:

$$\begin{cases} \min & \gamma \\ \text{s.t.} & P > 0 \\ & \begin{bmatrix} A^{\mathrm{T}}P + PA & PB & C^{\mathrm{T}} \\ B^{\mathrm{T}}P & -\gamma I & D^{\mathrm{T}} \\ C & D & -\gamma I \end{bmatrix} < 0, \end{cases} \tag{5.30}$$

or

$$\begin{cases} \min & \rho \\ \text{s.t.} & P > 0 \\ & \begin{bmatrix} A^{\mathrm{T}}P + PA & PB & C^{\mathrm{T}} \\ B^{\mathrm{T}}P & -\rho I & D^{\mathrm{T}} \\ C & D & -I \end{bmatrix} < 0, \end{cases} \tag{5.31}$$

with $\gamma = \sqrt{\rho}$. Optimization problems corresponding to the third condition of Theorem 5.3 and those in Theorem 5.4 can be presented similarly.

**Example 5.1**

(Yu 2002, p. 32) Consider the following second-order system:

$$G(s) = \frac{k}{s^2 + 2\zeta\omega s + \omega^2},$$

where
    $\zeta$ is the natural frequency
    $\omega$ is the damping factor

A state–space representation of this system is

$$\begin{cases} \dot{x}(t) = \begin{bmatrix} -2\zeta\omega & -\omega^2 \\ 1 & 0 \end{bmatrix} x(t) + \begin{bmatrix} 1 \\ 0 \end{bmatrix} u(t) \\ y(t) = \begin{bmatrix} 0 & k \end{bmatrix} x(t) \end{cases}.$$

When the system parameters are given, we can solve the corresponding LMI problem (5.30) with the MATLAB® function mincx in the MATLAB LMI toolbox.

Case I: $\zeta = 0.1$, $\omega = 1$, and $k = 1$
    The solution is obtained as

$$\gamma = 5.026, \quad P = \begin{bmatrix} 1.005 & 0.101 \\ 0.101 & 1.005 \end{bmatrix} > 0.$$

Case II: $\zeta = 1$, $\omega = 3$, and $k = 20$
    The solution is obtained as

$$\gamma = 2.222, \quad P = \begin{bmatrix} 13.334 & 19.985 \\ 19.985 & 120.007 \end{bmatrix} > 0.$$

## 5.2.2 *Deduced Conditions*

Using Theorem 5.3 and the projection lemma (Lemma 2.11), we can further give another pair of alternative conditions for $\|G(s)\|_\infty < \gamma$.

**Theorem 5.5**     Given an arbitrarily positive scalar $\gamma$, the transfer function matrix

$$G(s) = C(sI - A)^{-1} B + D \tag{5.32}$$

satisfies

$$\|G(s)\|_\infty < \gamma \tag{5.33}$$

if and only if there exist a symmetric matrix $X > 0$, and a matrix $\Omega$, such that

$$\Theta + \Phi^{\mathrm{T}}\Omega\Psi + \Psi^{\mathrm{T}}\Omega^{\mathrm{T}}\Phi < 0 \qquad (5.34)$$

with

$$\begin{cases} \Theta = \begin{bmatrix} 0 & X & 0 & 0 & 0 \\ X & -X & 0 & 0 & 0 \\ 0 & 0 & -\gamma I_m & 0 & 0 \\ 0 & 0 & 0 & -X & 0 \\ 0 & 0 & 0 & 0 & \gamma I_r \end{bmatrix} \\[3em] \Phi = \begin{bmatrix} -I_n & A^{\mathrm{T}} & C^{\mathrm{T}} & I_n & 0 \\ 0 & B^{\mathrm{T}} & D^{\mathrm{T}} & 0 & -\gamma I_r \end{bmatrix} \\[1.5em] \Psi = \begin{bmatrix} I_n & 0 & 0 & 0 & 0 \\ 0 & 0 & 0 & 0 & I_r \end{bmatrix}. \end{cases} \qquad (5.35)$$

*Proof*  According to the first condition in Theorem 5.3, (5.33) is satisfied if and only if there exists $P > 0$, such that

$$\begin{bmatrix} A^{\mathrm{T}}P + PA & PB & C^{\mathrm{T}} \\ B^{\mathrm{T}}P & -\gamma I & D^{\mathrm{T}} \\ C & D & -\gamma I \end{bmatrix} < 0, \qquad (5.36)$$

which can be turned, via exchanging the last two rows and the last two columns, into

$$\begin{bmatrix} A^{\mathrm{T}}P + PA & C^{\mathrm{T}} & PB \\ C & -\gamma I & D \\ B^{\mathrm{T}}P & D^{\mathrm{T}} & -\gamma I \end{bmatrix} < 0. \qquad (5.37)$$

Applying the Schur complement lemma to the LMI (5.37), we obtain

$$\begin{bmatrix} A^{\mathrm{T}}P + PA & C^{\mathrm{T}} \\ C & -\gamma I \end{bmatrix} + \frac{1}{\gamma}\begin{bmatrix} PB \\ D \end{bmatrix}\begin{bmatrix} B^{\mathrm{T}}P & D^{\mathrm{T}} \end{bmatrix} < 0,$$

which is equivalent to

$$\begin{bmatrix} A^{\mathrm{T}}P + PA + \frac{1}{\gamma}PBB^{\mathrm{T}}P & C^{\mathrm{T}} + \frac{1}{\gamma}PBD^{\mathrm{T}} \\ C + \frac{1}{\gamma}DB^{\mathrm{T}}P & -\gamma I + \frac{1}{\gamma}DD^{\mathrm{T}} \end{bmatrix} < 0.$$

Letting $X = P^{-1}$, and multiplying both sides of the aforementioned LMI by $\text{diag}(X, I_n)$, gives

$$\begin{bmatrix} XA^\mathrm{T} + AX + \frac{1}{\gamma}BB^\mathrm{T} & XC^\mathrm{T} + \frac{1}{\gamma}BD^\mathrm{T} \\ C + \frac{1}{\gamma}DB^\mathrm{T} & -\gamma + \frac{1}{\gamma}DD^\mathrm{T} \end{bmatrix} < 0,$$

which can clearly be written, according to Lemma 2.9, as

$$\begin{bmatrix} XA^\mathrm{T} + AX - X + \frac{1}{\gamma}BB^\mathrm{T} & XC^\mathrm{T} + \frac{1}{\gamma}BD^\mathrm{T} & X \\ CX + \frac{1}{\gamma}DB^\mathrm{T} & -\gamma I + \frac{1}{\gamma}DD^\mathrm{T} & 0 \\ X & 0 & -X \end{bmatrix} < 0. \tag{5.38}$$

On the other hand, with the matrices defined in (5.35), we can easily verify that

$$N_\Phi = \begin{bmatrix} A^\mathrm{T} & C^\mathrm{T} & I_n \\ I_n & 0 & 0 \\ 0 & I_m & 0 \\ 0 & 0 & I_n \\ \frac{1}{\gamma}B^\mathrm{T} & \frac{1}{\gamma}D^\mathrm{T} & 0 \end{bmatrix} \quad \text{and} \quad N_\Psi = \begin{bmatrix} 0 & 0 & 0 \\ I_n & 0 & 0 \\ 0 & I_m & 0 \\ 0 & 0 & I_n \\ 0 & 0 & 0 \end{bmatrix},$$

are the right orthogonal complements of $\Phi$ and $\Psi$, respectively. Furthermore, we can show that

$$N_\Phi^\mathrm{T} \Theta N_\Phi = \begin{bmatrix} XA^\mathrm{T} + AX - X + \frac{1}{\gamma}BB^\mathrm{T} & XC^\mathrm{T} + \frac{1}{\gamma}BD^\mathrm{T} & X \\ CX + \frac{1}{\gamma}DB^\mathrm{T} & -\gamma I + \frac{1}{\gamma}DD^\mathrm{T} & 0 \\ X & 0 & X \end{bmatrix},$$

and

$$N_\Psi^\mathrm{T} \Theta N_\Psi = \begin{bmatrix} -X & 0 & 0 \\ 0 & -\gamma I_m & 0 \\ 0 & 0 & -X \end{bmatrix}.$$

Therefore, the inequality (5.38) is equivalent to $N_\Phi^\mathrm{T} \Theta N_\Phi < 0$. Considering the definition of $X$, we also have $N_\Psi^\mathrm{T} \Theta N_\Psi < 0$. Then, by applying the projection lemma (Lemma 2.11), we know that there exists $P > 0$ satisfying (5.36) if and only if there exist a symmetric matrix $X > 0$ and a matrix $\Omega$ satisfying (5.34). Thus, the proof is completed. ∎

**Corollary 5.4**   Given a positive scalar $\gamma$, the transfer function matrix (5.32) satisfies (5.33) if there exist a symmetric matrix $X$, and a matrix $V$, such that

$$
\begin{bmatrix}
-V - V^T & V^T A^T + X & V^T C^T & V^T & 0 \\
AV + X & -X & 0 & 0 & B \\
CV & 0 & -\gamma I_m & 0 & D \\
V & 0 & 0 & -X & 0 \\
0 & B^T & D^T & 0 & -\gamma I_r
\end{bmatrix} < 0, \tag{5.39}
$$

or

$$
\begin{bmatrix}
-V - V^T & V^T A^T + X & V^T C^T & V^T & 0 \\
AV + X & -X & 0 & 0 & B \\
CV & 0 & -I_m & 0 & D \\
V & 0 & 0 & -X & 0 \\
0 & B^T & D^T & 0 & -\gamma^2 I_r
\end{bmatrix} < 0. \tag{5.40}
$$

*Proof*   Particularly choosing

$$
\Omega = \begin{bmatrix} V & 0 \\ 0 & I_r \end{bmatrix},
$$

gives

$$
\Theta + \Phi^T \Omega \Psi + \Psi^T \Omega^T \Phi
$$

$$
= \begin{bmatrix}
-V - V^T & V^T A^T + X & V^T C^T & V^T & 0 \\
AV + X & -X & 0 & 0 & B \\
CV & 0 & -\gamma I_m & 0 & D \\
V & 0 & 0 & -X & 0 \\
0 & B^T & D^T & 0 & -\gamma I_r
\end{bmatrix}.
$$

Therefore, the inequality (5.39) implies (5.34).

Further, the condition (5.40) can be shown to be equivalent to (5.39).   ■

Regarding solving the problem of minimizing $\gamma$ satisfying $\|G(s)\|_\infty < \gamma$ using Theorem 5.5 and its corollary, optimization problems similar to (5.30) and (5.31) can also be easily presented.

## 5.3 LMI Conditions for H₂ Index

Now let us focus on the rational matrix in the form of (5.2), but with $D = 0$. The problem we are interested in is to find criterion in terms of LMIs for the following $H_2$ norm condition:

$$||C (sI - A)^{-1} B||_2 < \gamma, \tag{5.41}$$

where $\gamma > 0$ is a given scalar.

### 5.3.1 Basic Conditions

**Theorem 5.6**    Let $A \in \mathbb{R}^{n \times n}$, $B \in \mathbb{R}^{n \times r}$, $C \in \mathbb{R}^{m \times n}$, and $\gamma > 0$ be a given scalar. Then, condition (5.41) holds if and only if one of the following statements holds true:

1. $\exists X > 0$, s.t.

$$AX + XA^T + BB^T < 0, \quad \text{trace}\left(CXC^T\right) < \gamma^2. \tag{5.42}$$

2. $\exists Y > 0$, s.t.

$$A^T Y + YA + C^T C < 0, \quad \text{trace}\left(B^T YB\right) < \gamma^2. \tag{5.43}$$

*Proof*    Sufficiency. Suppose that the two relations in (5.42) hold. Let

$$Q = AX + XA^T + BB^T < 0,$$

and recall the fact that the matrix $A$ is stable, we know that the Lyapunov matrix equation

$$AE + EA^T = Q,$$

has a unique solution $E > 0$. Note that

$$
\begin{aligned}
A (X - E) + (X - E) A^T + BB^T \\
= \left(AX + XA^T + BB^T\right) - (AE + EA^T) \\
= Q - Q \\
= 0,
\end{aligned}
$$

the matrix

$$P = X - E$$

thus satisfies the Lyapunov matrix equation

$$AP + PA^T + BB^T = 0. \tag{5.44}$$

It thus follows from Lemma 5.1 that

$$
\begin{aligned}
||G\,(s)||_2^2 &= \text{trace}\left(CPC^{\mathrm{T}}\right)\\
&= \text{trace}\left(CXC^{\mathrm{T}}\right) - \text{trace}\left(CEC^{\mathrm{T}}\right)\\
&\le \gamma^2 - \text{trace}\left(CEC^{\mathrm{T}}\right)\\
&< \gamma^2,
\end{aligned}
$$

which gives $||G\,(s)\,||_2 < \gamma$.

Necessity. Suppose $||G\,(s)\,||_2 < \gamma$ holds, and let $P$ be the solution to the Lyapunov matrix equation (5.44), then it follows from Lemma 5.1 again that

$$
\text{trace}\left(CPC^{\mathrm{T}}\right) = ||G\,(s)||_2^2 < \gamma^2. \tag{5.45}
$$

On the other hand, since $A$ is stable, there exists $P_0 > 0$ satisfying

$$
AP_0 + P_0 A^{\mathrm{T}} < 0.
$$

It clearly follows from (5.45) that there exists a sufficiently small scalar $\epsilon > 0$, such that the matrix

$$
X = P + \epsilon P_0 > 0
$$

satisfies

$$
\text{trace}\left(CXC^{\mathrm{T}}\right) < \gamma^2,
$$

meanwhile, it can be verified that

$$
\begin{aligned}
AX &+ XA^{\mathrm{T}} + BB^{\mathrm{T}}\\
&= AP + PA^{\mathrm{T}} + BB^{\mathrm{T}} + \epsilon\left(AP_0 + P_0 A^{\mathrm{T}}\right)\\
&= \epsilon\left(AP_0 + P_0 A^{\mathrm{T}}\right)\\
&< 0.
\end{aligned}
$$

The necessity is thus done. ∎

It is clearly seen from this theorem that condition (5.41) holds only if the matrix $A$ is stable.

Based on Theorem 5.6, the problem of solving a minimal $\gamma$ satisfying condition (5.41) can be equivalently converted into the following LMI problem

$$
\begin{cases}
\min & \rho\\
\text{s.t.} & AX + XA^{\mathrm{T}} + BB^{\mathrm{T}} < 0\\
& \text{trace}\left(CXC^{\mathrm{T}}\right) < \rho\\
& X > 0
\end{cases} \tag{5.46}
$$

or

$$\begin{cases} \min & \rho \\ \text{s.t.} & A^{\mathrm{T}}Y + YA + C^{\mathrm{T}}C < 0 \\ & \text{trace}\left(B^{\mathrm{T}}YB\right) < \rho \\ & Y > 0, \end{cases} \tag{5.47}$$

while the attenuation level $\gamma = \rho^{1/2}$.

**Example 5.2**

Consider Example 5.1 again. Our purpose is to solve the corresponding LMI problem (5.47) with the MATLAB function `mincx` in the MATLAB LMI toolbox. When $\zeta = 0.1$, $\omega = 1$, $k = 1$, the solution is obtained as

$$\rho = 2.500, \quad Y = \begin{bmatrix} 2.500 & 0.500 \\ 0.500 & 2.600 \end{bmatrix} > 0.$$

When $\zeta = 1$, $\omega = 3$, $k = 20$, the solution is obtained as

$$\rho = 3.703, \quad Y = \begin{bmatrix} 3.704 & 22.222 \\ 22.222 & 166.667 \end{bmatrix} > 0.$$

## 5.3.2 Deduced Conditions

Based on the aforementioned theorem, we can further give another pair of conditions for (5.41).

**Theorem 5.7**   Given a positive scalar $\gamma$, the transfer function matrix

$$G(s) = C\left(sI - A\right)^{-1}B$$

satisfies

$$||G(s)||_2 < \gamma \tag{5.48}$$

if and only if there exist symmetric matrices $Z$, $P$ and a matrix $V$, such that

$$\begin{cases} \text{trace}\left(Z\right) < \gamma^2 \\ \begin{bmatrix} -Z & C \\ C^{\mathrm{T}} & -P \end{bmatrix} < 0 \\ \begin{bmatrix} -\left(V + V^{\mathrm{T}}\right) & V^{\mathrm{T}}A + P & V^{\mathrm{T}}B & V^{\mathrm{T}} \\ A^{\mathrm{T}}V + P & -P & 0 & 0 \\ B^{\mathrm{T}}V & 0 & -I & 0 \\ V & 0 & 0 & -P \end{bmatrix} < 0, \end{cases}$$

or

$$\begin{cases} \operatorname{trace}(Z) < \gamma^2 \\ \begin{bmatrix} -Z & B^{\mathrm{T}} \\ B & -P \end{bmatrix} < 0 \\ \begin{bmatrix} -(V + V^{\mathrm{T}}) & V^{\mathrm{T}}A^{\mathrm{T}} + P & V^{\mathrm{T}}C^{\mathrm{T}} & V^{\mathrm{T}} \\ AV + P & -P & 0 & 0 \\ CV & 0 & -I & 0 \\ V & 0 & 0 & -P \end{bmatrix} < 0. \end{cases}$$

*Proof*    According to the first condition in Theorem 5.6 that (5.48) is satisfied if and only if there exists $X > 0$, such that

$$AX + XA^{\mathrm{T}} + BB^{\mathrm{T}} < 0, \quad \operatorname{trace}(CXC^{\mathrm{T}}) < \gamma^2. \tag{5.49}$$

It follows from Lemma 2.12 that the first inequality in (5.49) is equivalent to

$$\begin{bmatrix} BB^{\mathrm{T}} + X - (W^{\mathrm{T}} + W) & AX + W^{\mathrm{T}} \\ XA^{\mathrm{T}} + W & -X \end{bmatrix} < 0, \tag{5.50}$$

with $X$ being an arbitrary symmetric positive definite matrix. Letting

$$J = \operatorname{diag}(W^{-\mathrm{T}}, X^{-1}),$$

and pre- and postmultiplying both sides of (5.50) by $J$ and $J^{\mathrm{T}}$, respectively, yields

$$\begin{bmatrix} W^{-\mathrm{T}}BB^{\mathrm{T}}W^{-1} + W^{-\mathrm{T}}XW^{-1} - W^{-1} - W^{-\mathrm{T}} & W^{-\mathrm{T}}A + X^{-1} \\ A^{\mathrm{T}}W^{-1} + X^{-1} & -X^{-1} \end{bmatrix} < 0.$$

Further, denote $V = W^{-1}$, $P = X^{-1}$, we can write the LMI equivalently as follows:

$$\begin{bmatrix} V^{\mathrm{T}}BB^{\mathrm{T}}V + V^{\mathrm{T}}P^{-1}V - (V + V^{\mathrm{T}}) & V^{\mathrm{T}}A + P \\ A^{\mathrm{T}}V + P & -P \end{bmatrix} < 0. \tag{5.51}$$

Since

$$\begin{bmatrix} V^{\mathrm{T}}BB^{\mathrm{T}}V + V^{\mathrm{T}}P^{-1}V - (V + V^{\mathrm{T}}) & V^{\mathrm{T}}A + P \\ A^{\mathrm{T}}V + P & -P \end{bmatrix}$$

$$= \begin{bmatrix} -(V + V^{\mathrm{T}}) & V^{\mathrm{T}}A + P \\ A^{\mathrm{T}}V + P & -P \end{bmatrix} + \begin{bmatrix} V^{\mathrm{T}}BB^{\mathrm{T}}V + V^{\mathrm{T}}P^{-1}V & 0 \\ 0 & 0 \end{bmatrix}$$

$$= \begin{bmatrix} -(V + V^{\mathrm{T}}) & V^{\mathrm{T}}A + P \\ A^{\mathrm{T}}V + P & -P \end{bmatrix}$$

$$\quad + \begin{bmatrix} V^{\mathrm{T}}B & V^{\mathrm{T}} \\ 0 & 0 \end{bmatrix} \begin{bmatrix} I & 0 \\ 0 & P^{-1} \end{bmatrix} \begin{bmatrix} B^{\mathrm{T}}V & 0 \\ V & 0 \end{bmatrix},$$

according to Schur complement lemma, inequality (5.51) is equivalent to

$$\begin{bmatrix} -(V+V^{\mathrm{T}}) & V^{\mathrm{T}}A+P & V^{\mathrm{T}}B & V^{\mathrm{T}} \\ A^{\mathrm{T}}V+P & -P & 0 & 0 \\ B^{\mathrm{T}}V & 0 & -I & 0 \\ V & 0 & 0 & -P \end{bmatrix} < 0. \qquad (5.52)$$

By now, we have verified that the first inequality in (5.49) is equivalent to (5.52).

Next let us consider the second inequality in (5.49). In view of Lemma 2.13, trace $\left(CXC^{\mathrm{T}}\right) < \gamma^2$ is equivalent to

$$CXC^{\mathrm{T}} < Z, \quad \operatorname{trace}(Z) < \gamma^2,$$

which is clearly equivalent to

$$\begin{bmatrix} -Z & C \\ C^{\mathrm{T}} & -X^{-1} \end{bmatrix} < 0, \quad \operatorname{trace}(Z) < \gamma^2.$$

Recalling that $P = X^{-1}$, we obtain

$$\begin{bmatrix} -Z & C \\ C^{\mathrm{T}} & -P \end{bmatrix} < 0, \quad \operatorname{trace}(Z) < \gamma^2.$$

With this, the proof of the first conclusion is completed.

The second one can be verified similarly using the second condition in Theorem 5.6. ■

Regarding solving a minimal $\gamma$ satisfying condition (5.41) using Theorem 5.7, optimization problems similar to (5.46) and (5.47) can be similarly formulated.

## 5.4 Notes and References

Chapter 5 provides the LMI conditions for $H_\infty$ and $H_2$ indices. We have summarized these conditions in Tables 5.1 and 5.2, respectively.

The $H_\infty$ norm of a transfer function is the supremum of the maximum singular value of the frequency response of the system. For a stable system, the $H_\infty$ norm of the transfer function coincides with the $L_2$-induced norm of the input–output operator associated with the system (Theorem 5.1).

The squared $H_2$ norm of the transfer function coincides with the total "output energy" in the impulse responses of the system. The $H_2$ norm also has an interpretation that coincides with the $L_P$-induced norm of the input–output operator associated with the system (Theorem 5.2).

What we have given in this chapter about $H_\infty$ and $H_2$ indices are very basic. As a matter of fact, these concepts have generalizations in many directions. For example,

**Table 5.1  Criteria for $\|G(s)\|_\infty < \gamma$**

| | Criteria | Variables |
|---|---|---|
| 1 | $P > 0$ <br><br> $\begin{bmatrix} A^TP + PA & PB & C^T \\ B^TP & -\gamma I & D^T \\ C & D & -\gamma I \end{bmatrix} < 0$ | $P$ |
| 2 | $X > 0$ <br><br> $\begin{bmatrix} XA^T + AX & B & XC^T \\ B^T & -\gamma I & D^T \\ CX & D & -\gamma I \end{bmatrix} < 0$ | $X$ |
| 3 | $\begin{bmatrix} -V - V^T & V^TA^T + X & V^TC^T & V^T & 0 \\ AV + X & -X & 0 & 0 & B \\ CV & 0 & -\gamma I_m & 0 & D \\ V & 0 & 0 & -X & 0 \\ 0 & B^T & D^T & 0 & -\gamma I_r \end{bmatrix} < 0$ | $X, V$ |

Oliveira et al. (2002) extends $H_2$ and $H_\infty$ norm characterizations for discrete-time systems, and Zhou et al. (1995) generalizes robust $H_\infty$ performance to systems with structured uncertainties.

## Exercises

**5.1** Prove Corollary 5.3, that is, let $G(s) = C(sI - A)^{-1}B + D$, then the matrix $A$ is Hurwitz stable if $\|G(s)\|_\infty < \gamma$ for some $0 < \gamma < \infty$.

**5.2** (Scherer and Weiland 2000) Let

$$A = \begin{bmatrix} \sin \alpha & \cos \alpha \\ -\cos \alpha & \sin \alpha \end{bmatrix}, \quad 0 \le \alpha \le \pi.$$

Show that the solution $X$ of the Lyapunov equation $A^TX + XA + I = 0$ diverges in the sense that $\det(X) \to \infty$ whenever $\alpha \to 0$.

**Table 5.2  Criteria for $\|G(s)\|_2 < \gamma$**

| | Criteria | Variables |
|---|---|---|
| 1 | $X > 0$<br><br>$\mathrm{trace}\left(CXC^\mathsf{T}\right) < \gamma^2$<br><br>$AX + XA^\mathsf{T} + BB^\mathsf{T} < 0$ | $X$ |
| 2 | $Y > 0$<br><br>$\mathrm{trace}\left(B^\mathsf{T}YB\right) < \gamma^2$<br><br>$A^\mathsf{T}Y + YA + C^\mathsf{T}C < 0$ | $Y$ |
| 3 | $\mathrm{trace}\,(Z) < \gamma^2$<br><br>$\begin{bmatrix} -Z & C \\ C^\mathsf{T} & -P \end{bmatrix} < 0$<br><br>$\begin{bmatrix} -\left(V + V^\mathsf{T}\right) & V^\mathsf{T}A + P & V^\mathsf{T}B & V^\mathsf{T} \\ A^\mathsf{T}V + P & -P & 0 & 0 \\ B^\mathsf{T}V & 0 & -I & 0 \\ V & 0 & 0 & -P \end{bmatrix} < 0.$ | $P, Z, V$ |

**5.3** Consider the following system

$$\begin{cases} \dot{x} = Ax + Bu \\ y = Cx + Du, \end{cases}$$

where

$$A = \begin{bmatrix} -5 & 1 & 2 \\ 1 & -9 & 1 \\ -1 & -10 & -3 \end{bmatrix}, \quad B = \begin{bmatrix} 0 \\ 1 \\ 0 \end{bmatrix},$$

$$C = \begin{bmatrix} 1 & 0 & 0 \\ 0 & 0 & 1 \end{bmatrix}, \quad D = \begin{bmatrix} 1 \\ 1 \end{bmatrix}.$$

Find a minimal $\gamma$ such that the system transfer function

$$G(s) = C(sI - A)^{-1}B + D$$

satisfies $\|G(s)\|_\infty < \gamma$.

**5.4** Consider the system in Exercise 5.3 again. Find a minimal $\gamma$ such that the system transfer function

$$G(s) = C(sI - A)^{-1}B$$

satisfies $\|G(s)\|_2 < \gamma$.

**5.5** Consider a jet transport aircraft system studied by Liu et al. (2000a) and Grace et al. (1992), whose model is in the form of (5.15) with $D = 0$, and

$$A = \begin{bmatrix} -0.0558 & -0.9968 & 0.0802 & 0.0415 \\ 0.5980 & -0.1150 & -0.0318 & 0 \\ -3.0500 & 0.3880 & -0.4650 & 0 \\ 0 & 0.0805 & 1.0000 & 0 \end{bmatrix},$$

$$B = \begin{bmatrix} 0.0729 & 0.0001 \\ -4.7500 & 1.2300 \\ 1.5300 & 10.6300 \\ 0 & 0 \end{bmatrix}, \quad C = \begin{bmatrix} 0 & 1 & 0 & 0 \\ 0 & 0 & 0 & 1 \end{bmatrix}.$$

Find a minimal $\gamma$ such that the system transfer function

$$G(s) = C(sI - A)^{-1}B + D$$

satisfies

$$\|G(s)\|_\infty < \gamma.$$

**5.6** Consider the system in Exercise 5.5. Find a minimal $\rho$, such that $\|G(s)\|_2^2 < \rho$.

**5.7** Consider the transfer function

$$G(s) = C^T(sI - A)^{-1}B,$$

with

$$A = \begin{bmatrix} -1 & 0 \\ -2 & -1 \end{bmatrix}, \quad B = \begin{bmatrix} 1 \\ 0 \end{bmatrix}, \quad C = \begin{bmatrix} 1 & 1 \end{bmatrix}.$$

First find $\|G(s)\|_2$ using Lemma 5.1, and then compute it via solving (5.46) or (5.47), and compare the values of $\|G(s)\|_2$ obtained through the two different ways.

# *Chapter 6*

# Property Analysis

In this chapter, we examine certain properties of control systems, which include the stabilizability and detectability, dissipativity, passivity and positive-realness, nonexpansivity, and bounded-realness.

In the first two sections, we consider the stabilizability and detectability of the following linear system

$$\begin{cases} \rho x = Ax + Bu \\ y = Cx + Du, \end{cases} \tag{6.1}$$

where

- $\rho$ represents the differential operator (when the system is continuous-time), or the one step forward shift operator (when the system is discrete-time)
- $x \in \mathbb{R}^n$, $y \in \mathbb{R}^m$ and $u \in \mathbb{R}^r$ are the state vector, the output vector, and the input vector, respectively
- $A$, $B$, $C$, and $D$ are the system matrices of appropriate dimensions

## 6.1 Hurwitz Stabilizability and Detectability

Corresponding to the continuous-time systems and the discrete-time systems, respectively, we have the concepts of Hurwitz stabilizability and detectability and Schur stabilizability and detectability. In normal linear systems context, for example, Duan (2004a), Hurwitz stabilizability and detectability are simply termed as stabilizability and detectability.

### *6.1.1 Hurwitz Stabilizability*

#### *6.1.1.1 Definition and the PBH Criterion*

Hurwitz stabilizability is a very important concept in linear systems theory, which is defined as follows (Duan 2004a, 2010).

**Definition 6.1** The system (6.1), or the matrix pair $(A, B)$, is said to be Hurwitz stabilizable if there exists a real matrix $K$ such that $A + BK$ is Hurwitz stable.

For condition of Hurwitz stabilizability of a given matrix pair $(A, B)$, we have the following well-known PBH criterion (Duan 2004a, 2010).

**Theorem 6.1** The system (6.1) is Hurwitz stabilizable if and only if

$$\text{rank} \begin{bmatrix} sI - A & B \end{bmatrix} = n, \quad \forall s \in \mathbb{C}, \quad \text{Re}(s) \geq 0;$$

or

$$\text{rank} \begin{bmatrix} sI - A & B \end{bmatrix} = n, \quad \forall s \in \lambda(A), \quad \text{Re}(s) \geq 0.$$

The PBH criterion tells us that the system (6.1) is Hurwitz stabilizable if all its uncontrollable modes are Hurwitz stable.

#### *6.1.1.2 LMI Conditions*

In this section, we present a few LMI conditions for Hurwitz stabilizability.

**Theorem 6.2** The system (6.1), or the matrix pair $(A, B)$, is Hurwitz stabilizable if and only if there exist a symmetric positive definite matrix $P$ and a matrix $W$ satisfying

$$AP + PA^{\text{T}} + BW + W^{\text{T}}B^{\text{T}} < 0. \tag{6.2}$$

*Proof* Necessity: Following the definition of Hurwitz stabilizability and the well-known Lyapunov stability theory, we know that the first statement holds if and only if there exist a matrix $K$ and a matrix $P > 0$ satisfying

$$(A + BK) P + P (A + BK)^{\text{T}} < 0. \tag{6.3}$$

Letting

$$W = KP,$$

and substituting this into (6.3) gives (6.2).

Sufficiency: Suppose that there exist a symmetric positive definite matrix $P$ and a matrix $W$ satisfying (6.2). Then by letting

$$K = WP^{-1},$$

we obtain (6.3). With this we complete the proof. ■

The condition in the aforementioned theorem is clearly an LMI. Using the well-known projection lemma, we can further show the following result.

**Theorem 6.3**   System (6.1) is Hurwitz stabilizable if and only if there exists a symmetric positive definite matrix $P$ satisfying

$$N_b^{\mathrm{T}} \left( AP + PA^{\mathrm{T}} \right) N_b < 0, \tag{6.4}$$

with $N_b$ being the right orthogonal complement of $B^{\mathrm{T}}$.

*Proof*   It suffices to show the equivalence between (6.2) and (6.4). To do this, we make use of the projection lemma (Lemma 2.11).

Let

$$H = AP + PA^{\mathrm{T}}, \quad X = W, \quad Q = B^{\mathrm{T}}, \quad \hat{P} = I_n,$$

then the inequality (6.2) can be written as

$$H + \hat{P}^{\mathrm{T}} X^{\mathrm{T}} Q + Q^{\mathrm{T}} X \hat{P} < 0.$$

By Lemma 2.11, this is equivalent to

$$N_{\hat{p}}^{\mathrm{T}} H N_{\hat{p}} < 0, \quad N_q^{\mathrm{T}} H N_q < 0.$$

Since $\hat{P} = I_n$, $N_{\hat{p}}$ is of zero dimension, and thus the condition $N_{\hat{p}}^{\mathrm{T}} H N_{\hat{p}} < 0$ vanishes, while the second one is identical to (6.4). Thus, the equivalence directly follows from Lemma 2.11. ■

**Theorem 6.4**   System (6.1) is Hurwitz stabilizable if and only if there exists a symmetric positive definite matrix $P$ such that

$$AP + PA^{\mathrm{T}} < \gamma BB^{\mathrm{T}}, \tag{6.5}$$

for some scalar $\gamma > 0$.

*Proof*  Necessity: Let the system (6.1) be Hurwitz stabilizable, then it follows from theory of linear quadratic optimal control that, under the stabilizability condition, there exists a positive definite matrix $X$ satisfying the following Riccati matrix equation (Duan 2004a)

$$A^{\mathrm{T}}X + XA - \gamma XBB^{\mathrm{T}}X + I = 0.$$

Pre- and postmultiplying by $P = X^{-1}$ both sides of the aforementioned equation gives

$$PA^{\mathrm{T}} + AP - \gamma BB^{\mathrm{T}} = -PP < 0.$$

This shows that (6.5) holds.

Sufficiency: Now suppose that there exists a symmetric positive definite matrix $P$ satisfying (6.5), then by choosing

$$K = -\frac{\gamma}{2}B^{\mathrm{T}}P^{-1}, \tag{6.6}$$

we have

$$(A + BK)P + P(A + BK)^{\mathrm{T}} = AP + PA^{\mathrm{T}} - \gamma BB^{\mathrm{T}} < 0.$$

Therefore, the system (6.1) is stabilizable. The proof is then completed.  ■

**Remark 6.1**  It is suggested that in applications we can simply take $\gamma = 1$ in the inequality (6.5). In this case, (6.5) turns into

$$AP + PA^{\mathrm{T}} < BB^{\mathrm{T}}, \tag{6.7}$$

and the corresponding gain matrix is given by

$$K = -\frac{1}{2}B^{\mathrm{T}}P^{-1}. \tag{6.8}$$

This is not only because that this case is simple but also due to the fact that it would be in vain to adjust the parameter $\gamma$. In fact, once (6.5) has a solution $P$, (6.7) has a solution $P_0 = P/\gamma$. Thus, for arbitrary scalar $\gamma > 0$, there holds

$$K = -\frac{\gamma}{2}B^{\mathrm{T}}P^{-1} = -\frac{\gamma}{2}B^{\mathrm{T}}(\gamma P_0)^{-1} = -\frac{1}{2}B^{\mathrm{T}}P_0^{-1}.$$

This clearly shows that different choice of $\gamma$ really does not change the value of the feedback gain matrix $K$.

It is clear that the conditions (6.2), (6.4), and (6.5) in the three theorems are all LMIs. Compared with the second-rank condition in Theorem 6.1, the LMI conditions have an advantage in computation. For some high-order systems, the rank condition may not be reliable, while the LMI conditions have good numerical reliability.

**Example 6.1**

Duan (2003) and Kimura (1977) considered a linear system in the form of (6.1) with $D = 0$ and

$$
A = \begin{bmatrix}
-1 & 0 & 1 & 1 & 0 & 0 & -1 & 2 \\
0 & 0 & 1 & 0 & -3 & -1 & 0 & 0 \\
-1 & -2 & 1 & 0 & -3 & -1 & 0 & 0 \\
0 & 0 & 0 & 0 & 1 & 0 & 0 & 0 \\
0 & 0 & -1 & 0 & -3 & 0 & 4 & 2 \\
0 & 0 & 0 & 0 & 0 & 0 & 1 & 0 \\
0 & 0 & 0 & 0 & 0 & 0 & 0 & 1 \\
-1 & 2 & 0 & 0 & 1 & -1 & -1 & 2
\end{bmatrix},
\quad
B = \begin{bmatrix}
1 & 0 & 0 & 0 \\
0 & 0 & 0 & 0 \\
0 & 1 & 0 & 0 \\
0 & 0 & 0 & 0 \\
0 & 0 & 1 & 0 \\
0 & 0 & 0 & 0 \\
0 & 0 & 0 & 0 \\
0 & 0 & 0 & 0
\end{bmatrix},
$$

$$
C = \begin{bmatrix}
17 & 23 & 3 & 14 & -15 & 5 & 25 & 5 \\
3 & -3 & 2 & 3 & -2 & 2 & -3 & -6 \\
-2 & 1 & 0 & -1 & 2 & 0 & 1 & 4
\end{bmatrix}.
$$

Applying the function `feasp` in the MATLAB® LMI toolbox to the corresponding LMI problem (6.7) gives the following solution

$$
P = \begin{bmatrix}
75.8772 & -21.1790 & 40.3367 & -22.1019 \\
-21.1790 & 36.1906 & 11.6049 & 13.9642 \\
40.3367 & 11.6049 & 79.7923 & -4.5277 \\
-22.1019 & 13.9642 & -4.5277 & 12.3702 \\
8.6120 & 2.9424 & 17.8700 & -5.0391 \\
-28.3812 & 14.2712 & -21.5186 & 13.7992 \\
5.0920 & 15.7686 & 36.8485 & 2.4367 \\
24.3097 & -29.2853 & -7.3904 & -11.6437
\end{bmatrix}
$$

$$
\begin{bmatrix}
8.6120 & -28.3812 & 5.0920 & 24.3097 \\
2.9424 & 14.2712 & 15.7686 & -29.2853 \\
17.8700 & -21.5186 & 36.8485 & -7.3904 \\
-5.0391 & 13.7992 & 2.4367 & -11.6437 \\
12.0971 & -5.8385 & 5.1002 & -1.4904 \\
-5.8385 & 23.7600 & -10.5280 & -9.2869 \\
5.1002 & -10.5280 & 27.9910 & -16.0292 \\
-1.4904 & -9.2869 & -16.0292 & 26.3990
\end{bmatrix} > 0.
$$

Thus, the system is Hurwitz stabilizable. On the other hand, it can be obtained that

$$\lambda(A) = \{-4.1259, 2.0444 \pm 2.0378i, 0.0038 \pm 1.3685i, -0.5280$$
$$\pm 0.6694i, 0.0855\}$$

and it can also be easily verified that

$$\text{rank}\begin{bmatrix} sI - A & B \end{bmatrix} = 8,$$

for $s = 2.0444 \pm 2.0378i, 0.0038 \pm 1.3685i, 0.0855$. Thus, the conclusion is true following the PBH criterion.

## 6.1.2 Hurwitz Detectability

Hurwitz detectability is a dual concept of Hurwitz stabilizability and is defined as follows (Duan 2004a, 2010).

**Definition 6.2** The system (6.1), or the matrix pair $(A, C)$, is said to be Hurwitz detectable if there exists a real matrix $L$ such that $A + LC$ is Hurwitz stable.

Note that

$$\lambda(A) = \lambda(A^{\mathrm{T}}),$$

and

$$(A + LC)^{\mathrm{T}} = A^{\mathrm{T}} + C^{\mathrm{T}}K,$$

with $K = L^{\mathrm{T}}$, we know immediately that there exists an $L$ such that $(A + LC)$ is Hurwitz stable if and only if there exists a $K$ such that $A^{\mathrm{T}} + C^{\mathrm{T}}K$ is Hurwitz stable. This implies the following duality principle.

**Proposition 6.1** A linear system $(A, B, C)$ of appropriate dimensions is Hurwitz stabilizable if and only if its dual system $(A^{\mathrm{T}}, C^{\mathrm{T}}, B^{\mathrm{T}})$ is Hurwitz detectable.

It follows from Theorems 6.1 through 6.4, and the duality principle, we immediately have the following results for Hurwitz detectability.

**Theorem 6.5** The system (6.1) is Hurwitz detectable if and only if one of the following condition holds:

1. All the unobservable modes of system (6.1) are stable, that is,

$$\text{rank} \begin{bmatrix} sI - A \\ C \end{bmatrix} = n, \quad \forall s, \quad \text{Re}(s) \geq 0. \tag{6.9}$$

2. There exist a symmetric positive definite matrix $P$ and a matrix $W$ satisfying

$$A^T P + PA + W^T C + C^T W < 0. \tag{6.10}$$

3. There exists a symmetric positive definite matrix $P$ satisfying

$$N_c^T \left( A^T P + PA \right) N_c < 0, \tag{6.11}$$

with $N_c$ being the right orthogonal complement of $C$.

4. There exists a symmetric positive definite matrix $P$ such that

$$A^T P + PA < \gamma C^T C, \tag{6.12}$$

for some scalar $\gamma > 0$

**Example 6.2**

Consider a jet transport aircraft system studied by Liu et al. (2000a) and Grace et al. (1992), whose model is in the form of (6.1) with $D = 0$ and

$$A = \begin{bmatrix} -0.0558 & -0.9968 & 0.0802 & 0.0415 \\ 0.5980 & -0.1150 & -0.0318 & 0.0000 \\ -3.0500 & 0.3880 & -0.4650 & 0.0000 \\ 0.0000 & 0.0805 & 1.0000 & 0.0000 \end{bmatrix},$$

$$B = \begin{bmatrix} 0.0729 & 0.0001 \\ -4.7500 & 1.2300 \\ 1.5300 & 10.6300 \\ 0.0000 & 0.0000 \end{bmatrix}, \quad C = \begin{bmatrix} 0 & 1 & 0 & 0 \\ 0 & 0 & 0 & 1 \end{bmatrix}.$$

Applying the MATLAB function `feasp` in the MATLAB LMI toolbox to the corresponding LMI problem (6.12), with $\gamma = 1$, gives the following solution

$$P = \begin{bmatrix} 0.9581 & -0.2349 & 0.0996 & 0.1333 \\ -0.2349 & 1.2111 & -0.0202 & 0.0314 \\ 0.0996 & -0.0202 & 0.1933 & -0.1444 \\ 0.1333 & 0.0314 & -0.1444 & 0.3248 \end{bmatrix} > 0.$$

Thus, the system is Hurwitz detectable.

# 6.2 Schur Stabilizability and Detectability

Corresponding to Schur stability of discrete-time linear systems, we have the concepts of Schur stabilizability and detectability. They are a pair of very important concepts in linear systems theory.

## 6.2.1 Schur Stabilizability

### 6.2.1.1 Definition and the PBH Criterion

First, let us investigate Schur stabilizability, which is defined as follows.

**Definition 6.3**    The system (6.1), or the matrix pair $(A, B)$, is said to be Schur stabilizable if there exists a real matrix $K$ such that $A + BK$ is Schur stable.

For condition for Schur stabilizability, we have the following well-known PBH criterion.

**Theorem 6.6**    The system (6.1) is Schur stabilizable if and only if

$$\text{rank} \begin{bmatrix} sI - A & B \end{bmatrix} = n, \quad \forall s \in \mathbb{C}, \quad |s| \geq 1,$$

or

$$\text{rank} \begin{bmatrix} sI - A & B \end{bmatrix} = n, \quad \forall s \in \lambda(A), \quad |s| \geq 1.$$

The aforementioned result states that all the uncontrollable modes of a Schur stabilizable system (6.1) must be Schur stable.

### 6.2.1.2 LMI Conditions

In this section, we will present three LMI conditions for Schur stabilizability of system (6.1).

**Theorem 6.7**    The system (6.1) is Schur stabilizable if and only if there exist a symmetric matrix $P$ and a matrix $W$ satisfying

$$\begin{bmatrix} -P & AP + BW \\ PA^{\mathrm{T}} + W^{\mathrm{T}} B^{\mathrm{T}} & -P \end{bmatrix} < 0. \tag{6.13}$$

*Proof*    Necessity: Following Proposition 4.2, we know that the system (6.1) is Schur stabilizable if and only if there exist a matrix $K$ and a symmetric matrix $P$ satisfying

$$\begin{bmatrix} -P & (A + BK) P \\ P (A + BK)^{\mathrm{T}} & -P \end{bmatrix} < 0. \tag{6.14}$$

Letting

$$W = KP,$$

and substituting it into (6.14) gives (6.13).

Sufficiency: Suppose that there exist a symmetric matrix $P$ and a matrix $W$ satisfying (6.13). Then by letting

$$K = WP^{-1},$$

we obtain (6.14). Thus, the system (6.1) is Schur stabilizable. With this we complete the proof of the theorem. ∎

Corresponding to Theorem 6.3, the following theorem further gives another LMI condition for Schur stabilizability of linear systems.

**Theorem 6.8** System (6.1) is Schur stabilizable if and only if there exists a symmetric matrix $P$ satisfying

$$\begin{bmatrix} -N_b^{\mathrm{T}} P N_b & N_b^{\mathrm{T}} AP \\ PA^{\mathrm{T}} N_b & -P \end{bmatrix} < 0, \tag{6.15}$$

with $N_b$ being the right orthogonal complement of $B^{\mathrm{T}}$.

*Proof* It suffices to show the equivalence between (6.13) and (6.15). To do this we make use of the projection lemma (Lemma 2.11). Corresponding to the notations in Lemma 2.11, here we take

$$H = \begin{bmatrix} -P & AP \\ PA^{\mathrm{T}} & -P \end{bmatrix}, \quad X = \begin{bmatrix} 0 & W \end{bmatrix}, \quad Q = \begin{bmatrix} B^{\mathrm{T}} & 0 \end{bmatrix}, \quad \hat{P} = I_{2n},$$

then the inequality (6.13) can be written as

$$H + \hat{P}^{\mathrm{T}} X^{\mathrm{T}} Q + Q^{\mathrm{T}} X \hat{P} < 0.$$

By Lemma 2.11, this is equivalent to

$$N_{\hat{p}}^{\mathrm{T}} H N_{\hat{p}} < 0, \quad N_q^{\mathrm{T}} H N_q < 0.$$

Let $N_b$ be the right orthogonal complement of $B^{\mathrm{T}}$, then it is easy to verify that

$$N_q = \begin{bmatrix} N_b & 0 \\ 0 & I_n \end{bmatrix}$$

is the right orthogonal complement of $Q$. Since $\hat{P} = I_{2n}$, $N_{\hat{P}}$ is of zero dimension, and thus does not exist. Correspondingly, the condition $N_{\hat{P}}^{\mathrm{T}} HN_{\hat{P}} < 0$ vanishes, while the second one is identical with

$$N_q^{\mathrm{T}} \begin{bmatrix} -P & AP \\ PA^{\mathrm{T}} & -P \end{bmatrix} N_q < 0,$$

which is equivalent to (6.15) in view of the format of $N_q$. Therefore, it follows from Lemma 2.11 that (6.13) and (6.15) are equivalent. The proof is done. ■

Corresponding to Theorem 6.4, we have the following theorem that gives a different LMI condition for Schur stabilizability of linear systems.

**Theorem 6.9**    System (6.1) is Schur stabilizable if and only if there exists a symmetric matrix $P$ such that

$$\begin{bmatrix} -P & PA^{\mathrm{T}} \\ AP & -P - \gamma BB^{\mathrm{T}} \end{bmatrix} < 0, \quad \gamma \geq 1. \tag{6.16}$$

*Proof*    Necessity: Let the system (6.1) be Schur stabilizable, it then follows from the Riccati equation theory for discrete-time linear systems that, under the Schur stabilizability condition, there exists a positive definite matrix $X$ satisfying the following Riccati matrix equation

$$A^{\mathrm{T}} \left( X - XB \left( \frac{1}{\gamma} I + B^{\mathrm{T}} XB \right)^{-1} B^{\mathrm{T}} X \right) A - X + I = 0.$$

Pre- and postmultiplying by $P = X^{-1}$ both sides of the aforementioned equation gives

$$PA^{\mathrm{T}} \left( X - XB \left( \frac{1}{\gamma} I + B^{\mathrm{T}} XB \right)^{-1} B^{\mathrm{T}} X \right) AP - P = -PP < 0. \tag{6.17}$$

Using Corollary 2.1, we have

$$X - XB \left( \frac{1}{\gamma} I + B^{\mathrm{T}} XB \right)^{-1} B^{\mathrm{T}} X = \left( X^{-1} + \gamma BB^{\mathrm{T}} \right)^{-1}. \tag{6.18}$$

Substituting this relation together with $P = X^{-1}$ into (6.17) yields

$$PA^{\mathrm{T}} \left( P + \gamma BB^{\mathrm{T}} \right)^{-1} AP - P < 0, \tag{6.19}$$

which can be easily shown to be equivalent to (6.16) with the help of the Schur complement lemma.

Sufficiency: Now suppose that there exists a symmetric positive definite matrix $P$ satisfying (6.16). Choosing

$$K = -\left(\frac{2}{\gamma}I + B^{\mathrm{T}}P^{-1}B\right)^{-1} B^{\mathrm{T}}P^{-1}A$$

$$= -\left(\frac{2}{\gamma}I + B^{\mathrm{T}}XB\right)^{-1} B^{\mathrm{T}}XA, \quad X = P^{-1}, \quad \gamma \geq 1, \qquad (6.20)$$

then for this gain $K$, we need only to show

$$(A + BK)^{\mathrm{T}} X (A + BK) - X < 0. \qquad (6.21)$$

First, with the help of Corollary 2.1, again, we have

$$A + BK$$

$$= A - B\left(\frac{2}{\gamma} + B^{\mathrm{T}}P^{-1}B\right)^{-1} B^{\mathrm{T}}P^{-1}A$$

$$= P\left(P^{-1} - P^{-1}B\left(\frac{2}{\gamma} + B^{\mathrm{T}}P^{-1}B\right)^{-1} B^{\mathrm{T}}P^{-1}\right) A$$

$$= P\left(P + \frac{\gamma}{2}BB^{\mathrm{T}}\right)^{-1} A,$$

thus

$$(A + BK)^{\mathrm{T}} P^{-1} (A + BK) - P^{-1}$$

$$= A^{\mathrm{T}}\left(P + \frac{\gamma}{2}BB^{\mathrm{T}}\right)^{-1} P \left(P + \frac{\gamma}{2}BB^{\mathrm{T}}\right)^{-1} A - P^{-1}$$

$$= A^{\mathrm{T}}\left(P + \gamma BB^{\mathrm{T}} + \frac{\gamma^2}{4}BB^{\mathrm{T}}P^{-1}BB^{\mathrm{T}}\right)^{-1} A - P^{-1}. \qquad (6.22)$$

Second, noticing $P > 0$, $\gamma \geq 1$, and $BB^{\mathrm{T}} \geq 0$, we clearly have

$$\frac{\gamma^2}{4}BB^{\mathrm{T}}P^{-1}BB^{\mathrm{T}} \geq 0,$$

hence

$$P + \gamma BB^{\mathrm{T}} + \frac{\gamma^2}{4}BB^{\mathrm{T}}P^{-1}BB^{\mathrm{T}} \geq P + BB^{\mathrm{T}}.$$

With the help of the aforementioned relation and (6.16), we further have

$$
\begin{bmatrix} -P & PA^T \\ AP & -P - \gamma BB^T - \frac{\gamma^2}{4} BB^T P^{-1} BB^T \end{bmatrix} \le \begin{bmatrix} -P & PA^T \\ AP & -P - \gamma BB^T \end{bmatrix} < 0,
$$

which is equivalent, according to the Schur complement lemma, to

$$
PA^T \left( P + \gamma BB^T + \frac{\gamma^2}{4} BB^T P^{-1} BB^T \right)^{-1} AP - P < 0.
$$

Pre- and postmultiplying both sides of the aforementioned inequality by $P^{-1}$ yields

$$
A^T \left( P + \gamma BB^T + \frac{\gamma^2}{4} BB^T P^{-1} BB^T \right)^{-1} A - P^{-1} < 0. \qquad (6.23)
$$

Combining (6.22) with (6.23) yields (6.21). The proof is then completed. ∎

**Remark 6.2**    It is suggested that in applications the scalar $\gamma$ in the inequality (6.16) be simply chosen to be 1. In this case, the inequality turns out to be

$$
\begin{bmatrix} -P & PA^T \\ AP & -P - BB^T \end{bmatrix} < 0, \qquad (6.24)
$$

and the corresponding gain matrix is given by

$$
K = -\left( 2I + B^T P^{-1} B \right)^{-1} B^T P^{-1} A.
$$

This is not only because that this case is simple but also due to the fact that it is in vain to adjust the parameter $\gamma$. In fact, once (6.24) has a solution $P$, (6.16) has a solution $P_0 = P/\gamma$. This fact gives

$$
\begin{aligned}
K &= -\left( \frac{2}{\gamma} I + B^T P^{-1} B \right)^{-1} B^T P^{-1} A \\
&= -\left( \frac{2}{\gamma} I + B^T (\gamma P_0)^{-1} B \right)^{-1} B^T (\gamma P_0)^{-1} A \\
&= -\left( 2I + B^T P_0^{-1} B \right)^{-1} B^T P_0^{-1} A.
\end{aligned}
$$

This clearly states that the gain matrix $K$, although has a relation to $\gamma$ in expression, is invariant with $\gamma$ in value.

It is clear that the conditions given in the aforementioned three theorems are all LMIs. Compared with the second-rank condition in Theorem 6.6, the LMI conditions have an advantage in computation. For some high-order systems, the rank condition may not be reliable, while the LMI conditions have good numerical reliability.

**Example 6.3**

The following discrete-time linear system

$$\begin{cases} x(k+1) = \begin{bmatrix} 0 & 1 & 0 \\ 1 & 1 & 0 \\ -1 & 0 & 0 \end{bmatrix} x(k) + \begin{bmatrix} 0 \\ 1 \\ 0 \end{bmatrix} u(k) \\ y(k) = \begin{bmatrix} 1 & 0 & 0 \\ 0 & 0 & 1 \end{bmatrix} x(k) \end{cases} \tag{6.25}$$

has been considered by Duan et al. (2003). For this system, applying the MATLAB function `feasp` in the MATLAB LMI toolbox to the corresponding LMI problem (6.24) yields

$$P = \begin{bmatrix} 0.4088 & -0.1085 & 0.0000 \\ -0.1085 & 0.2421 & 0.0000 \\ 0.0000 & 0.0000 & 1.1178 \end{bmatrix} > 0.$$

Thus, the system is Schur stabilizable. In fact, it can be obtained that

$$\lambda(A) = \{0, 1.6180, -0.6180\}$$

and it can also be easily verified that

$$\text{rank} \begin{bmatrix} sI - A & B \end{bmatrix} = 3, \quad s = 1.6180.$$

Thus, the conclusion is true following the well-known PBH criterion.

## 6.2.2 Schur Detectability

Schur detectability is a dual concept of Schur stabilizability and is defined as follows.

**Definition 6.4** The system (6.1), or the matrix pair $(A, C)$, is said to be Schur detectable if there exists a real matrix $L$ such that $A + LC$ is Schur stable.

Following Theorems 6.7, 6.8, and the duality principle, we immediately have the following results for Schur detectability.

**Theorem 6.10** The system (6.1) is Schur detectable if and only if one of the following conditions holds:

1. All the unobservable modes of system (6.1) are Schur stable, that is,

$$\text{rank} \begin{bmatrix} sI - A \\ C \end{bmatrix} = n, \quad \forall s \in \mathbb{C}, \quad |s| \geq 1.$$

2. There exist a symmetric matrix $P$ and a matrix $W$ satisfying

$$\begin{bmatrix} -P & A^T P + C^T W^T \\ PA + WC & -P \end{bmatrix} < 0.$$

3. There exists a symmetric matrix $P$ satisfying

$$\begin{bmatrix} -N_c^T P N_c & N_c^T A^T P \\ PAN_c & -P \end{bmatrix} < 0,$$

with $N_c$ being the right orthogonal complement of $C$.

4. There exists a symmetric matrix $P$ such that

$$\begin{bmatrix} -P & PA \\ A^T P & -P - \gamma C^T C \end{bmatrix} < 0, \quad \gamma \geq 1. \tag{6.26}$$

**Example 6.4**

Consider the discrete-time linear system (6.25) studied in Example 6.3. For this system, applying the MATLAB function feasp in the MATLAB LMI toolbox to the corresponding LMI problem (6.26) yields

$$P = \begin{bmatrix} 3.4186 & -2.2241 & -0.3139 \\ -2.2241 & 1.9483 & 1.0138 \\ -0.3139 & 1.0138 & 2.3788 \end{bmatrix} > 0.$$

Thus, the system is Schur detectable.

## 6.3 Dissipativity

The concept of dissipativity is related to the linear system of the following form:

$$\begin{cases} \dot{x} = Ax + Bu \\ y = Cx + Du, \end{cases} \tag{6.27}$$

where

- $x \in \mathbb{R}^n$, $u \in \mathbb{R}^r$, and $y \in \mathbb{R}^m$ are the system state vector, input vector, and the output vector, respectively
- $A$, $B$, $C$, and $D$ are the system coefficient matrices of appropriate dimensions. Obviously, the transfer function of the system is $G(s) = C(sI - A)^{-1}B + D$

For strictness, in this section, the control input $u(t)$ is restricted to be piece-wise continuous vector functions defined on $[0, \infty)$.

## 6.3.1 Definition

The definition of dissipativity is associated with a so-called supply function. Consider a general quadratic supply function

$$s(u, y) = \begin{bmatrix} y \\ u \end{bmatrix}^{\mathrm{T}} Q \begin{bmatrix} y \\ u \end{bmatrix}, \qquad (6.28)$$

where $Q$ is a real symmetric matrix of dimension $(m + r)$. We emphasize that the matrix $Q$ is not assumed to be symmetric positive or negative definite.

**Definition 6.5**    The linear system (6.27) with $x(0) = 0$ is said to be dissipative with the supply function $s(u, y)$ given in (6.28) if

$$\int_0^T s(u, y) \, dt \geq 0 \qquad (6.29)$$

holds for all $T \geq 0$, all piece-wise continuous input $u(t)$ and the corresponding output $y(t)$.

The following proposition gives an equivalent definition for dissipativity of the linear system (6.27).

**Proposition 6.2**    The linear system (6.27) with $x(0) = 0$ is dissipative with the supply function (6.28) if and only if there exists a matrix $P \geq 0$, such that

$$\int_{t_0}^{t_1} s(u, y) \, dt \geq x^{\mathrm{T}}(t_1) Px(t_1) - x^{\mathrm{T}}(t_0) Px(t_0) \qquad (6.30)$$

holds for all $t_0 \leq t_1$.

*Proof* Sufficiency: Let (6.30) holds all $t_0 \le t_1$. Choosing $t_0 = 0$, $t_1 = T$ in (6.30) we can immediately see that (6.30) implies (6.29) if $x(0) = 0$.

Necessity: Let the linear system (6.27) be dissipative with the supply function (6.28), then (6.29) holds. Further, noting that the supply function $s(u, y)$ is quadratic, we know that it suits for the optimal cost of a linear quadratic optimization problem. Let the optimal solution to the optimal control problem

$$\begin{cases} \min_u \int_{t_0}^{t_1} s(u, y) dt \\ \text{s.t. } \dot{x} = Ax + Bu, \ x(0) = 0 \\ \quad\quad y = Cx + Du \end{cases}$$

be denoted by

$$u^*(t) = Kx(t),$$

then it is well known from optimal control theory that there exists a matrix $P > 0$ such that

$$\min_u \int_{t_0}^{t_1} s(u, y) \, dt$$

$$= \int_{t_0}^{t_1} s(u^*, y^*) \, dt$$

$$= \frac{1}{2} x^{\mathrm{T}}(t_0) Px(t_0) - \frac{1}{2} x^{\mathrm{T}}(t_1) Px(t_1), \tag{6.31}$$

where $y^*$ is the optimal system output corresponding to the optimal control $u^*$.

Since (6.29) holds, we also have

$$\min_u \int_{t_0}^{t_1} s(u, y) \, dt \ge 0.$$

Otherwise, we can construct

$$\tilde{u}(t) = \begin{cases} u(t), & t \ge t_0 \\ 0, & 0 \le t < t_0, \end{cases}$$

which gives, in view of $x(0) = 0$, the system output

$$\tilde{y}(t) = \begin{cases} y(t), & t \ge t_0 \\ 0, & 0 \le t < t_0, \end{cases}$$

and hence

$$\int_0^{t_1} s(\tilde{u}, \tilde{y}) \, dt = \int_{t_0}^{t_1} s(\tilde{u}, \tilde{y}) \, dt < 0.$$

This clearly contradicts with the assumption (6.29).

Combining this with (6.31) gives

$$x^{\mathrm{T}}(t_0)Px(t_0) - x^{\mathrm{T}}(t_1)Px(t_1) \geq 0.$$

Hence,

$$x^{\mathrm{T}}(t_1)Px(t_1) - x^{\mathrm{T}}(t_0)Px(t_0) \leq 0. \tag{6.32}$$

Finally, combining (6.29) with (6.32) yields (6.30). The proof is then completed. ■

For convenience, (6.30) is often called the dissipativity inequality.

## 6.3.2 Equivalent Conditions

### 6.3.2.1 Frequency Domain Condition

The following gives a necessary and sufficient condition for dissipativity of linear system (6.27) in terms of its transfer function.

**Theorem 6.11**  Suppose that the linear system (6.27) is controllable. Then the linear system (6.27) is dissipative with the supply function defined by (6.28) if and only if

$$\begin{bmatrix} G(s) \\ I \end{bmatrix}^{\mathrm{H}} Q \begin{bmatrix} G(s) \\ I \end{bmatrix} \geq 0, \quad \forall s \in \mathbb{C}, \quad \mathrm{Re}(s) > 0. \tag{6.33}$$

The proof of this theorem is rather lengthy and is given in Appendix A.

### 6.3.2.2 Matrix Inequality Condition

For dissipativity of the system (6.27), we can also derive a criterion based on a matrix inequality in terms of the system coefficients. This derivation requires the following preliminary result.

**Lemma 6.1** Suppose that the linear system (6.27), with $x(0) = 0$, is controllable, and

$$\begin{bmatrix} x(t) \\ u(t) \end{bmatrix}^{\mathrm{T}} \Phi \begin{bmatrix} x(t) \\ u(t) \end{bmatrix} \geq 0, \quad \forall t \geq 0, \tag{6.34}$$

holds for some $\Phi \in \mathbb{S}^{(n+r)}$ and arbitrary piece-wise continuous input $u(t)$ and the corresponding state $x(t)$. Then $\Phi \geq 0$.

*Proof* It follows from the definition of positive definiteness of matrices, it is sufficient to show that, for arbitrary $z \in \mathbb{R}^{n+r}$, there exist a time moment $t_1$, and a pair of input vector $\tilde{u}(t)$ and state vector $\tilde{x}(t)$ of the linear system (6.27) satisfying

$$\begin{bmatrix} \tilde{x}(t_1) \\ \tilde{u}(t_1) \end{bmatrix} = z. \tag{6.35}$$

Partition the vector $z$ as

$$z = \begin{bmatrix} z_1 \\ z_2 \end{bmatrix}, \quad z_1 \in \mathbb{R}^n, \quad z_2 \in \mathbb{R}^r.$$

Then, by the controllability assumption, there exists an input $u(t)$ which drives $x(t)$ to $z_1$, that is,

$$x(t_1) = \int_0^{t_1} e^{A(t-\tau)} Bu(\tau)\, d\tau = z_1. \tag{6.36}$$

Define a new piece-wisely continuous input as

$$\tilde{u}(t) = \begin{cases} u(t), & 0 \leq t < t_1 \\ z_2, & t_1 \leq t. \end{cases}$$

Then, we have $\tilde{u}(t_1) = z_2$. While the corresponding state vector at this time moment $t_1$ is

$$\tilde{x}(t_1) = \int_0^{t_1} e^{A(t-\tau)} B\tilde{u}(\tau)\, d\tau = \int_0^{t_1} e^{A(t-\tau)} Bu(\tau)\, d\tau = z_1.$$

Therefore, we have found a pair of input vector $\tilde{u}(t)$ and state vector $\tilde{x}(t)$ of the linear system (6.27) satisfying (6.35). The proof is done. ■

**Theorem 6.12**    Suppose that the linear system (6.27) is controllable. Then, the system (6.27) is dissipative with the supply function defined by (6.28) if and only if there exists a matrix $P \geq 0$ such that

$$\begin{bmatrix} A^{\mathrm{T}}P + PA & PB \\ B^{\mathrm{T}}P & 0 \end{bmatrix} - \begin{bmatrix} C & D \\ 0 & I \end{bmatrix}^{\mathrm{T}} Q \begin{bmatrix} C & D \\ 0 & I \end{bmatrix} \leq 0. \tag{6.37}$$

*Proof*    Sufficiency: Let the linear system (6.27) be dissipative, then the dissipativity inequality (6.30) holds, which clearly can be rewritten as

$$\int_{t_0}^{t_1} \left( s(u, y) - \frac{\mathrm{d}}{\mathrm{d}t} \left[ x(t)^{\mathrm{T}} P x(t) \right] \right) \mathrm{d}t \geq 0. \tag{6.38}$$

Using the system equations, we have

$$s(u, y) = \begin{bmatrix} y(t) \\ u(t) \end{bmatrix}^{\mathrm{T}} Q \begin{bmatrix} y(t) \\ u(t) \end{bmatrix}$$

$$= \begin{bmatrix} Cx(t) + Du(t) \\ u(t) \end{bmatrix}^{\mathrm{T}} Q \begin{bmatrix} Cx(t) + Du(t) \\ u(t) \end{bmatrix}$$

$$= \begin{bmatrix} x(t) \\ u(t) \end{bmatrix}^{\mathrm{T}} \begin{bmatrix} C & D \\ 0 & I \end{bmatrix}^{\mathrm{T}} Q \begin{bmatrix} C & D \\ 0 & I \end{bmatrix} \begin{bmatrix} x(t) \\ u(t) \end{bmatrix},$$

and

$$\frac{\mathrm{d}}{\mathrm{d}t} \left[ x(t)^{\mathrm{T}} P x(t) \right] = \dot{x}(t)^{\mathrm{T}} P x(t) + x(t)^{\mathrm{T}} P \dot{x}(t)$$

$$= (Ax(t) + Bu(t))^{\mathrm{T}} P x(t) + x(t)^{\mathrm{T}} P (Ax(t) + Bu(t))$$

$$= x(t)^{\mathrm{T}} \left( A^{\mathrm{T}}P + PA \right) x(t) + u(t)^{\mathrm{T}} B^{\mathrm{T}} P x(t) + x(t)^{\mathrm{T}} P B u(t)$$

$$= \begin{bmatrix} x(t) \\ u(t) \end{bmatrix}^{\mathrm{T}} \begin{bmatrix} A^{\mathrm{T}}P + PA & PB \\ B^{\mathrm{T}}P & 0 \end{bmatrix} \begin{bmatrix} x(t) \\ u(t) \end{bmatrix}.$$

Thus, we obtain

$$s(u, y) - \frac{\mathrm{d}}{\mathrm{d}t} \left[ x(t)^{\mathrm{T}} P x(t) \right] = \begin{bmatrix} x(t) \\ u(t) \end{bmatrix}^{\mathrm{T}} F(P) \begin{bmatrix} x(t) \\ u(t) \end{bmatrix}, \tag{6.39}$$

where

$$F(P) = - \begin{bmatrix} A^{\mathrm{T}}P + PA & PB \\ B^{\mathrm{T}}P & 0 \end{bmatrix} + \begin{bmatrix} C & D \\ 0 & I \end{bmatrix}^{\mathrm{T}} Q \begin{bmatrix} C & D \\ 0 & I \end{bmatrix}.$$

Therefore, (6.38) is equivalent to

$$\int_{t_0}^{t_1} \begin{bmatrix} x(t) \\ u(t) \end{bmatrix}^T F(P) \begin{bmatrix} x(t) \\ u(t) \end{bmatrix} dt \geq 0. \tag{6.40}$$

Further, in view of the arbitrariness of $t_0$ and $t_1$, (6.40) is clear seen to be equivalent to

$$\begin{bmatrix} x(t) \\ u(t) \end{bmatrix}^T F(P) \begin{bmatrix} x(t) \\ u(t) \end{bmatrix} > 0, \quad \forall t \geq 0.$$

Since the aforementioned inequality holds for all input $u(t)$, and also the system is controllable, by Lemma 6.1 we know that (6.40) is equivalent to $F(P) \geq 0$. Therefore, we have shown that (6.30) implies (6.37).

Necessity: This is actually a reverse process of the sufficiency part. Suppose that there exists a matrix $P \in \mathbb{S}^n$ such that (6.37) holds, then we have (6.40), which is equivalent to (6.38) in view of (6.39). Hence we further have the dissipativity inequality (6.30). Therefore, the system (6.27) is dissipative with the supply function $s(u, y)$. Thus, the conclusion is true in view of Proposition 6.2. ■

**Example 6.5**

Consider an advanced (CCV-type) fighter aircraft system (Syrmos and Lewis 1993, Duan 2003) in the form of (6.27) with the following coefficients:

$$A = \begin{bmatrix} -1.3410 & 0.9933 & 0 & -0.1689 & -0.2518 \\ 43.2230 & -0.8693 & 0 & -17.2510 & -1.5766 \\ 1.3410 & 0.0067 & 0 & 0.1689 & 0.2518 \\ 0 & 0 & 0 & -20.0000 & 0 \\ 0 & 0 & 0 & 0 & -20.0000 \end{bmatrix}, \quad B = \begin{bmatrix} 0 & 0 \\ 0 & 0 \\ 0 & 0 \\ 20 & 0 \\ 0 & 20 \end{bmatrix},$$

$$C = \begin{bmatrix} I_3 & 0 \end{bmatrix}, \quad D = 0.$$

For this system, we choose

$$Q = \begin{bmatrix} -3 & 2 & 1 & 0 & -1 \\ 2 & 1 & 1 & 0 & 0 \\ 1 & 1 & 2 & -1 & 0 \\ 0 & 0 & -1 & 1 & 0 \\ -1 & 0 & 0 & 0 & 1 \end{bmatrix}.$$

Since

$$\lambda(Q) = \{-4.0539, 0.0470, 1.0000, 1.4961, 3.5108\}$$

contains both positive and negative values, the matrix $Q$ is neither positive definite nor negative definite.

Solving the LMI in (6.37) with the MATLAB function `feasp` yields a positive definite solution

$$P = \begin{bmatrix} 2.7598 & -0.0841 & 0.7253 & -0.0315 & -0.0408 \\ -0.0841 & 0.0359 & 0.0301 & 0.0135 & -0.0029 \\ 0.7253 & 0.0301 & 1.3723 & -0.0553 & -0.0077 \\ -0.0315 & 0.0135 & -0.0553 & 0.0735 & -0.0009 \\ -0.0408 & -0.0029 & -0.0077 & -0.0009 & 0.0637 \end{bmatrix} > 0.$$

Therefore, the system is dissipative with respect to this particular supply function.

## 6.4 Passivity and Positive-Realness

In the proceeding section, we have introduced for the linear system (6.27) the concept of dissipativity and also given some equivalent conditions. In this section, we further require that the dimension of the input vector and that of the output vector of system (6.27) are the same, that is, $m = r$.

It is seen from the definition that dissipativity varies with the supply function

$$s_Q(u, y) = \begin{bmatrix} y \\ u \end{bmatrix}^T Q \begin{bmatrix} y \\ u \end{bmatrix}.$$

Specially choosing

$$Q = Q_P = \begin{bmatrix} 0 & I \\ I & 0 \end{bmatrix}, \tag{6.41}$$

leads to the concept of passivity.

### 6.4.1 Definitions

With the choice of $Q$ matrix given in (6.41), the supply function becomes

$$s_P(u, y) = 2u^T(t)y(t).$$

In this case, the dissipativity condition (6.29) turns into

$$\int_0^T u^T(t)y(t)\,dt \geq 0. \tag{6.42}$$

**Definition 6.6** The linear system (6.27), with the same number of input and output variables, is called passive if (6.42) holds for arbitrary $T \geq 0$, arbitrary input $u(t)$, and the corresponding solution $y(t)$ of (6.27) with $x(0) = 0$.

Obviously, passivity describes a property of a linear system in the time domain. While the following definition, which has a close relation to the concept of passivity, describes a property of a linear system in the frequency domain.

**Definition 6.7**   The transfer function matrix

$$G(s) = C\,(sI - A)^{-1}\,B + D$$

of system (6.27) is called positive-real if it is square and satisfies

$$G^{H}(s) + G(s) \geq 0, \quad \forall s \in \mathbb{C}, \quad \text{Re}(s) > 0. \tag{6.43}$$

## 6.4.2 Positive-Real Lemma

Passivity and positive-realness are two very important concepts in linear control systems theory. The following well-known positive-real lemma reveals the relation between passivity and positive-realness.

**Theorem 6.13**   Let the linear system (6.27) be controllable. Then, the system (6.27) is passive if and only if its transfer function is positive-real.

*Proof*   When the $Q$ matrix is specially taken as in (6.41), the inequality (6.33) reduces to (6.43). Thus, we can easily obtain the conclusion from the equivalence of (6.29) and (6.33) in Theorem 6.11 (see Figure 6.1). ■

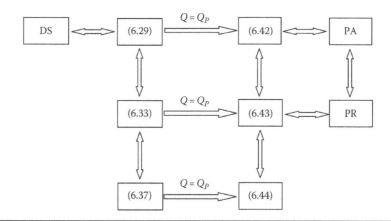

**Figure 6.1   Relations among dissipativity (DS), passivity (PA), and positive-realness (PR).**

The aforementioned theorem tells us that passivity and positive-realness in fact describe the same property of a linear system, one gives the time-domain feature of this property, while the other gives the frequency-domain feature.

## 6.4.3 LMI Condition

The following theorem gives an LMI condition for passivity of linear systems.

**Theorem 6.14**    Let the linear system (6.27) be controllable. Then, the system (6.27) is passive if and only if there exists a matrix $P > 0$, such that

$$\begin{bmatrix} A^{\mathrm{T}} P + PA & PB - C^{\mathrm{T}} \\ B^{\mathrm{T}} P - C & -D^{\mathrm{T}} - D \end{bmatrix} \leq 0. \tag{6.44}$$

*Proof*    It is known from earlier text that, when the $Q$ matrix is specially chosen as in (6.41), the dissipativity of the linear system (6.27) becomes the passivity of the system. Meanwhile, the inequality (6.37) can be easily seen to reduce to (6.44). Therefore, we can easily obtain the conclusion from Theorem 6.12 (see Figure 6.1). ■

It is obvious that (6.44) implies $A^{\mathrm{T}} P + PA \leq 0$, thus the following conclusion immediately follows.

**Corollary 6.1**    The linear system (6.27) is passive only if the matrix $A$ is stable or critically stable.

Using the Schur complement lemma, we can also get the following corollary.

**Corollary 6.2**    Let the linear system (6.27) be controllable, and $D + D^{\mathrm{T}} > 0$. Then, the linear system (6.27) is passive if and only if there exists a matrix $P \in \mathbb{S}^n$ satisfying the following Riccati matrix inequality:

$$A^{\mathrm{T}} P + PA + \left(PB - C^{\mathrm{T}}\right) \left(D + D^{\mathrm{T}}\right)^{-1} \left(B^{\mathrm{T}} P - C\right) \leq 0.$$

**Example 6.6**

Consider a continuous-time linear system in the form of (6.27) with

$$A = \begin{bmatrix} -1 & 0 \\ 1 & -1 \end{bmatrix}, \quad B = \begin{bmatrix} -4 \\ 1 \end{bmatrix},$$

$$C = \begin{bmatrix} -100 & 10 \end{bmatrix}, \quad D = 1.$$

Applying the MATLAB function `feasp` in the MATLAB LMI toolbox to the corresponding LMI problem in (6.44) with these parameters gives the following solution

$$P = \begin{bmatrix} 26.0514 & 1.8531 \\ 1.8531 & 19.9938 \end{bmatrix} > 0.$$

By Theorem 6.13, this linear system is passive.

## 6.5 Nonexpansivity and Bounded-Realness

In this section, we also consider the controllable linear system (6.27) with $m = r$.
As in Section 6.4, specially choosing

$$Q = Q_N = \begin{bmatrix} -I & 0 \\ 0 & I \end{bmatrix}, \tag{6.45}$$

in the definition of dissipativity, or equally, in the supply function

$$s_Q(u, y) = \begin{bmatrix} y \\ u \end{bmatrix}^T Q \begin{bmatrix} y \\ u \end{bmatrix},$$

leads to the concept of nonexpansivity.

### 6.5.1 Definitions

With the choice of $Q$ matrix given in (6.45), the supply function becomes to

$$s_N(u, y) = u^T(t)u(t) - y^T(t)y(t).$$

In this case, the dissipativity condition (6.29) turns into

$$\int_0^T y^T(t)y(t)\, dt \leq \int_0^T u^T(t)u(t)\, dt. \tag{6.46}$$

**Definition 6.8**  The linear system (6.27) is called nonexpansive if (6.46) holds for arbitrary $T \geq 0$, arbitrary input $u(t)$ and the corresponding solution $y(t)$ of system (6.27) with $x(0) = 0$.

Clearly, nonexpansivity is a property of linear systems in the time domain. The following definition describes a property of a linear system in the frequency domain.

**Definition 6.9**   The transfer function matrix

$$G(s) = C\,(sI - A)^{-1}\,B + D$$

of the system (6.27) is called bounded-real if it satisfies

$$G^{\mathrm{H}}(s)G(s) \leq I, \quad \forall s \in \mathbb{C}, \quad \mathrm{Re}(s) > 0. \tag{6.47}$$

The following proposition gives another interpretation for the concept of bounded-realness.

**Proposition 6.3**   The bounded-real condition (6.47) is equivalent to

$$||G(s)||_\infty \leq 1. \tag{6.48}$$

*Proof*   It is obvious that (6.47) is equivalent to

$$\sigma_{\max}\,(G(s)) \leq 1, \quad \forall s \in \mathbb{C}, \quad \mathrm{Re}(s) > 0. \tag{6.49}$$

Thus, by Proposition 5.1, the aforementioned condition is equivalent to

$$||G(s)||_\infty = \sup\{\sigma_{\max}\,(G(s)) \mid \mathrm{Re}(s) > 0\} \leq 1.$$

With this we complete the proof.   ■

## 6.5.2 Bounded-Real Lemma

Like passivity and positive-realness, nonexpansivity and bounded-realness are also a pair of concepts describing a property of a linear system in the time-domain and the frequency-domain, respectively. The following well-known bounded-real lemma reveals this relation.

**Theorem 6.15**   Let the linear system (6.27) be controllable. Then, the system (6.27) is nonexpansive if and only if its transfer function is bounded-real.

*Proof*   When the $Q$ matrix is specially taken as in (6.45), the inequality (6.33) reduces to (6.47). Thus, we can easily obtain the conclusion from the equivalence of (6.29) and (6.33) in Theorem 6.11 (see Figure 6.2).   ■

The aforementioned theorem tells us that nonexpansivity and bounded-realness in fact describe a same property of a linear system, one gives the time-domain feature of this property, while the other gives the frequency-domain feature.

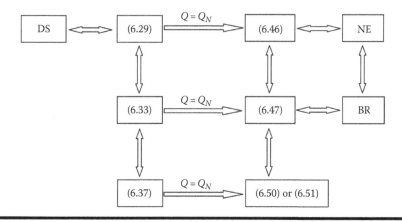

**Figure 6.2 Relations among dissipativity (DS), nonexpansivity (NE), and bounded-realness (BR).**

### 6.5.3 LMI Conditions

Proposition 6.3 has linked the concept of bounded-realness of a linear system with the $H_\infty$ performance of the system. Using such a relation and Theorem 5.3, we can immediately get the LMI condition for bounded-realness of a linear system.

**Theorem 6.16**  Let the linear system (6.27) be controllable. Then, the transfer function of system (6.27) is bounded-real if and only if there exists a matrix $P > 0$ satisfying one of the following LMIs:

$$\begin{bmatrix} A^\mathrm{T}P + PA & PB & C^\mathrm{T} \\ B^\mathrm{T}P & -I & D^\mathrm{T} \\ C & D & -I \end{bmatrix} < 0, \tag{6.50}$$

and

$$\begin{bmatrix} PA + A^\mathrm{T}P + C^\mathrm{T}C & PB + C^\mathrm{T}D \\ B^\mathrm{T}P + D^\mathrm{T}C & D^\mathrm{T}D - I \end{bmatrix} < 0. \tag{6.51}$$

*Proof*  This directly follows from Proposition 6.3 and Theorem 5.3, where the scalar $\gamma$ is simply set to 1. ■

**Remark 6.3**  In fact, as a slight generalization of Proposition 6.3, we can easily show that the transfer function matrix of the linear system (6.27), with $x(0) = 0$, satisfies $||G(s)||_\infty < \gamma$ for some $\gamma > 0$ if and only if the system input and output admit the following relation

$$\int_0^T y^{\mathrm{T}}(t)y(t)\,\mathrm{d}t \le \gamma^2 \int_0^T u^{\mathrm{T}}(t)u(t)\,\mathrm{d}t \tag{6.52}$$

for arbitrary $T > 0$. Such an equivalent relation clearly reveals the meaning of $H_\infty$ indices in the time domain.

The following is an immediate corollary of Theorem 6.16.

**Corollary 6.3**   The linear system (6.27) is nonexpansive only if the matrix $A$ is stable.

Further, using Schur complement lemma, we can also get the following corollary.

**Corollary 6.4**   Let the linear system (6.27) be controllable and $I - D^{\mathrm{T}}D > 0$. Then the transfer function of system (6.27) is bounded-real if and only if there exists a matrix $P > 0$ satisfying the following Riccati matrix inequality:

$$A^{\mathrm{T}}P + PA + C^{\mathrm{T}}C + \left(PB + C^{\mathrm{T}}D\right)\left(I - D^{\mathrm{T}}D\right)^{-1}\left(PB + C^{\mathrm{T}}D\right)^{\mathrm{T}} \le 0.$$

**Example 6.7**

Consider the linear system in Example 6.6 again. Applying the MATLAB function `feasp` in the MATLAB LMI toolbox to the corresponding LMI problem in (6.51), we find that this LMI constraint is infeasible. By Theorem 6.15, the system is not nonexpansive. In fact, solving the corresponding $H_\infty$ optimization problem, we actually obtain $\gamma = 372.2820 \gg 1$.

To finish this section, we mention that, for simplicity and also for emphasizing the relationship between dissipativity and nonexpansivity (bounded realness), we have chosen to prove Theorem 6.15, the bounded-real lemma, using Theorem 6.11. While this bounded-real lemma can also be proven directly. A direct proof is provided in Appendix A .

## 6.6 Notes and References

This chapter investigates certain properties of linear systems, which include stabilizability, detectability, dissipativity, passivity and positive-realness, nonexpansivity, and bounded-realness.

**Table 6.1  Criteria for Stabilizability of (A, B)**

| | Criteria | Variables |
|---|---|---|
| Hurwitz | $P > 0$<br><br>$AP + PA^\mathrm{T} + BW + W^\mathrm{T}B^\mathrm{T} < 0$ | $P, W$ |
| | $P > 0$<br><br>$N_b^\mathrm{T}\left(AP + PA^\mathrm{T}\right)N_b < 0$ | $P$ |
| | $P > 0$<br><br>$AP + PA^\mathrm{T} - BB^\mathrm{T} < 0$ | $P$ |
| Schur | $\begin{bmatrix} -P & AP + BW \\ PA^\mathrm{T} + W^\mathrm{T}B^\mathrm{T} & -P \end{bmatrix} < 0$ | $P, W$ |
| | $\begin{bmatrix} -N_b^\mathrm{T}PN_b & N_b^\mathrm{T}AP \\ PA^\mathrm{T}N_b & -P \end{bmatrix} < 0$ | $P$ |
| | $\begin{bmatrix} -P & PA^\mathrm{T} \\ AP & -P - BB^\mathrm{T} \end{bmatrix} < 0$ | $P$ |

The LMI conditions of stabilizability and detectability of the linear system $(A, B, C)$ are summarized in Tables 6.1 and 6.2, respectively, while those for passivity (positive-realness) and nonexpansivity (bounded-realness) of the linear system $(A, B, C, D)$ are given in Table 6.3.

Again, what we have presented in this chapter are very basic. For generalization of the concepts of detectability to descriptor linear systems, one can refer to Duan (2010) for detectability and observability of stochastic systems, to Li et al. (2009, 2010) for positive-realness of descriptor linear systems, and to Zhou et al. (2010b) and Yang et al. (2007).

As we have seen in this chapter that passivity and nonexpansivity are both special cases of dissipativity, which is defined for the linear system $(A, B, C, D)$, with a transfer function

$$G(s) = C\,(sI - A)^{-1}\,B + D.$$

Conditions for dissipativity of system $(A, B, C, D)$ with respect to the following supply function

$$s(u, y) = \begin{bmatrix} y \\ u \end{bmatrix}^\mathrm{T} Q \begin{bmatrix} y \\ u \end{bmatrix}, \tag{6.53}$$

**Table 6.2   Criteria for Detectability of $(A, B, C)$**

| | Criteria | Variables |
|---|---|---|
| Hurwitz | $P > 0$ <br><br> $A^\mathsf{T}P + PA + W^\mathsf{T}C + C^\mathsf{T}W < 0$ | $P, W$ |
| | $P > 0$ <br><br> $N_c^\mathsf{T}\left(A^\mathsf{T}P + PA\right)N_c < 0$ | $P$ |
| | $P > 0$ <br><br> $A^\mathsf{T}P + PA - C^\mathsf{T}C < 0$ | $P$ |
| Schur | $\begin{bmatrix} -P & A^\mathsf{T}P + C^\mathsf{T}W^\mathsf{T} \\ PA + WC & -P \end{bmatrix} < 0$ | $P, W$ |
| | $\begin{bmatrix} -N_c^\mathsf{T}PN_c & N_c^\mathsf{T}A^\mathsf{T}P \\ PAN_c & -P \end{bmatrix} < 0$ | $P$ |
| | $\begin{bmatrix} -P & PA \\ A^\mathsf{T}P & -P - C^\mathsf{T}C \end{bmatrix} < 0$ | $P$ |

are composed of the following:

1. $\int_0^T s(u, y)\,\mathrm{d}t \geq 0, \quad \forall T > 0.$
2. $\int_{t_0}^{t_1} s(u, y)\,\mathrm{d}t \geq x^\mathsf{T}(t_1)\,Px(t_1) - x^\mathsf{T}(t_0)\,Px(t_0), \quad \forall t_1 \geq t_0 > 0.$
3. $\begin{bmatrix} G(s) \\ I \end{bmatrix}^\mathsf{H} Q \begin{bmatrix} G(s) \\ I \end{bmatrix} \geq 0, \forall s \in \mathbb{C}, \; \mathrm{Re}(s) > 0.$
4. $\begin{bmatrix} A^\mathsf{T}P + PA & PB \\ B^\mathsf{T}P & 0 \end{bmatrix} - \begin{bmatrix} C & D \\ 0 & I \end{bmatrix}^\mathsf{T} Q \begin{bmatrix} C & D \\ 0 & I \end{bmatrix} \leq 0, \quad P > 0.$

We have seen that nonexpansivity is equivalent to $||G(s)||_\infty \leq 1$. This gives a clear meaning of the $H_\infty$ index. Also, it tells us that the $H_\infty$ norm of a stable linear system admits an interpretation in terms of dissipativity of the system with respect to a specific quadratic supply function.

Most of the results for dissipativity of linear systems presented in this chapter are closely related to those in Willems (1971b) and Scherer and Weiland (2000). These include result in Proposition 6.2 and those in Theorems 6.11 and 6.12. However, the proofs given in Willems (1971b) and Scherer and Weiland (2000) are

**Table 6.3    Criteria for Passivity and Nonexpansivity**

| *Problems* | *Criteria* | | |
|---|---|---|---|
| Passivity or positive-realness | $\int_0^T u^{\mathsf{T}}(t)y(t)dt \geq 0, \quad \forall T \geq 0$ | | |
| | $G^{\mathsf{H}}(s) + G(s) \geq 0, \quad \forall s \in \mathbb{C}, \ \mathrm{Re}(s) > 0$ | | |
| | $\exists P > 0$, s.t. $\begin{bmatrix} A^{\mathsf{T}}P + PA & PB - C^{\mathsf{T}} \\ B^{\mathsf{T}}P - C & -D^{\mathsf{T}} - D \end{bmatrix} < 0$ | | |
| Nonexpansivity or bounded-realness | $\int_0^T y^{\mathsf{T}}(t)y(t)\,dt \leq \int_0^T u^{\mathsf{T}}(t)u(t)\,dt, \quad \forall T \geq 0$ | | |
| | $G^{\mathsf{H}}(s)G(s) \leq I|$ | | |
| | $|G(s)\|_\infty \leq 1$ | | |
| | $\exists P > 0$, s.t. $\begin{bmatrix} A^{\mathsf{T}}P + PA & PB & C^{\mathsf{T}} \\ B^{\mathsf{T}}P & -I & D^{\mathsf{T}} \\ C & D & -I \end{bmatrix} < 0$ | | |
| | $\exists P > 0$, s.t. $\begin{bmatrix} A^{\mathsf{T}}P + PA + C^{\mathsf{T}}C & PB + C^{\mathsf{T}}D \\ B^{\mathsf{T}}P + D^{\mathsf{T}}C & D^{\mathsf{T}}D - I \end{bmatrix} \leq 0$ | | |

very brief and provide only some outlines. For the sake of most of the audience, we have provided detailed proofs of these results in Section 6.3.

Finally, we mention that the concept of dissipativity has been extended to other types of systems, for example, neural networks with mixed time-varying delays (Cao et al. 2006), and smart dampers systems (Johnson and Erkus 2007).

# Exercises

**6.1** Prove Corollary 6.1, that is, the linear system

$$\begin{cases} \dot{x} = Ax + Bu \\ y = Cx + Du \end{cases}$$

is passive only if the matrix $A$ is stable.

**6.2** Verify the Hurwitz stabilizability of the following linear system using LMI technique:

$$\dot{x} = \begin{bmatrix} 0 & 1 & 2 \\ 0 & 1 & 0 \\ 1 & 1 & 1 \end{bmatrix} x + \begin{bmatrix} 0 & 1 \\ 1 & 0 \\ 0 & 1 \end{bmatrix} u.$$

**6.3** Verify using LMI technique the Hurwitz detectability of the following system

$$\begin{cases} \dot{x} = \begin{bmatrix} -2 & 1 & 0 \\ 0 & -2 & 0 \\ 0 & 0 & -2 \end{bmatrix} x \\ y = \begin{bmatrix} 1 & 0 & 4 \\ 2 & 0 & 8 \end{bmatrix} x. \end{cases}$$

**6.4** Consider the following linear system (Duan 2003)

$$\begin{cases} \dot{x} = Ax + Bu \\ y = Cx \end{cases}$$

with

$$A = \begin{bmatrix} 0.5 & 0 & 0 \\ 0 & -2 & 10 \\ 0 & 1 & -2 \end{bmatrix}, \quad B = \begin{bmatrix} 1 & 0 \\ -2 & 2 \\ 0 & 1 \end{bmatrix}, \quad C = \begin{bmatrix} 1 & 0 & 0 \\ 0 & 0 & 1 \end{bmatrix}.$$

Determine whether or not this system is Hurwitz stabilizable and detectable.

**6.5** Consider the following linear system

$$G(s) = \frac{1}{s^2 + 3s + 2}. \tag{6.54}$$

1. Determine whether or not this system is passive.
2. Determine whether or not this system is bounded-real.

**6.6** Consider the jet transport aircraft system in Example 6.2.
1. Determine whether or not this system is passive.
2. Determine whether or not this system is bounded-real.

**6.7** Prove Theorem 6.16 using Theorem 6.12.

# CONTROL SYSTEMS DESIGN

# Chapter 7

# Feedback Stabilization

In the previous three chapters, we have studied several types of analysis problems, while from this chapter on we will look at certain control systems design problems. We start with in this chapter the problem of state feedback stabilization.

The results in the first five sections are closely related to the stability analysis results covered in Chapter 4. In Chapter 4, we have given various analysis conditions for different types of stability. But in this chapter, we will first apply those conditions to the state feedback stabilization system, and then, through a variable substitution, we can turn the design conditions for stabilization into LMIs, and meanwhile get the expression of the gain matrix. Such a procedure is shown in Figure 7.1.

## 7.1 State Feedback Stabilization

In this section, we examine the problem of state feedback stabilization. The purpose is to present a solution to the problem based on LMIs.

### 7.1.1 Case of Continuous-Time Systems

The problem of state feedback stabilization for continuous-time systems can be stated as follows.

**Problem 7.1**    For the following continuous-time linear system

$$\dot{x} = Ax + Bu, \tag{7.1}$$

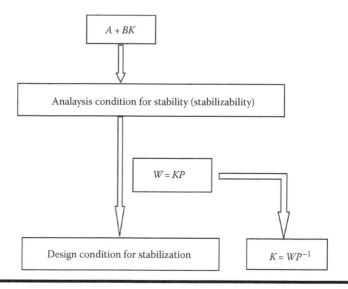

**Figure 7.1  Basic procedure for stabilization design.**

design a state feedback control law $u = Kx$, such that the closed-loop system

$$\dot{x} = (A + BK)\,x$$

is asymptotically stable.

By Lyapunov stability theory, the problem has a solution if and only if there exists a symmetric positive definite matrix $P$, such that

$$(A + BK)\,P + P\,(A + BK)^{\mathrm{T}} < 0.$$

Thus, the following result immediately follows from Theorems 6.2 and 6.4 and their proof.

**Theorem 7.1**   Consider the state feedback stabilization problem stated earlier.

1. The problem has a solution if and only if there exist a symmetric positive definite matrix $P$ and a matrix $W$ satisfying

$$AP + PA^{\mathrm{T}} + BW + W^{\mathrm{T}}B^{\mathrm{T}} < 0. \tag{7.2}$$

In this case, a solution to the problem is given by

$$K = WP^{-1}.$$

2. The problem has a solution if and only if there exists a symmetric positive definite matrix $P$ satisfying

$$AP + PA^{\mathrm{T}} - BB^{\mathrm{T}} < 0. \tag{7.3}$$

In this case, a solution to the problem is given by

$$K = -\frac{1}{2}B^{\mathrm{T}}P^{-1}.$$

**Remark 7.1** In Remark 6.1, we point out that the inequality (7.3) can be replaced by the LMI

$$AP + PA^{\mathrm{T}} < \gamma BB^{\mathrm{T}}, \quad \gamma > 0.$$

In this case, the corresponding gain matrix is given by

$$K = -\frac{\gamma}{2}B^{\mathrm{T}}P^{-1},$$

which is actually invariant with the scalar $\gamma$.

**Example 7.1**

Consider the following benchmark system, which has been introduced in Problem 3.10,

$$\begin{cases} \dot{x}(t) = \begin{bmatrix} 20 & 0 & 0 & \cdots & 0 \\ 20 & 19 & 0 & \cdots & 0 \\ 0 & 20 & 18 & \ddots & \vdots \\ \vdots & \ddots & \ddots & \ddots & 0 \\ 0 & \cdots & 0 & 20 & 1 \end{bmatrix} x(t) + \begin{bmatrix} 1 \\ 0 \\ 0 \\ \vdots \\ 0 \end{bmatrix} u(t) \\ y = \begin{bmatrix} 0 & 0 & \cdots & 0 & 1 \end{bmatrix} x. \end{cases}$$

Applying the MATLAB® function `feasp` in the MATLAB LMI toolbox to the corresponding LMI problem (7.2), we obtain the parameters $W$ and $P > 0$, and the stabilizing feedback gain is obtained via

$$K = \begin{bmatrix} k_1 & k_2 & \cdots & k_{20} \end{bmatrix} = WP^{-1}.$$

Entries in the gain matrix $K$ are given in Table 7.1.

**Table 7.1  Entries of Gain Matrix K**

| j | $k_{3j+1}$ | $k_{3j+2}$ | $k_{3j+3}$ |
|---|---|---|---|
| 0 | −591.4762 | −10314.5551 | −92979.44209 |
| 1 | −540566.8807 | −2209854.2130 | −6688509.0488 |
| 2 | −15474002.5449 | −27952786.7039 | −39997637.3117 |
| 3 | −45768927.4074 | −42125136.3660 | −31262321.5973 |
| 4 | −18691639.2974 | −8961636.0608 | −3411864.4338 |
| 5 | −1013690.8962 | −228097.5187 | −36841.7712 |
| 6 | −3836.6356 | −196.2084 | |

By using the MATLAB function `eig`, the set of closed-loop eigenvalues are obtained as

$$\lambda (A + BK) = \{-144.1706 \pm 238.9446i, -14.5740 \pm 58.0656i,$$
$$- 5.4157 \pm 33.3797i, -2.7601 \pm 23.2861i,$$
$$- 1.4620 \pm 18.0273i, -1.6736 \pm 15.3718i,$$
$$- 3.1287 \pm 11.5781i, -5.3042 \pm 8.5417i,$$
$$- 6.8075 \pm 2.03473i, -5.4416 \pm 4.6807i\}, \qquad (7.4)$$

which is clearly located on the left-half complex plane.

## 7.1.2  Case of Discrete-Time Systems

The problem of state feedback stabilization of discrete-time linear systems can be stated as follows.

**Problem 7.2**   For the following discrete-time linear system

$$x (k + 1) = Ax (k) + Bu (k),$$

design a state feedback control law

$$u (k) = Kx (k),$$

such that the closed-loop system

$$x (k + 1) = (A + BK) x (k)$$

is asymptotically stable.

By Lyapunov stability theory, the problem has a solution if and only if there exists a symmetric positive definite matrix $P$, such that

$$(A + BK) P (A + BK)^{\mathrm{T}} - P < 0.$$

Thus, the following result immediately follows from Theorems 6.7 and 6.9 and their proofs.

**Theorem 7.2**   Consider the state feedback stabilization problem stated earlier.

1. The problem has a solution if and only if there exist a symmetric matrix $P$ and a matrix $W$ satisfying

$$\begin{bmatrix} -P & AP + BW \\ PA^{\mathrm{T}} + W^{\mathrm{T}} B^{\mathrm{T}} & -P \end{bmatrix} < 0. \tag{7.5}$$

   In this case, a feedback gain matrix is given by

$$K = WP^{-1}.$$

2. The problem has a solution if and only if there exists a symmetric matrix $P$ satisfying

$$\begin{bmatrix} -P & PA^{\mathrm{T}} \\ AP & -P - BB^{\mathrm{T}} \end{bmatrix} < 0. \tag{7.6}$$

   In this case, a feedback gain matrix is given by

$$K = -(2I + B^{\mathrm{T}} P^{-1} B)^{-1} B^{\mathrm{T}} P^{-1} A.$$

**Remark 7.2**   In Remark 6.2, we point out that the inequality (7.6) can be replaced by the following alternate LMI condition

$$\begin{bmatrix} -P & PA^{\mathrm{T}} \\ AP & -P - \gamma BB^{\mathrm{T}} \end{bmatrix} < 0, \quad \gamma > 0.$$

In this case, the corresponding gain matrix is given by

$$K = -\left(\frac{2}{\gamma}I + B^{\mathrm{T}} P^{-1} B\right)^{-1} B^{\mathrm{T}} P^{-1} A,$$

which is actually invariant with the scalar $\gamma$.

**Example 7.2**

Consider the discrete-time linear system in Example 6.3 again. For this system, applying the MATLAB function `feasp` in the MATLAB LMI toolbox to the corresponding LMI problem (7.5) yields

$$P = \begin{bmatrix} 48.2702 & -0.0000 & 0.0000 \\ -0.0000 & 26.5366 & -0.0000 \\ 0.0000 & -0.0000 & 85.6691 \end{bmatrix} > 0,$$

$$W = \begin{bmatrix} -48.2702 & -26.5366 & 0 \end{bmatrix}.$$

Thus, a stabilizing feedback gain is obtained as

$$K = WP^{-1} = \begin{bmatrix} -1.0000 & -1.0000 & -0.0000 \end{bmatrix}.$$

It can be verified using the MATLAB function `eig` that the closed-loop eigenvalues are $(-0.1046 \pm 0.1813i) \times 10^{-5}$ and $2.093 \times 10^{-6}$, which are all stable in the discrete-time sense.

## 7.2 $\mathbb{D}$-Stabilization

In this section, we examine several problems of $\mathbb{D}$-stabilization of the following system

$$\dot{x} = Ax + Bu, \tag{7.7}$$

where $x \in \mathbb{R}^n$ and $u \in \mathbb{R}^r$ are the state vector and the input vector, respectively.

### 7.2.1 $\mathbb{H}_{(\alpha,\beta)}$-Stabilization

The problem of $\mathbb{H}_{(\alpha,\beta)}$-stabilization of the linear system (7.7) is stated as follows.

**Problem 7.3**   For the linear system (7.7), design a state feedback control law $u = Kx$ such that the closed-loop system

$$\dot{x} = (A + BK)x$$

is $\mathbb{H}_{(\alpha,\beta)}$ stable.

It follows from Proposition 4.3 that the problem has a solution if and only if there exists a symmetric positive definite matrix $P$ satisfying

$$\begin{cases} (A + BK)P + P(A + BK)^{\mathrm{T}} + 2\alpha P < 0 \\ -(A + BK)P - P(A + BK)^{\mathrm{T}} - 2\beta P < 0. \end{cases}$$

Based on this observation, the following result can be immediately obtained.

**Theorem 7.3**    Problem 7.3 has a solution if and only if there exist a symmetric positive definite matrix $P$ and a matrix $W$ satisfying

$$\begin{cases} AP + PA^{\mathrm{T}} + BW + W^{\mathrm{T}}B^{\mathrm{T}} + 2\alpha P < 0 \\ -AP - PA^{\mathrm{T}} - BW - W^{\mathrm{T}}B^{\mathrm{T}} - 2\beta P < 0. \end{cases} \tag{7.8}$$

In this case, a solution to the problem is given by

$$K = WP^{-1}.$$

**Example 7.3**

Consider the benchmark system in Example 7.1 again. Let the associated region be

$$\mathbb{H}_{(1,3)} = \{x + yj \mid -3 < x < -1 < 0\}.$$

By using the MATLAB function `feasp` in the MATLAB LMI toolbox, the corresponding LMI problem (7.8) is solved. Based on the obtained parameter matrices $W$ and $P$, the feedback gain is obtained by

$$K = \begin{bmatrix} k_1 & k_2 & \cdots & k_{20} \end{bmatrix} = WP^{-1}.$$

Entries in the gain matrix $K$ are given in Table 7.2.

**Table 7.2    Entries of Gain Matrix $K$**

| $j$ | $k_{3j+1}$ | $k_{3j+2}$ | $k_{3j+3}$ |
|---|---|---|---|
| 0 | −247.2992 | −3641.8033 | −27720.8477 |
| 1 | −150480.7997 | −588726.8296 | −1753787.5424 |
| 2 | −4043652.1159 | −7366762.0268 | −10691261.2443 |
| 3 | −12474312.3964 | −11723931.5461 | −8899960.6584 |
| 4 | −5433676.2155 | −2659386.8988 | −1028572.6855 |
| 5 | −310360.7401 | −70269.3239 | −11477.0865 |
| 6 | −1184.2997 | −61.5274 | |

By using the MATLAB function `eig`, the closed-loop eigenvalues are obtained as

$$\lambda\,(A+BK) = \{-2.0033 \pm 196.2688i, -1.9091 \pm 64.4927i,$$
$$-1.7565 \pm 37.7301i, -1.6236 \pm 26.1631i,$$
$$-1.5371 \pm 19.8647i, -1.6132 \pm 16.3433i,$$
$$-2.0503 \pm 13.8705i, -2.0603 \pm 10.6065i,$$
$$-2.0475 \pm 6.5857i, -2.0487 \pm 2.2323i\},$$

which are clearly located in the region $\mathbb{H}_{(1,3)}$.

### 7.2.2 $\mathbb{D}_{(q,r)}$-Stabilization

The problem of $\mathbb{D}_{(q,r)}$-stabilization is stated as follows.

**Problem 7.4** Let

$$\mathbb{D} = \mathbb{D}_{(q,r)} = \left\{ x + jy \,\middle|\, (x+q)^2 + y^2 < r^2 \right\}$$

be a domain on the complex plane. Design a state feedback control law $u = Kx$ for the linear system

$$\rho x = Ax + Bu,$$

such that the closed-loop system

$$\rho x = (A + BK)\,x$$

is $\mathbb{D}$-stable. Here $\rho$ represents either the differential operator (for continuous-time system case) or the one-step forward operator (for discrete-time system case).

It follows from Proposition 4.4 that the problem has a solution if and only if there exists a symmetric matrix $P$ satisfying the following LMI:

$$\begin{bmatrix} -rP & qP + (A+BK)\,P \\ qP + P\,(A+BK)^{\mathrm{T}} & -rP \end{bmatrix} < 0.$$

Denoting

$$W = KP,$$

we can convert the above inequality into the following equivalent one:

$$\begin{bmatrix} -rP & qP + AP + BW \\ qP + PA^{\mathrm{T}} + W^{\mathrm{T}}B^{\mathrm{T}} & -rP \end{bmatrix} < 0. \tag{7.9}$$

From the aforementioned deduction, we have the following result.

**Theorem 7.4**   Problem 7.4 has a solution if and only if there exists a symmetric positive definite matrix $P$ satisfying the LMI (7.9). In this case, a feedback gain matrix is given by

$$K = WP^{-1}.$$

**Example 7.4**

Consider the following linear system

$$\dot{x}(t) = \begin{bmatrix} -5 & 1 & 0 \\ 0 & 1 & 1 \\ 1 & 1 & 1 \end{bmatrix} x(t) + \begin{bmatrix} 0 & 0 \\ 0 & 1 \\ 1 & 0 \end{bmatrix} u(t).$$

This system has been used in Duan et al. (1990, 2001) and Duan and Chen (1995). Let the parameters associated with the interested region $\mathbb{D}_{(q,r)}$ be $q = 2$ and $r = 1$, that is,

$$\mathbb{D}_{(2,1)} = \left\{ x + jy \,\middle|\, (x+2)^2 + y^2 < 1 \right\}.$$

Using the MATLAB function `feasp` in the MATLAB LMI toolbox, the corresponding LMI problem (7.9) is solved, and the parameter matrices are obtained as

$$P = \begin{bmatrix} 2.2026 & 5.7346 & -0.0000 \\ 5.7346 & 16.2965 & -0.0000 \\ -0.0000 & -0.0000 & 11.4349 \end{bmatrix} > 0,$$

and

$$W = \begin{bmatrix} -7.9372 & -22.0312 & -34.3047 \\ -19.5344 & -51.3145 & -11.4349 \end{bmatrix}.$$

Hence, a feedback gain is given by

$$K = WP^{-1} = \begin{bmatrix} -1.0000 & -1.0000 & -3.0000 \\ -8.0032 & -0.3325 & -1.0000 \end{bmatrix}.$$

It can be verified that the corresponding closed-loop poles are $-2.3302$, $-2.0023$, and $-2.0000$, which are indeed located in the region $\mathbb{D}_{(2,1)}$.

### 7.2.3 General $\mathbb{D}$-Stabilization

The problem of $\mathbb{D}$-stabilization, with $\mathbb{D}$ being an LMI region, is stated as follows.

**Problem 7.5** Let $\mathbb{D} = \mathbb{D}_{(L,M)}$ be an LMI region, whose characteristic function is

$$F_{\mathbb{D}} = L + sM + \bar{s}M^{\mathrm{T}}.$$

Design a state feedback control law $u = Kx$ for the linear system

$$\rho x = Ax + Bu, \tag{7.10}$$

such that the closed-loop system

$$\rho x = (A + BK)x$$

is $\mathbb{D}_{(L,M)}$-stable.

It follows from Theorem 4.1 that the matrix $(A + BK)$ is $\mathbb{D}_{(L,M)}$-stable if and only if there exists a symmetric positive definite matrix $P$ such that

$$R_{\mathbb{D}}(A, P) = L \otimes P + M \otimes ((A + BK)P) + M^{\mathrm{T}} \otimes ((A + BK)P)^{\mathrm{T}} < 0,$$

where $\otimes$ represents the Kronecker product. Let

$$W = KP,$$

the aforementioned inequality can be rewritten as

$$L \otimes P + M \otimes AP + M^{\mathrm{T}} \otimes PA^{\mathrm{T}} + M \otimes BW + M^{\mathrm{T}} \otimes W^{\mathrm{T}}B^{\mathrm{T}} < 0. \tag{7.11}$$

With this deduction, we now have the following theorem.

**Theorem 7.5** The aforementioned $\mathbb{D}_{(L,M)}$-stabilization problem has a solution if and only if there exist a symmetric positive definite matrix $P$ and a matrix $W$ satisfying the LMI (7.11). In this case, a solution to the problem is given by

$$K = WP^{-1}.$$

**Example 7.5**

Consider the LMI region $\mathbb{S}(\alpha, r, \theta)$ given in (4.17). In Example 4.8, we have shown that $(A + BK)$ is $\mathbb{S}(\alpha, r, \theta)$-stable if and only if there exists a symmetric positive definite matrix $P$, such that

$$2\alpha P + (A + BK)P + P(A + BK)^{\mathrm{T}} < 0,$$

$$\begin{bmatrix} -rP & (A+BK)\,P \\ P\,(A+BK)^{\mathsf{T}} & -rP \end{bmatrix} < 0,$$

$$\begin{bmatrix} \left((A+BK)\,P + P\,(A+BK)^{\mathsf{T}}\right)\sin\theta & \left((A+BK)\,P - P\,(A+BK)^{\mathsf{T}}\right)\cos\theta \\ \left(P\,(A+BK)^{\mathsf{T}} - (A+BK)\,P\right)\cos\theta & \left((A+BK)\,P + P\,(A+BK)^{\mathsf{T}}\right)\sin\theta \end{bmatrix} < 0.$$

Through introducing the relation

$$W = KP,$$

we immediately have the conclusion that the LMI region stabilization problem in the case of $\mathbb{S}(\alpha, r, \theta)$ has a solution if and only if there exist a symmetric matrix $P$ and a matrix $W$, satisfying

$$2\alpha P + AP + PA^{\mathsf{T}} + BW + W^{\mathsf{T}}B^{\mathsf{T}} < 0,$$

$$\begin{bmatrix} -rP & AP + BW \\ PA^{\mathsf{T}} + W^{\mathsf{T}}B^{\mathsf{T}} & -rP \end{bmatrix} < 0,$$

$$\begin{bmatrix} \left(AP + PA^{\mathsf{T}} + BW + W^{\mathsf{T}}B^{\mathsf{T}}\right)\sin\theta & \left(AP - PA^{\mathsf{T}} + BW - W^{\mathsf{T}}B^{\mathsf{T}}\right)\cos\theta \\ \left(PA^{\mathsf{T}} - AP + W^{\mathsf{T}}B^{\mathsf{T}} - BW\right)\cos\theta & \left(AP + PA^{\mathsf{T}} + BW + W^{\mathsf{T}}B^{\mathsf{T}}\right)\sin\theta \end{bmatrix} < 0.$$

In this case, the feedback gain is given by

$$K = WP^{-1}.$$

## Example 7.6

Consider the following linear system, which has been studied by Duan (2003):

$$\dot{x}(t) = \begin{bmatrix} -0.5 & 0 & 0 \\ 0 & -2 & 10 \\ 0 & 1 & -2 \end{bmatrix} x(t) + \begin{bmatrix} 1 & 0 \\ -2 & 2 \\ 0 & 1 \end{bmatrix} u(t).$$

Let the matrices associated with the characteristic function of the LMI region $\mathbb{D}$ be

$$L = \begin{bmatrix} -1 & 2 \\ 2 & -1 \end{bmatrix}, \quad M = \frac{1}{4}\begin{bmatrix} 0 & 3 \\ 1 & 0 \end{bmatrix}.$$

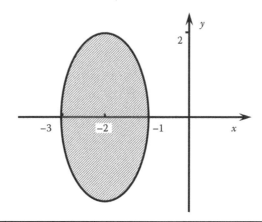

**Figure 7.2   The LMI region in Example 7.6.**

Thus,

$$\mathbb{D} = \left\{ s \, \middle| \, s \in \mathbb{C}, \; L + sM + \bar{s}M^\mathsf{T} < 0 \right\}$$

$$= \left\{ s \, \middle| \, s \in \mathbb{C}, \; \begin{bmatrix} -1 & 2 \\ 2 & -1 \end{bmatrix} + \begin{bmatrix} 0 & \frac{3}{4}s \\ \frac{1}{4}s & 0 \end{bmatrix} + \begin{bmatrix} 0 & \frac{1}{4}\bar{s} \\ \frac{3}{4}\bar{s} & 0 \end{bmatrix} < 0 \right\}$$

$$= \left\{ s \, \middle| \, s \in \mathbb{C}, \; \begin{bmatrix} -1 & 2 + \frac{3}{4}s + \frac{1}{4}\bar{s} \\ 2 + \frac{1}{4}s + \frac{3}{4}\bar{s} & -1 \end{bmatrix} < 0 \right\}$$

$$= \left\{ x + yj \, \middle| \, x, y \in \mathbb{R}, \; (x+2)^2 + \frac{1}{4}y^2 < 1 \right\},$$

which is obviously an ellipsoid (see Figure 7.2).

With this system and the aforementioned given region, the corresponding LMI is

$$\begin{bmatrix} -P & 2P + \frac{3}{4}AP + \frac{1}{4}PA^\mathsf{T} + \frac{3}{4}BW + \frac{1}{4}W^\mathsf{T}B^\mathsf{T} \\ * & -P \end{bmatrix} < 0,$$

where "*" denotes the symmetric entry in the matrix. Using the MATLAB function `feasp` in the MATLAB LMI toolbox, the aforementioned LMI is solved, and the parameter matrices are obtained as

$$P = \begin{bmatrix} 65.3429 & 7.9921 & -14.5394 \\ 7.9921 & 47.8959 & 6.0223 \\ -14.5394 & 6.0223 & 12.8421 \end{bmatrix} > 0,$$

and

$$W = \begin{bmatrix} -101.5266 & -20.7431 & 51.3169 \\ -26.7148 & -51.3939 & -9.0775 \end{bmatrix}.$$

Hence, a feedback gain is given by

$$K = WP^{-1} = \begin{bmatrix} -0.6521 & -0.7799 & 3.6234 \\ -0.4885 & -0.8853 & -0.8448 \end{bmatrix}.$$

Using the MATLAB command eig, the corresponding closed-loop poles are computed to be $-2.1039 \pm 1.0842i$ and $-2.0000$, which are indeed located in the LMI region $\mathbb{D}$.

## 7.3 Quadratic Stabilization

Quadratic stabilization deals with the stabilization of a family of systems. It arises from the problem of robust control.

### 7.3.1 Family of Systems

This is concerned with the stabilization of the following uncertain system

$$\rho x = A(\delta(t))x + B(\delta(t))u, \tag{7.12}$$

with $\rho$ representing the differential operator, or the one step forward shift operator. As stated in the beginning of Section 4.3, in the former case, the system is continuous-time, and $t$ is a continuous variable; while in the latter case, the system is discrete-time, and now $t$ is a discrete variable.

The coefficient matrices of the system take the forms of

$$A(\delta(t)) = A_0 + \Delta A(\delta(t)), \quad B(\delta(t)) = B_0 + \Delta B(\delta(t)),$$

with

$$\Delta A(\delta(t)) = \delta_1(t)A_1 + \delta_2(t)A_2 + \cdots + \delta_k(t)A_k,$$

$$\Delta B(\delta(t)) = \delta_1(t)B_1 + \delta_2(t)B_2 + \cdots + \delta_k(t)B_k,$$

where

- $A_0 \in \mathbb{R}^{n \times n}$ and $B_0 \in \mathbb{R}^{n \times r}$ are the nominal system coefficient matrices
- $\Delta A(\delta(t))$ and $\Delta B(\delta(t))$ are the parameter matrix perturbations
- $A_i \in \mathbb{R}^{n \times n}$, $B_i \in \mathbb{R}^{n \times r}$, $i = 1, 2, \ldots, k$, are known matrices that represent the perturbation directions

- $\delta_i(t)$, $i = 1, 2, \ldots, k$, are arbitrary time functions, which represent the uncertain parameters
- $\delta(t) = \begin{bmatrix} \delta_1(t) & \delta_2(t) & \cdots & \delta_k(t) \end{bmatrix}$ is the uncertain parameter vector, which is taken from a compact set $\Delta$, that is,

$$\delta(t) = \begin{bmatrix} \delta_1(t) & \delta_2(t) & \cdots & \delta_k(t) \end{bmatrix} \in \Delta.$$

Clearly, the uncertain system (7.12) can be written as

$$\rho x = A(\delta(t))x + B(\delta(t))u, \quad \delta(t) \in \Delta. \tag{7.13}$$

## 7.3.2 Quadratic Hurwitz Stabilization

First, let us treat the continuous-time case, and investigate the quadratic Hurwitz stabilization of the uncertain continuous-time system (7.13).

### 7.3.2.1 Problem Formulation

The problem of quadratic Hurwitz stabilization can be stated as follows.

**Problem 7.6**   Find a fixed state feedback control law $u = Kx$ for the uncertain system (7.13) such that the set of closed-loop systems

$$\dot{x} = A_c(\delta(t))x, \quad \delta(t) \in \Delta, \tag{7.14}$$

with

$$A_c(\delta(t)) = A(\delta(t)) + B(\delta(t))K,$$

is quadratically Hurwitz stable, that is, there exists a symmetric positive definite matrix $P$, such that

$$A_c(\delta(t))P + PA_c^{\mathrm{T}}(\delta(t)) < 0, \quad \forall \delta(t) \in \Delta. \tag{7.15}$$

In the following of this section, we seek solutions to the aforementioned quadratic Hurwitz stabilization problem for the cases of interval systems and polytopic systems.

### 7.3.2.2 Case of Interval Systems

As in Section 4.3, we first consider the case of $\Delta = \Delta_I$, with

$$\Delta_I = \left\{ \delta(t) = \begin{bmatrix} \delta_1(t) & \delta_2(t) & \cdots & \delta_k(t) \end{bmatrix} \mid \delta_i(t) \in [\delta_i^-, \delta_i^+], \quad i = 1, 2, \ldots, k \right\}. \tag{7.16}$$

In this case, the system becomes an interval system, and the set of the extremes of $\Delta_I$ is

$$\Delta_E = \left\{ \delta = \begin{bmatrix} \delta_1 & \delta_2 & \cdots & \delta_k \end{bmatrix} \mid \delta_i = \delta_i^- \text{ or } \delta_i^+, \quad i = 1, 2, \ldots, k \right\}.$$

It follows from Corollary 4.1 that the system (7.14), with $\delta(t) \in \Delta = \Delta_I$, is quadratically Hurwitz stable if and only if there exists a symmetric positive definite matrix $P$, such that

$$A_c(\delta)P + PA_c^{\mathrm{T}}(\delta) < 0, \quad \delta_i = \delta_i^- \text{ or } \delta_i^+, \tag{7.17}$$

that is,

$$[A(\delta) + B(\delta)K]P + P[A(\delta) + B(\delta)K]^{\mathrm{T}} < 0, \quad \delta_i = \delta_i^- \text{ or } \delta_i^+. \tag{7.18}$$

Following the typical treatment of this type of inequalities, we immediately have the following theorem.

**Theorem 7.6**    With $\Delta = \Delta_I$ given by (7.16), the aforementioned quadratic Hurwitz stabilization problem (Problem 7.6) has a solution if and only if there exist a symmetric positive definite matrix $P$ and a matrix $W$ satisfying the following set of LMIs:

$$A(\delta)P + PA^{\mathrm{T}}(\delta) + B(\delta)W + W^{\mathrm{T}}B^{\mathrm{T}}(\delta) < 0, \quad \delta_i = \delta_i^- \text{ or } \delta_i^+. \tag{7.19}$$

In this case, a solution to the problem is given by

$$K = WP^{-1}.$$

When the matrix $B$ does not have perturbations, that is, $B_i = 0, i = 1, 2, \ldots, k$, following Theorem 6.4 and its proof we further have the following result.

**Theorem 7.7**    When the matrix $B$ does not have perturbations, and $\Delta = \Delta_I$ is given by (7.16), the aforementioned quadratic Hurwitz stabilization problem (Problem 7.6) has a solution if and only if there exists a symmetric positive definite matrix $P$ satisfying the following set of LMIs

$$A(\delta)P + PA^{\mathrm{T}}(\delta) - B_0 B_0^{\mathrm{T}} < 0, \quad \delta_i = \delta_i^- \text{ or } \delta_i^+. \tag{7.20}$$

In this case, a solution to the problem is given by

$$K = -\frac{1}{2}B_0^{\mathrm{T}}P^{-1}.$$

**Example 7.7**

Consider the following continuous-time linear system:

$$\dot{x}(t) = A(\delta(t))x(t) + B(\delta(t))u(t),$$

with

$$A(\delta(t)) = A_0 + \delta_1(t)A_1, \quad B(\delta(t)) = B_0 + \delta_1(t)B_1,$$

$$A_0 = \begin{bmatrix} -3 & 0 & 1 \\ 0 & -3 & 0 \\ 1 & 0 & -4 \end{bmatrix}, \quad A_1 = \begin{bmatrix} 1 & 0 & 1 \\ 0 & -1 & 0 \\ 0 & 1 & 1 \end{bmatrix},$$

$$B_0 = \begin{bmatrix} 1 & 1 \\ 3 & 4 \\ 5 & 1 \end{bmatrix}, \quad B_1 = \begin{bmatrix} 1 & 0 \\ 0 & -2 \\ 0 & 1 \end{bmatrix},$$

$$\delta_1(t) \in [-0.1, 0.1].$$

Using the MATLAB function feasp in the MATLAB LMI toolbox, the corresponding LMI problem (7.19) is solved and the two parameter matrices are obtained as

$$P = \begin{bmatrix} 0.3429 & 0.1269 & 0.1949 \\ 0.1269 & 0.9457 & 0.1456 \\ 0.1949 & 0.1456 & 1.1505 \end{bmatrix} > 0,$$

$$W = \begin{bmatrix} -0.0438 & 0.2574 & 0.8922 \\ 0.0476 & 0.3169 & -0.8434 \end{bmatrix}.$$

Thus, the feedback gain matrix can be obtained as

$$K = WP^{-1} = \begin{bmatrix} -0.7059 & 0.2336 & 0.8655 \\ 0.4812 & 0.4037 & -0.8657 \end{bmatrix}.$$

### 7.3.2.3 Case of Polytopic Systems

As in Section 4.3, let us consider the case of $\Delta = \Delta_P$, with

$$\Delta_P = \left\{ \delta(t) = \begin{bmatrix} \delta_1(t) & \delta_2(t) & \cdots & \delta_k(t) \end{bmatrix} \middle| \sum_{i=1}^{k} \delta_i(t) = 1, \, \delta_i(t) \geq 0, \right.$$

$$\left. i = 1, 2, \ldots, k \right\}, \tag{7.21}$$

and in this case the set of extremes of $\Delta$ is

$$\Delta_E = \left\{ \begin{pmatrix} 1 & 0 & \cdots & 0 \end{pmatrix}, \begin{pmatrix} 0 & 1 & \cdots & 0 \end{pmatrix}, \ldots, \begin{pmatrix} 0 & 0 & \cdots & 1 \end{pmatrix} \right\}.$$

It follows from Corollary 4.2 that the system (7.14), with $\delta(t) \in \Delta = \Delta_P$, is quadratically Hurwitz stable if and only if there exists a symmetric positive definite matrix $P$, such that

$$[(A_0 + A_i) + (B_0 + B_i)\,K]\,P + P\,[(A_0 + A_i) + (B_0 + B_i)\,K]^{\mathrm{T}} < 0,$$
$$i = 1, 2, \ldots, k.$$

Following the typical treatment of this type of inequalities, we immediately have the following theorem.

**Theorem 7.8**  With $\Delta = \Delta_P$ given by (7.21), the aforementioned quadratic Hurwitz stabilization problem (Problem 7.6) has a solution if and only if there exist a symmetric positive definite matrix $P$ and a matrix $W$ satisfying the following set of LMIs:

$$(A_0 + A_i)\,P + P\,(A_0 + A_i)^{\mathrm{T}} + (B_0 + B_i)\,W + W^{\mathrm{T}}\,(B_0 + B_i)^{\mathrm{T}} < 0,$$
$$i = 1, 2, \ldots, k. \tag{7.22}$$

In this case, a solution to the problem is given by

$$K = WP^{-1}.$$

When the matrix $B$ does not have perturbations, that is, $B_i = 0$, $i = 1, 2, \ldots, k$, following Theorem 6.4 and its proof we further have the following result.

**Theorem 7.9**  When the matrix $B$ does not have perturbations, and $\Delta = \Delta_P$ is given by (7.21), the aforementioned quadratic Hurwitz stabilization problem (Problem 7.6) has a solution if and only if there exists a symmetric positive definite matrix $P$ satisfying the following set of LMIs

$$(A_0 + A_i)\,P + P\,(A_0 + A_i)^{\mathrm{T}} - B_0 B_0^{\mathrm{T}} < 0, \quad i = 1, 2, \ldots, k.$$

In this case, a solution to the problem is given by

$$K = -\frac{1}{2}B_0^T P^{-1}.$$

**Example 7.8**

Consider the following continuous-time linear system with parameter uncertainties:

$$\dot{x}(t) = A(\delta(t))x(t) + B(\delta(t))u(t),$$

where

$$A(\delta(t)) = A_0 + \delta_1(t)A_1 + \delta_2(t)A_2,$$

$$B(\delta(t)) = B_0 + \delta_1(t)B_1 + \delta_2(t)B_2,$$

with

$$A_0 = \begin{bmatrix} -1 & 2 \\ 2 & -3 \end{bmatrix}, \quad A_1 = \begin{bmatrix} 2 & -1 \\ -1.5 & 2 \end{bmatrix}, \quad A_2 = \begin{bmatrix} 1 & 2 \\ 2 & 1 \end{bmatrix},$$

$$B_0 = \begin{bmatrix} 1 \\ 3 \end{bmatrix}, \quad B_1 = \begin{bmatrix} 1 \\ 0 \end{bmatrix}, \quad B_2 = \begin{bmatrix} 0 \\ 1 \end{bmatrix},$$

and let us assume that the set of perturbation parameters $\delta_i(t)$, $i = 1, 2$, form a polytope in the form of (7.21), with $k = 2$.

Using the MATLAB function feasp in the MATLAB LMI toolbox, the corresponding LMI problem (7.22) is solved and the two parameter matrices are obtained as

$$P = \begin{bmatrix} 8.3116 & -1.1787 \\ -1.1787 & 22.0326 \end{bmatrix} > 0,$$

and

$$W = \begin{bmatrix} -20.8876 & -0.0551 \end{bmatrix}.$$

Thus, the robust feedback control gain matrix can be computed as

$$K = WP^{-1} = \begin{bmatrix} -2.5326 & -0.1380 \end{bmatrix}.$$

## 7.3.3 *Quadratic Schur Stabilization*

### 7.3.3.1 *Problem Formulation*

The problem of quadratic Schur stabilization can be stated as follows.

**Problem 7.7** Find a fixed state feedback control law $u = Kx$ for the uncertain system (7.13) such that the set of closed-loop systems (7.14) is quadratically Schur stable, that is, there exists a symmetric positive definite matrix $P$, such that

$$A_c(\delta(t))PA_c(\delta(t))^T - P < 0, \quad \forall \delta(t) \in \Delta. \tag{7.23}$$

In the following of this section, we seek solutions to the aforementioned quadratic Schur stabilization problem for the cases of interval systems and polytopic systems.

### 7.3.3.2 Case of Interval Systems

It follows from Corollary 4.3 that the system (7.14), with $\delta(t) \in \Delta = \Delta_I$, is quadratically Schur stable if and only if there exists a symmetric matrix $P$, such that

$$\begin{bmatrix} -P & A_c(\delta)P \\ PA_c(\delta)^T & -P \end{bmatrix} < 0, \quad \delta_i = \delta_i^- \text{ or } \delta_i^+,$$

that is,

$$\begin{bmatrix} -P & [A(\delta) + B(\delta)K]P \\ P[A(\delta) + B(\delta)K]^T & -P \end{bmatrix} < 0, \quad \delta_i = \delta_i^- \text{ or } \delta_i^+. \tag{7.24}$$

Following the typical treatment of this type of inequalities, we immediately have the following theorem.

**Theorem 7.10** With $\Delta = \Delta_I$ given by (7.16), the aforementioned quadratic Schur stabilization problem (Problem 7.7) has a solution if and only if there exist a symmetric matrix $P$ and a matrix $W$ satisfying the following set of LMIs:

$$\begin{bmatrix} -P & A(\delta)P + B(\delta)W \\ PA(\delta)^T + W^T B^T(\delta) & -P \end{bmatrix} < 0, \quad \delta_i = \delta_i^- \text{ or } \delta_i^+. \tag{7.25}$$

In this case, a solution to the problem is given by

$$K = WP^{-1}.$$

**Example 7.9**

Consider the following uncertain discrete-time linear system

$$x(k + 1) = A(\delta(k))x(k) + B(\delta(k))u(k),$$

with

$$A(\delta(k)) = A_0 + \delta_1(k)A_1, \quad B(\delta(k)) = B_0 + \delta_1(k)B_1,$$

$$A_0 = \begin{bmatrix} 2 & 1 & 0 \\ 1 & -2 & 0 \\ 0 & -1 & 4 \end{bmatrix}, \quad A_1 = \begin{bmatrix} 0 & 0 & 1 \\ 1 & -1 & 0 \\ 1 & 1 & 0 \end{bmatrix},$$

$$B_0 = \begin{bmatrix} 2 & 1 \\ 0 & 2 \\ -1 & 3 \end{bmatrix}, \quad B_1 = \begin{bmatrix} 0 & 1 \\ -1 & 2 \\ 0 & 1 \end{bmatrix},$$

$$\delta_1(k) \in [-0.1, 0.1].$$

Using the MATLAB function `feasp` in the MATLAB LMI toolbox to the corresponding LMI problem (7.25), we obtain

$$P = \begin{bmatrix} 0.9160 & -0.1146 & 0.2421 \\ -0.1146 & 0.9031 & -0.5255 \\ 0.2421 & -0.5255 & 0.5795 \end{bmatrix} > 0,$$

$$W = \begin{bmatrix} -0.5803 & -0.6959 & 0.4364 \\ -0.5596 & 0.8327 & -0.7482 \end{bmatrix}.$$

Thus, the robustly stabilizing feedback gain matrix can be obtained to be

$$K = WP^{-1} = \begin{bmatrix} -0.8730 & -0.4891 & 0.6743 \\ -0.3611 & 0.4505 & -0.7317 \end{bmatrix}.$$

### 7.3.3.3 Case of Polytopic Systems

It follows from Corollary 4.4 that the system (7.14), with $\delta(t) \in \Delta = \Delta_P$, is quadratically Schur stable if and only if there exists a symmetric matrix $P$, such that

$$\begin{bmatrix} -P & [(A_0 + A_i) + (B_0 + B_i)K]P \\ P[(A_0 + A_i) + (B_0 + B_i)K]^{\mathrm{T}} & -P \end{bmatrix} < 0,$$

$$i = 1, 2, \ldots, k.$$

Following the typical treatment of this type of inequalities, we immediately have the following theorem.

**Theorem 7.11**  With $\Delta = \Delta_P$ given by (7.21), the aforementioned quadratic Schur stabilization problem (Problem 7.7) has a solution if and only if there exist a symmetric matrix $P$ and a matrix $W$ satisfying the following set of LMIs:

$$\begin{bmatrix} -P & (A_0 + A_i)\,P + (B_0 + B_i)\,W \\ P\,(A_0 + A_i)^{\mathrm{T}} + W^{\mathrm{T}}\,(B_0 + B_i)^{\mathrm{T}} & -P \end{bmatrix} < 0,$$

$$i = 1, 2, \ldots, k. \tag{7.26}$$

In this case, a solution to the problem is given by

$$K = WP^{-1}.$$

**Example 7.10**

Consider the following discrete-time linear system with parameter uncertainties:

$$x(k+1) = A(\delta(k))x(k) + B(\delta(k))u(k),$$

where

$$A(\delta(k)) = A_0 + \delta_1(k)A_1 + \delta_2(k)A_2,$$

$$B(\delta(k)) = B_0 + \delta_1(k)B_1 + \delta_2(k)B_2,$$

$$A_0 = \begin{bmatrix} 1 & 4 \\ 2 & -3 \end{bmatrix}, \quad A_1 = \begin{bmatrix} 0 & -1.5 \\ -1 & 0 \end{bmatrix}, \quad A_2 = \begin{bmatrix} -1 & 2 \\ 2 & 1 \end{bmatrix},$$

$$B_0 = \begin{bmatrix} 1 & 0 \\ 0 & 1 \end{bmatrix}, \quad B_1 = \begin{bmatrix} 0 & 0 \\ 0 & 1 \end{bmatrix}, \quad B_2 = \begin{bmatrix} 0 & -1 \\ 1 & 0 \end{bmatrix},$$

and assume that the set of perturbation parameters $\delta_i(k)$, $i = 1, 2$, forms a polytope in the form of (4.27).

Using the MATLAB function `feasp` in the MATLAB LMI toolbox to the corresponding LMI problem (7.26), we obtain

$$P = \begin{bmatrix} 5.4590 & 3.0723 \\ 3.0723 & 1.9068 \end{bmatrix} > 0,$$

and

$$W = \begin{bmatrix} -15.8760 & -8.6388 \\ 1.1703 & 1.1635 \end{bmatrix}.$$

Thus, the robust feedback control gain matrix can be obtained to be

$$K = WP^{-1} = \begin{bmatrix} -3.8466 & 1.6672 \\ -1.3848 & 2.8415 \end{bmatrix}.$$

## 7.4 Quadratic $\mathbb{D}$-Stabilization

In the aforementioned section, we have treated quadratic Hurwitz and Schur stabilization using LMI techniques. In this section, we further investigate the problem of quadratic $\mathbb{D}$-stabilization.

### 7.4.1 Problem Formulation

The problem of quadratic $\mathbb{D}$-stabilization can be stated as follows.

**Problem 7.8**  Let $\mathbb{D} = \mathbb{D}_{(L,M)}$ be an LMI region, whose characteristic function is

$$F_{\mathbb{D}} = L + sM + \bar{s}M^{\mathrm{T}}.$$

Find a fixed state feedback control law $u = Kx$ for the uncertain system (7.13), such that the set of closed-loop systems (7.14) is quadratically $\mathbb{D}_{(L,M)}$-stable, that is, there exists a symmetric positive definite matrix $P$, such that

$$L \otimes P + M \otimes (A_c(\delta(t))P) + M^{\mathrm{T}} \otimes (A_c(\delta(t))P)^{\mathrm{T}} < 0, \quad \forall \delta(t) \in \Delta. \quad (7.27)$$

In the following of this section, we seek a solution to the aforementioned quadratic $\mathbb{D}_{(L,M)}$-stabilization problem based on Theorem 4.7.

### 7.4.2 Solution

It follows from Theorem 4.7 that the system (7.14), with $\delta(t) \in \Delta$, is quadratically $\mathbb{D}_{(L,M)}$-stable if and only if there exists a symmetric positive definite matrix $P$, such that

$$L \otimes P + M \otimes (A_c(\delta(t))P) + M^{\mathrm{T}} \otimes (A_c(\delta(t))P)^{\mathrm{T}} < 0, \quad \forall \delta \in \Delta_E,$$

that is

$$L \otimes P + M \otimes ((A(\delta(t)) + B(\delta(t))K)P)$$
$$+ M^{\mathrm{T}} \otimes ((A(\delta(t)) + B(\delta(t))K)P)^{\mathrm{T}} < 0, \quad \forall \delta \in \Delta_E. \quad (7.28)$$

Following the typical treatment of this type of inequalities, we immediately have the following theorem.

**Theorem 7.12**  Let $\Delta \subset \mathbb{R}^k$ be a compact convex set. Then, the aforementioned quadratic $\mathbb{D}_{(L,M)}$-stabilization problem (Problem 7.8) has a solution if and only if there exist a symmetric positive definite matrix $P$ and a matrix $W$ satisfying the following LMI:

$$L \otimes P + M \otimes (A(\delta(t))P) + M^T \otimes (A(\delta(t))P)^T$$
$$+ M \otimes (B(\delta(t))W) + M^T \otimes (B(\delta(t))W)^T < 0, \quad \forall \delta(t) \in \Delta_E. \quad (7.29)$$

In this case, a solution to the problem is given by

$$K = WP^{-1}.$$

Consider the specific LMI region

$$\mathbb{S}(\alpha, r, \theta) = \left\{ (x, y) \mid x < -\alpha < 0, \, |x + jy| < r, \, |y| < -x \tan \theta \right\}. \quad (7.30)$$

It is seen in Example 4.8 that the characteristic function of the LMI region $\mathbb{S}(\alpha, r, \theta)$ is

$$F_{\mathbb{S}(\alpha, r, \theta)}(s) = L + sM + \bar{s}M^T,$$

where

$$L = \mathrm{diag}(L_1, L_2, L_3), \quad M = \mathrm{diag}(M_1, M_2, M_3),$$

with

$$L_1 = 2\alpha, \quad M_1 = 1,$$

$$L_2 = \begin{bmatrix} -r & 0 \\ 0 & -r \end{bmatrix}, \quad M_2 = \begin{bmatrix} 0 & 1 \\ 0 & 0 \end{bmatrix},$$

$$L_3 = 0, \quad M_3 = \begin{bmatrix} \sin\theta & \cos\theta \\ -\cos\theta & \sin\theta \end{bmatrix}.$$

Therefore, the following corollary follows from Theorem 7.12.

**Corollary 7.1** The quadratic $\mathbb{S}(\alpha, r, \theta)$-stabilization problem has a solution if and only if there exist a symmetric positive definite matrix $P$ and a matrix $W$ satisfying the following set of LMIs:

$$A(\delta(t))P + PA^T(\delta(t)) + B(\delta(t))W + W^TB^T(\delta(t)) + 2\alpha P < 0,$$
$$\forall \delta(t) \in \Delta_E, \quad (7.31)$$

$$\begin{bmatrix} -rP & A\left(\delta\left(t\right)\right)P + B\left(\delta\left(t\right)\right)W \\ PA^{\mathrm{T}}\left(\delta\left(t\right)\right) + W^{\mathrm{T}}B^{\mathrm{T}}\left(\delta\left(t\right)\right) & -rP \end{bmatrix} < 0, \quad \forall \delta\left(t\right) \in \Delta_E,$$

(7.32)

and

$$\begin{bmatrix} \Omega_1\left(\delta\left(t\right)\right)\sin\theta & \Omega_2\left(\delta\left(t\right)\right)\cos\theta \\ \Omega_2^{\mathrm{T}}\left(\delta\left(t\right)\right)\cos\theta & \Omega_1\left(\delta\left(t\right)\right)\sin\theta \end{bmatrix} < 0, \quad \forall \delta\left(t\right) \in \Delta_E,$$

(7.33)

with

$$\Omega_1\left(\delta\left(t\right)\right) = A\left(\delta\left(t\right)\right)P + PA^{\mathrm{T}}\left(\delta\left(t\right)\right) + B\left(\delta\left(t\right)\right)W + W^{\mathrm{T}}B^{\mathrm{T}}\left(\delta\left(t\right)\right),$$

$$\Omega_2\left(\delta\left(t\right)\right) = A\left(\delta\left(t\right)\right)P - PA^{\mathrm{T}}\left(\delta\left(t\right)\right) + B\left(\delta\left(t\right)\right)W - W^{\mathrm{T}}B^{\mathrm{T}}\left(\delta\left(t\right)\right).$$

In this case, a solution to the problem is given by

$$K = WP^{-1}.$$

## 7.4.3 Special Cases

### 7.4.3.1 Case of Interval Systems

Now let us consider the case of $\delta(t) \in \Delta = \Delta_I$, with

$$\Delta_I = \left\{ \delta(t) \,\middle|\, \delta_i(t) \in [\delta_i^-, \, \delta_i^+], \quad i = 1, 2, \ldots, k \right\}.$$

(7.34)

In this case, we have

$$\Delta_E = \left\{ \delta = \begin{bmatrix} \delta_1 & \delta_2 & \cdots & \delta_k \end{bmatrix} \,\middle|\, \delta_i = \delta_i^- \text{ or } \delta_i^+, \quad i = 1, 2, \ldots, k \right\}.$$

With this observation, we immediately have, following from Theorem 7.12, the following result.

**Theorem 7.13** With $\Delta = \Delta_I$ given by (7.34), the aforementioned quadratic $\mathbb{D}_{(L,M)}$-stabilization problem (Problem 7.8) has a solution if and only if there exist a symmetric positive definite matrix $P$ and a matrix $W$ satisfying the following set of LMIs:

$$L \otimes P + M \otimes (A(\delta)P) + M^{\mathrm{T}} \otimes (A(\delta)P)^{\mathrm{T}}$$

$$+ M \otimes (B(\delta)W) + M^{\mathrm{T}} \otimes (B(\delta)W)^{\mathrm{T}} < 0,$$

$$\delta_i = \delta_i^- \text{ or } \delta_i^+, \quad i = 1, 2, \ldots, k.$$

(7.35)

In this case, a solution to the problem is given by

$$K = WP^{-1}.$$

## 7.4.3.2 Case of Polytopic Systems

In the case of $\delta(t) \in \Delta = \Delta_P$, with

$$\Delta_P = \left\{ \delta(t) \,\middle|\, \sum_{i=1}^{k} \delta_i(t) = 1, \; \delta_i(t) \geq 0, \; i = 1, 2, \ldots, k \right\}, \tag{7.36}$$

we have

$$\Delta_E = \left\{ e_1^{\mathrm{T}}, e_2^{\mathrm{T}}, \ldots, e_k^{\mathrm{T}} \right\} \subset \mathbb{R}^k, \tag{7.37}$$

where $e_i \in \mathbb{R}^k$ is the vector with the $i$th element being 1 while all the others being zero. Noticing that

$$A(e_i) = A_0 + A_i, \; B(e_i) = B_0 + B_i, \quad i = 1, 2, \ldots, k,$$

following from Theorem 7.12, we immediately have the following result.

**Theorem 7.14**   With $\Delta = \Delta_P$ given by (7.36), the aforementioned quadratic $\mathbb{D}_{(L,M)}$-stabilization problem (Problem 7.8) has a solution if and only if there exist a symmetric positive definite matrix $P$ and a matrix $W$ satisfying the following LMI:

$$L \otimes P + M \otimes ((A_0 + A_i) P) + M^{\mathrm{T}} \otimes ((A_0 + A_i) P)^{\mathrm{T}} + M \otimes ((B_0 + B_i) W)$$

$$+ M^{\mathrm{T}} \otimes ((B_0 + B_i) W)^{\mathrm{T}} < 0, \quad i = 1, 2, \ldots, k. \tag{7.38}$$

In this case, a solution to the problem is given by

$$K = W P^{-1}.$$

### Example 7.11

Consider a parameter-dependent interval system that represents a simplified and linearized longitudinal dynamic model of the flight control system of a supersonic cruise missile. The state-space representation of the system is as follows:

$$\begin{cases} \dot{x}(t) = A(\delta) \, x(t) + B(\delta) \, u(t) \\ y(t) = C x(t), \end{cases} \tag{7.39}$$

where

$$A(\delta(t)) = A_0 + \sum_{i=1}^{4} \delta_i(t) A_i, \quad B(\delta(t)) = B_0 + \sum_{i=1}^{4} \delta_i(t) B_i,$$

with

$$A_0 = \begin{bmatrix} -5 & 0 & 0 \\ 0 & 0 & 1 \\ 0 & 0 & 0 \end{bmatrix}, \quad A_1 = \begin{bmatrix} -1 & 1 & 0 \\ 0 & 0 & 0 \\ 0 & 0 & 0 \end{bmatrix},$$

$$A_2 = \begin{bmatrix} 0 & 0 & 0 \\ 0 & 0 & 0 \\ 1 & -1 & 0 \end{bmatrix}, \quad A_3 = A_4 = 0,$$

$$B_0 = \begin{bmatrix} 2 & 1 & 3 \end{bmatrix}^T, \quad B_1 = B_2 = 0, \quad B_3 = \begin{bmatrix} 1 & 0 & 0 \end{bmatrix}^T,$$

$$B_4 = \begin{bmatrix} 0 & 0 & -1 \end{bmatrix}, \quad C = \begin{bmatrix} 0 & 1 & 0 \end{bmatrix},$$

and $\delta_i(t)$, $i = 1, 2, 3, 4$, being the uncertain variables that take values in the following regular polyhedron:

$$\Delta_I = \{-0.1 \le \delta_1 \le 0.2, \ 0.08 \le \delta_2 \le 0.1, \ 0.03 \le \delta_3 \le 0.05,$$
$$-0.35 \le \delta_4 \le 0.35\}.$$

Using the conclusion of Corollary 7.1, a quadratically $\mathbb{S}(\alpha, r, \theta)$-stabilizing state feedback control law for the quadratically stabilizable parameter-dependent interval system can be obtained with a given level of performance. For example, we need to design a state feedback control law $u = Kx$ that meets the following specifications:

1. The closed-loop system is quadratically stable.
2. The overshoot $\sigma$ of the unit step response of the nominal system is less than 2%.
3. The setting time $t_s$ of the unit step response of the nominal system is lower than 0.5s.

To meet these conditions, we have chosen, for the LMI region $\mathbb{S}(\alpha, r, \theta)$, the following parameters

$$\alpha = 2.5, \quad r = 40, \quad \theta = \pi/15.6 \deg.$$

By using the MATLAB function feasp in the MATLAB LMI toolbox, a feasible solution of (7.31) through (7.33) is obtained as follows:

$$P = \begin{bmatrix} 0.0571 & 0.0028 & -0.0088 \\ 0.0028 & 0.0010 & 0.0026 \\ -0.0088 & 0.0026 & 0.0388 \end{bmatrix} > 0,$$

$$W = \begin{bmatrix} -0.0619 & -0.0291 & -0.1222 \end{bmatrix}.$$

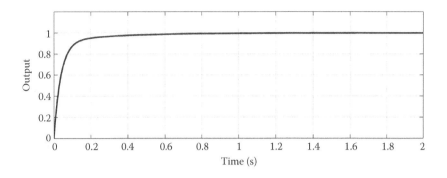

**Figure 7.3    Unit step response, Example 7.11.**

Furthermore, the quadratically $\mathbb{S}(\alpha, r, \theta)$-stabilizing state feedback matrix is

$$K = WP^{-1} = \begin{bmatrix} -0.1022 & -24.9468 & -1.5032 \end{bmatrix},$$

and the corresponding unit step response of the nominal closed-loop system is shown in Figure 7.3. More importantly, due to the $\mathbb{S}(\alpha, r, \theta)$-stabilization, the aforementioned specifications are perfectly achieved with $\sigma = 0$ and $t_s = 0.1909s$.

## 7.5  Insensitive Region Design

In this section, we investigate two types of robust stabilization problems, namely, the insensitive strip region design problem and the insensitive disk region design problem. The nature of these designs is to place the closed-loop eigenvalues in a particular region on the complex plane, and this region has an inner boundary that is insensitive to perturbations to the system parameter matrices. The design is based on the sensitivity theory of matrix eigenvalues.

### 7.5.1  Sensitivity of Matrix Eigenvalues

Let $A, \Delta A \in \mathbb{R}^{n \times n}$, and $B$ be a matrix that is resulted in by adding perturbation $\Delta A$ to $A$, that is,

$$B = A + \Delta A.$$

In this case, we are interested in the problem of how big would the eigenvalues of $A$ vary when it is perturbed. To do this, we need first to define a measure. Let

$$\lambda_i(A) = \lambda_i, \quad \lambda_i(B) = \mu_i, \quad i = 1, 2, \ldots, n.$$

Obviously, the following index

$$S_A(B) = \max_{1 \le j \le n} \left\{ \min_{1 \le i \le n} |\lambda_i - \mu_j| \right\}$$

is a natural one.

For simplicity, let us focus on one type of important matrices.

**Definition 7.1**   A matrix $A \in \mathbb{R}^{n \times n}$ is said to be nondefective if there exists a nonsingular matrix $V \in \mathbb{C}^{n \times n}$ such that

$$VAV^{-1} = \Lambda = \operatorname{diag}(\lambda_1, \lambda_2, \dots, \lambda_n). \qquad (7.40)$$

The following is the well-known Bauer–Fike theorem (Bauer and Fike 1960), which gives an estimated bound of $S_A(B)$.

**Theorem 7.15**   Let $A \in \mathbb{R}^{n \times n}$ be nondefective, and $V \in \mathbb{C}^{n \times n}$ be a nonsingular matrix satisfying (7.40). Then,

$$S_A(A + \Delta A) = k_A(V) \, \|\Delta A\|_2 \,,$$

with

$$k_A(V) = \|V\|_2 \, \|V^{-1}\|_2. \qquad (7.41)$$

It follows from this theorem that the eigenvalues of matrix $A$ are insensitive to perturbations in its elements if $k_A(V)$ is small. Regarding this index, we have the following proposition.

**Proposition 7.1**   Let $A$ be nondefective, and $k_A(V)$ be defined by (7.40) and (7.41). Then, $k_A(V) \ge 1$, and the equality holds if and only if $A$ is symmetric.

*Proof*   Note that

$$\|V\|_2 = \sigma_{\max}(V),$$

$$\|V^{-1}\|_2 = \sigma_{\max}(V^{-1}) = \frac{1}{\sigma_{\min}(V)},$$

we thus have

$$k_A(V) = \|V\|_2 \, \|V^{-1}\|_2 = \frac{\sigma_{\max}(V)}{\sigma_{\min}(V)} \ge 1. \qquad (7.42)$$

When $A$ is symmetric, it is well known that there exists a unitary matrix $V$ satisfying (7.40). For such a $V$, we have $V^H V = I$, and hence

$$\sigma_i^2(V) = \lambda_i(V^H V) = 1, \quad i = 1, 2, \ldots, n.$$

Thus,

$$k_A(V) = \frac{\sigma_{\max}(V)}{\sigma_{\min}(V)} = 1.$$

Conversely, let $k_A(V) = 1$, then it follows from (7.42) that $\sigma_{\max}(V) = \sigma_{\min}(V)$, hence

$$\sigma_{\max}(V) = \sigma_{\min}(V) = \sigma_i(V) = \sigma, \quad i = 1, 2, \ldots, n.$$

Therefore, it follows from the well-known singular value decomposition that there exist a pair of unitary matrices $U_1$ and $U_2$ such that

$$V = \sigma U_1 U_2.$$

In this case, we have from (7.40)

$$A = V^{-1} \Lambda V$$

$$= (\sigma U_1 U_2)^{-1} \Lambda (\sigma U_1 U_2)$$

$$= \left( \frac{1}{\sigma} U_2^T U_1^T \right) \Lambda (\sigma U_1 U_2)$$

$$= U_2^T U_1^T \Lambda U_1 U_2,$$

which is obviously symmetric. ∎

The aforementioned proposition tells us that symmetric matrices have the smallest eigenvalue sensitivities in terms of the measure $k_A(V)$.

## 7.5.2 Insensitive Strip Region Design

In this section, we consider the control of the following continuous-time linear system

$$\begin{cases} \dot{x} = Ax + Bu \\ y = Cx, \end{cases} \tag{7.43}$$

where $x \in \mathbb{R}^n$, $y \in \mathbb{R}^m$, and $u \in \mathbb{R}^r$ are the state vector, the output vector, and the input vector, respectively.

## 7.5.2.1 Problem

The insensitive strip region design problem is stated as follows.

**Problem 7.9**   Consider the continuous-time linear system (7.43). For given real scalars $\gamma_1$ and $\gamma_2$, design an output feedback control law

$$u = Ky,$$

such that

$$\gamma_1 < \lambda_i \left( A_c^s \right) < \gamma_2, \quad i = 1, 2, \ldots, n, \tag{7.44}$$

where

$$A_c^s \triangleq \frac{1}{2} \left( A_c \right)_s = \frac{(A + BKC)^\mathrm{T} + (A + BKC)}{2}.$$

The motivation of the aforementioned problem lies in the following lemma.

**Lemma 7.1**   Let $K$ be a solution to the aforementioned problem, then

$$\gamma_1 < \alpha_1 \leq \mathrm{Re} \left( \lambda_i \left( A + BKC \right) \right) \leq \alpha_2 < \gamma_2, \quad i = 1, 2, \ldots, n,$$

where

$$\alpha_1 = \lambda_{\min} \left( A_c^s \right), \quad \alpha_2 = \lambda_{\max} \left( A_c^s \right).$$

*Proof*   Note that for an arbitrary matrix $L \in \mathbb{R}^{n \times n}$ there holds

$$\min_i \lambda_i \left( \frac{L + L^\mathrm{T}}{2} \right) \leq \mathrm{Re} \left( \lambda_i \left( L \right) \right) \leq \max_i \lambda_i \left( \frac{L + L^\mathrm{T}}{2} \right), \quad i = 1, 2, \ldots, n.$$

Therefore, when condition (7.44) is satisfied, the conclusion clearly holds.   ■

The aforementioned lemma tells us that the eigenvalues of the matrix $(A + BKC)$ are within a strip (see Figure 7.4). More precisely, we have

$$\lambda_i \left( A + BKC \right) \in \mathbb{H}_{(-\alpha_2, -\alpha_1)} \subset \mathbb{H}_{(-\gamma_2, -\gamma_1)}, \quad i = 1, 2, \ldots, n,$$

where, by definition,

$$\mathbb{H}_{(-\alpha_2, -\alpha_1)} = \{ s \mid s \in \mathbb{C}, \ \alpha_1 < \mathrm{Re} \left( s \right) < \alpha_2 \},$$

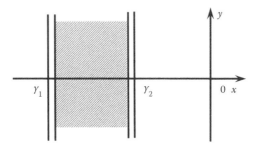

**Figure 7.4    The Strip field.**

and

$$\mathbb{H}_{(-\gamma_2,-\gamma_1)} = \{s \mid s \in \mathbb{C}, \ \gamma_1 < \mathrm{Re}\,(s) < \gamma_2\}.$$

Therefore, the matrix $A + BKC$ is $\mathbb{H}_{(-\gamma_2,-\gamma_1)}$-stable. Particularly, when $\gamma_2 = 0$, $\gamma_1 = -\infty$, the matrix $A + BKC$ is Hurwitz stable, or the closed-loop system

$$\dot{x} = (A + BKC)\,x$$

is stable in the continuous-time sense.

The importance of this problem also lies in the robustness issues. We point out that the closed-loop eigenvalues are not easy to get out of the desired region $\mathbb{H}_{(-\gamma_2,-\gamma_1)}$ since they are first bounded by $\mathbb{H}_{(-\alpha_2,-\alpha_1)} \subset \mathbb{H}_{(-\gamma_2,-\gamma_1)}$. Note that the boundaries $\alpha_1$ and $\alpha_2$ of $\mathbb{H}_{(-\alpha_2,-\alpha_1)}$ are the minimal and maximal eigenvalues of a symmetric matrix, and it follows from Proposition 7.1 that the eigenvalues of symmetric matrices are insensitive to parameter perturbations, the boundaries of $\mathbb{H}_{(-\alpha_2,-\alpha_1)}$ are eventually insensitive to perturbations in the system matrices as well as in the gain matrix $K$. They thus prevent the closed-loop eigenvalues from getting out of the desired region. Therefore, this design is robust in the sense that the system has a strip eigenvalue region that is insensitive to small parameter perturbations in the system matrices.

### 7.5.2.2 Solution

The solution to the problem is straightforward. The condition (7.44) is clearly equivalent to

$$2\gamma_1 I < (A + BKC)^{\mathrm{T}} + (A + BKC) < 2\gamma_2 I. \tag{7.45}$$

Therefore, the problem is converted into one of finding a matrix $K$ satisfying the aforementioned LMIs.

In practical applications, we can set $\gamma_1$ to be $-\infty$, and solve the following optimization problem:

$$\begin{cases} \min & \gamma \\ \text{s.t.} & (A + BKC)^{\mathrm{T}} + (A + BKC) < \gamma I. \end{cases}$$

If a solution with a nonpositive $\gamma$ is obtained, the problem gives a stabilizing output feedback controller in the Hurwitz stability sense. When the level $\gamma$ is positive but very small, we need to check $\alpha_2 = \lambda_{\max}\left(A_c^s\right)$. As long as $\alpha_2$ is negative, the problem provides a solution which robustly stabilizes the system in the Hurwitz stability sense.

**Example 7.12**

Consider the following continuous-time linear system:

$$\begin{cases} \dot{x}(t) = \begin{bmatrix} -0.5 & 0 & 1 \\ 0 & -2 & 0 \\ 0 & 0 & -5 \end{bmatrix} x(t) + \begin{bmatrix} 5 & 3 \\ 6 & 5 \\ 7 & 6 \end{bmatrix} u(t) \\ y(t) = \begin{bmatrix} 2 & -2 & 1 \end{bmatrix} x(t), \end{cases}$$

and choose real scalars $\gamma_1 = -5$, $\gamma_2 = -1$. By using the MATLAB function `feasp` in the MATLAB LMI toolbox, the corresponding LMI problem (7.45) is solved and a solution is found to be

$$K = \begin{bmatrix} -0.6811 \\ 0.8503 \end{bmatrix}.$$

With this solution and the help of the MATLAB function `eig`, we can find the set of closed-loop eigenvalues as

$$\lambda(A + BKC) = \{-4.6256, -1.4836, -3.0956\}.$$

It is obvious that

$$\alpha_1 = \lambda_{\min}\left(A_c^s\right) = -4.8107, \quad \alpha_2 = \lambda_{\max}\left(A_c^s\right) = -1.2436,$$

and

$$\lambda(A + BKC) \subset \mathbb{H}_{(-\alpha_2, -\alpha_1)} \subset \mathbb{H}_{(-\gamma_2, -\gamma_1)}.$$

## 7.5.3 Insensitive Disk Region Design

In this section, we consider the control of the following linear system

$$\begin{cases} \rho x = Ax + Bu \\ y = Cx, \end{cases} \tag{7.46}$$

where $x \in \mathbb{R}^n$, $y \in \mathbb{R}^m$, and $u \in \mathbb{R}^r$ are the state vector, the output vector and the input vector, respectively. Again, the symbol $\rho$ represents the differential operator (in the continuous-time case), or the one-step shift forward operator, that is, $\rho x(k) = x(k+1)$ (in the discrete-time case).

### 7.5.3.1 Problem

In this section, we are going to solve the following insensitive disk region design problem.

**Problem 7.10**   Consider the linear system (7.46). For given positive scalars $\gamma$ and $q$, design an output feedback control law $u = Ky$, such that

$$\eta = ||A + BKC + qI||_2 < \gamma. \tag{7.47}$$

Recalling the definition, we have

$$\mathbb{D}_{(q,\eta)} = \left\{s \mid s \in \mathbb{C}, \ |s+q| < \eta\right\}$$
$$= \left\{x + jy \mid x, y \in \mathbb{R}, \ (x+q)^2 + y^2 < \eta^2\right\},$$

and

$$\mathbb{D}_{(q,r)} = \left\{s \mid s \in \mathbb{C}, \ |s+q| < \gamma\right\}$$
$$= \left\{x + jy \mid x, y \in \mathbb{R}, \ (x+q)^2 + y^2 < \gamma^2\right\}.$$

The motivation of this problem lies in the following lemma.

**Lemma 7.2**   Let $K$ be a solution to the aforementioned problem, then

$$\lambda_i (A + BKC) \in \mathbb{D}_{(q,\eta)} \subset \mathbb{D}_{(q,r)}, \quad i = 1, 2, \ldots, n. \tag{7.48}$$

*Proof*   Note that for an arbitrary matrix $L \in \mathbb{R}^{n \times n}$ there holds

$$|\lambda_i (L)| \leq ||L||_2, \quad i = 1, 2, \ldots, n.$$

Therefore, when condition (7.47) is satisfied, we have

$$
\begin{aligned}
&\left|\operatorname{Re}\left(\lambda_i\left(A+BKC\right)\right)+q+j\operatorname{Im}\left(\lambda_i\left(A+BKC\right)\right)\right|\\
&=\left|\lambda_i\left(A+BKC\right)+q\right|\\
&=\left|\lambda_i\left(A+BKC+qI\right)\right|\\
&\leq\eta\\
&<\gamma,\quad i=1,2,\ldots,n.
\end{aligned}
$$

This is equivalent to condition (7.48). ■

The closed-loop system is clearly given by

$$
\rho x=(A+BKC)\,x.
$$

The aforementioned lemma tells us that, with this type of design, the closed-loop eigenvalues are located within the circle $\mathbb{D}_{(q,\eta)}\subset\mathbb{D}_{(q,r)}$ (see Figure 7.5). Therefore, the matrix $A+BKC$ is $\mathbb{D}_{(q,r)}$-stable. Particularly,

1. When $q=0$, $r=1$, the matrix $A+BKC$ is Schur stable, or the system (7.46) is stable in the discrete-time sense
2. When $q>0$, $r\leq q$, the matrix $A+BKC$ is Hurwitz stable, or the system (7.46) is stable in the continuous-time sense

Like the case for insensitive strip design, the importance of this insensitive disk design problem also lies in the robustness issues. We point out that the closed-loop eigenvalues are not easy to get out of the desired region $\mathbb{D}_{(q,r)}$ since they are first bounded by $\mathbb{D}_{(q,\eta)}\subset\mathbb{D}_{(q,r)}$. Note that the boundary $\eta$ of $\mathbb{D}_{(q,\eta)}$ is in fact an eigenvalue of a symmetric matrix, and it follows from Proposition 7.1 that the eigenvalues of symmetric matrices are insensitive to parameter perturbations, the

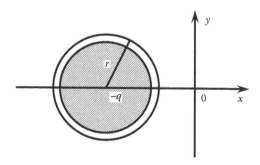

**Figure 7.5   The Disk field.**

boundary of $\mathbb{D}_{(q,\eta)}$ is eventually insensitive to perturbations in the system matrices as well as in the gain matrix $K$. They thus prevent the closed-loop eigenvalues from getting out of the desired region. Therefore, this design is robust in the sense that the system has a disk eigenvalue region that is insensitive to small parameter perturbations in the system matrices.

### 7.5.3.2 Solution

The condition (7.47) is clearly equivalent to

$$(A + BKC + qI)^{\mathrm{T}} (A + BKC + qI) < \gamma^2 I,$$

which can be equivalently converted into the following LMI with the help of the Schur complement lemma:

$$\begin{bmatrix} -\gamma I & (A + BKC + qI) \\ (A + BKC + qI)^{\mathrm{T}} & -\gamma I \end{bmatrix} < 0.$$

Thus, the problem is turned into an LMI feasibility problem.

In some applications, we may choose to solve the optimization problem

$$\begin{cases} \min & \gamma \\ \text{s.t.} & \begin{bmatrix} -\gamma I & (A + BKC + qI) \\ (A + BKC + qI)^{\mathrm{T}} & -\gamma I \end{bmatrix} < 0. \end{cases} \tag{7.49}$$

Particularly, for the case of Schur stabilization, we choose to solve the aforementioned optimization problem with $q = 0$. If a solution with $\gamma \leq 1$ is obtained, the problem gives a stabilizing output feedback controller in the Schur stability sense. When the level $\gamma$ is greater than 1 but very close to 1, we need to check $\eta = ||A + BKC + qI||_2$. As long as $\eta \leq 1$, the problem provides a solution that robustly stabilizes the system in the Schur stability sense.

**Remark 7.3**    This insensitive disk region design problem was first proposed and solved by Duan (1991), which provides for this problem a necessary and sufficient algebraic condition and also gives an efficient way of finding the gain matrix $K$.

**Example 7.13**

Consider the following continuous-time linear system

$$\begin{cases} \dot{x}(t) = \begin{bmatrix} -1 & 3 & 2 \\ 0 & 1 & 0 \\ 1 & 2 & -1 \end{bmatrix} x(t) + \begin{bmatrix} 1 & 0 \\ 2 & 3 \\ 1 & 1 \end{bmatrix} u(t) \\ y(t) = \begin{bmatrix} 1 & 1 & 0 \end{bmatrix} x(t). \end{cases}$$

For the case of $q = 4$, using the MATLAB function `mincx` in the MATLAB LMI toolbox, we solve the minimization problem (7.49) and obtain the solution

$$K = \begin{bmatrix} -2.7535 \\ 1.0334 \end{bmatrix}, \quad \gamma = 3.8888.$$

With this solution, it is computed that

$$\eta = ||A + BKC + qI||_2 = 3.8886 < \gamma.$$

Further, with the MATLAB function `eig` we find the closed-loop eigenvalues to be $-3.0958$ and $1.5322 \pm 0.7994i$, which are indeed all located within the desired region $\mathbb{D}_{(q,\eta)}$.

# 7.6 Robust Stabilization of Second-Order Systems

In the proceeding section, we investigated stabilization of first-order linear systems by making the symmetric part of the closed-loop system matrix positive definite. In this section, we further investigate using a similar approach the robust stabilization of second-order linear systems in the following form:

$$M\ddot{x} + D\dot{x} + Kx = Bu, \tag{7.50}$$

where

$x \in \mathbb{R}^n$ and $u \in \mathbb{R}^r$ are the state vector and the control vector, respectively
$M$, $D$, $K$, and $B$ are the system coefficient matrices of appropriate dimensions

In practical applications, the matrices $M$, $D$, and $K$ are called the mass matrix, the structural damping matrix, and the stiffness matrix, respectively.

Second-order linear systems capture the dynamic behavior of many natural phenomena, and hence have applications in many fields, such as vibration and structure analysis, spacecraft control and robotics control, and have attracted much attention (Duan and Liu 2002, Duan 2004b, Wang et al. 2005, Wu and Duan 2005). In many applications, control of a second-order linear system is realized by converting the system into a first-order one, while in this section, a direct approach is proposed, which maintains the second-order system frame.

Before getting on the general robust stabilization problem, we first investigate the problem of stability of second-order systems.

## 7.6.1 Stabilization

### 7.6.1.1 Problem Formulation

Due to usage consistency of notations, here we rewrite (7.50) and propose the following model for second-order linear system with state and state derivative observations:

$$\begin{cases} A_2\ddot{x} + A_1\dot{x} + A_0 x = Bu \\ y_d = C_d\dot{x} \\ y_p = C_p x, \end{cases} \tag{7.51}$$

where

- $x \in \mathbb{R}^n$ and $u \in \mathbb{R}^r$ are the state vector and the control vector, respectively
- $y_d \in \mathbb{R}^{m_d}$ and $y_p \in \mathbb{R}^{m_p}$ are the derivative output vector and the proportional output vector, respectively
- $A_2$, $A_1$, $A_0$, $B$, $C_d$, and $C_p$ are the system coefficient matrices of appropriate dimensions

In many applications, the following assumption on the system matrices is required:

**Assumption 7.1** $A_2, A_0 \in \mathbb{S}^n$ and $A_2 > 0$.

For the second-order linear system (7.51), choosing the following control law

$$\begin{aligned} u &= K_p y_p + K_d y_d \\ &= K_p C_p \dot{x} + K_d C_d x, \end{aligned} \tag{7.52}$$

with $K_p \in \mathbb{R}^{r \times m_p}$, $K_d \in \mathbb{R}^{r \times m_d}$, we obtain the closed-loop system as follows:

$$A_2\ddot{x} + \left(A_1 - BK_p C_p\right)\dot{x} + \left(A_0 - BK_d C_d\right)x = 0. \tag{7.53}$$

With the aforementioned assumption, the problem of state feedback stabilization for the continuous-time second-order system (7.51) can be stated as follows.

**Problem 7.11** For the continuous-time second-order linear system (7.51), with Assumption 7.1 being satisfied, design a state feedback control law (7.52) such that the closed-loop system (7.53) is Hurwitz stable.

### 7.6.1.2 Solution

To solve the aforementioned stabilization problem, we first introduce a stability result for second-order linear systems.

**Lemma 7.3** (Duan et al. 1989) The second-order linear system (7.51) is Hurwitz stable if

$$A_2 > 0, \quad A_1 + A_1^{\mathrm{T}} > 0, \quad A_0 > 0. \tag{7.54}$$

For a proof of this lemma, refer to Duan et al. (1989) or Appendix A.6.

With the help of the aforementioned lemma, we can easily obtain the following conclusion about solution to Problem 7.11.

**Theorem 7.16**   Problem 7.11 has a solution if there exist matrices $K_p \in \mathbb{R}^{r \times m_p}$ and $K_d \in \mathbb{R}^{r \times m_d}$ satisfying the LMIs

$$A_0 - BK_dC_d > 0 \qquad (7.55)$$

and

$$\left(A_1 - BK_pC_p\right) + \left(A_1 - BK_pC_p\right)^{\mathrm{T}} > 0. \qquad (7.56)$$

In many practical applications, the $B$ and $C_d$ matrices take the following special forms

$$B = \begin{bmatrix} I_r \\ 0 \end{bmatrix}, \quad C_d = \begin{bmatrix} I_{m_d} & 0 \end{bmatrix}. \qquad (7.57)$$

Correspondingly, we can partition the matrix $A_0$ as follows:

$$A_0 = \begin{bmatrix} A_{11}^0 & A_{12}^0 \\ \left(A_{12}^0\right)^{\mathrm{T}} & A_{22}^0 \end{bmatrix}, \quad A_{11}^0 \in \mathbb{S}^{\min\{r,m_d\}}. \qquad (7.58)$$

The following corollary states that in this case (7.55) has an analytical solution and hence is not needed to be solved via the MATLAB LMI toolbox.

**Corollary 7.2**   Let $B$ and $C_d$ be in the forms of (7.57), and the matrix $A_0$ be partitioned as in (7.58). Then, (7.55) has a solution with respect to $K_d$ if and only if $A_{22}^0 > 0$, and in this case, a general solution is given by

$$K_d = \begin{bmatrix} K_{d0} & 0 \\ 0 & 0 \end{bmatrix} \in \mathbb{R}^{r \times m_d}, \qquad (7.59)$$

with $K_{d0} \in \mathbb{S}^{\min\{r,m_d\}}$ satisfying

$$K_{d0} < A_{11}^0 - A_{12}^0 \left(A_{22}^0\right)^{-1} \left(A_{12}^0\right)^{\mathrm{T}}. \qquad (7.60)$$

*Proof*   When $B$ and $C_d$ are in the forms of (7.57), we have

$$BK_dC_d = \begin{bmatrix} K_d & 0 \\ 0 & 0 \end{bmatrix} \in \mathbb{S}^n.$$

With matrix $A_0$ being partitioned as in (7.58), and the matrix $K_d$ being given by (7.59), we further have

$$A_0 - BK_dC_d = \begin{bmatrix} A^0_{11} - K_{d0} & A^0_{12} \\ \left(A^0_{12}\right)^{\mathrm{T}} & A^0_{22} \end{bmatrix}.$$

Applying Schur complement lemma, we know that there exists a $K_d$ such that $A_0 - BK_dC_d > 0$ if and only if $A^0_{22} > 0$ and

$$A^0_{11} - K_{d0} - A^0_{12}\left(A^0_{22}\right)^{-1}\left(A^0_{12}\right)^{\mathrm{T}} > 0. \qquad (7.61)$$

Since (7.61) can be satisfied by choosing $K_{d0}$ as in (7.60), inequality (7.55) has a solution with respect to $K_d$ if and only if $A^0_{22} > 0$, and in this case, a general solution is given by (7.59) and (7.60). ■

## 7.6.2 Robust Stabilization

### 7.6.2.1 Problem Formulation

Consider the following uncertain second-order linear system

$$\begin{cases} (A_2 + \Delta A_2)\ddot{x} + (A_1 + \Delta A_1)\dot{x} + (A_0 + \Delta A_0)x = Bu \\ y_d = C_d\dot{x} \\ y_p = C_px, \end{cases} \qquad (7.62)$$

where

- $x \in \mathbb{R}^n$, $u \in \mathbb{R}^r$, $y_d \in \mathbb{R}^{m_d}$, $y_p \in \mathbb{R}^{m_p}$ and $A_2$, $A_1$, $A_0$, $B$, $C_d$, $C_p$ are as stated in the last subsection
- $\Delta A_2$, $\Delta A_1$ and $\Delta A_0$ are the perturbations of matrices $A_2$, $A_1$ and $A_0$, respectively

The perturbations are bounded and satisfy

$$\|\Delta A_2\|_2 \le \epsilon_2, \quad \|\Delta A_1\|_2 \le \epsilon_1, \quad \|\Delta A_0\|_2 \le \epsilon_0,$$

or

$$\max\left\{\left|\Delta a_{2ij}\right|\right\} \le \varepsilon_2, \quad \max\left\{\left|\Delta a_{1ij}\right|\right\} \le \varepsilon_1, \quad \max\left\{\left|\Delta a_{0ij}\right|\right\} \le \varepsilon_0,$$

where

- $\epsilon_2, \epsilon_1, \epsilon_0$, and $\varepsilon_2, \varepsilon_1, \varepsilon_0$ are two sets of given positive scalars
- $\Delta a_{2ij}$, $\Delta a_{1ij}$ and $\Delta a_{0ij}$ are the $i$-th row and $j$-th column elements of matrices $\Delta A_2$, $\Delta A_1$ and $\Delta A_0$, respectively

Furthermore, these perturbation notations satisfy the following basic assumption:

**Assumption 7.2**   $\Delta A_2, \Delta A_0 \in \mathbb{S}^n$ and $A_2 + \Delta A_2 > 0$.

**Problem 7.12**   For the uncertain continuous-time linear second-order system (7.62), with Assumptions 7.1 and 7.2 satisfied, design a state feedback control law (7.52) such that the closed-loop system

$$(A_2 + \Delta A_2)\ddot{x} + \left(A_1 - BK_pC_p + \Delta A_1\right)\dot{x} + (A_0 - BK_dC_d + \Delta A_0)x = 0, \tag{7.63}$$

is Hurwitz stable.

## 7.6.2.2 Solution

To solve the aforementioned robust stabilization problem, we first introduce the following lemma about robust stability of second-order linear systems.

**Lemma 7.4**   (Duan and Huang 1990) Let $A_2 > 0$, $A_1 + A_1^T > 0$ and $A_0 > 0$. Then, the following conclusions hold.

1. System (7.62) is Hurwitz stable if

$$\lambda_{\min}(A_2) > \|\Delta A_2\|_2,$$
$$\lambda_{\min}(A_1 + A_1^T) > \|\Delta A_1\|_2,$$
$$\lambda_{\min}(A_0) > \|\Delta A_0\|_2.$$

2. System (7.62) is Hurwitz stable if

$$\lambda_{\min}(A_2) > \sqrt{l_2}\max\left\{\left|\Delta a_{2ij}\right|\right\},$$
$$\lambda_{\min}(A_1 + A_1^T) > \sqrt{l_1}\max\left\{\left|\Delta a_{1ij}\right|\right\},$$
$$\lambda_{\min}(A_0) > \sqrt{l_0}\max\left\{\left|\Delta a_{0ij}\right|\right\},$$

   where $l_2$, $l_1$, $l_0$ are the numbers of nonzero elements in matrices $\Delta A_2$, $\Delta A_1$, $\Delta A_0$, respectively.

For a proof of the aforementioned lemma, refer to Duan and Huang (1990) or Appendix A.6.

With the help of this lemma, we can easily obtain the following conclusion about solution to Problem 7.12.

**Theorem 7.17**   Problem 7.12 has a solution if one of the following conditions is satisfied:

1. There exist matrices $K_p \in \mathbb{R}^{r \times m_p}$ and $K_d \in \mathbb{R}^{r \times m_d}$ satisfying

$$A_0 - BK_d C_d > \epsilon_0 I, \tag{7.64}$$

$$\left(A_1 - BK_p C_p\right) + \left(A_1 - BK_p C_p\right)^{\mathrm{T}} > \epsilon_1 I. \tag{7.65}$$

2. There exist matrices $K_p \in \mathbb{R}^{r \times m_p}$ and $K_d \in \mathbb{R}^{r \times m_d}$ satisfying

$$A_0 - BK_d C_d > \varepsilon_0 \sqrt{l_0} I, \tag{7.66}$$

$$\left(A_1 - BK_p C_p\right) + \left(A_1 - BK_p C_p\right)^{\mathrm{T}} > \varepsilon_1 \sqrt{l_1} I. \tag{7.67}$$

Parallel to Corollary 7.2, we have the following result (proof omitted).

**Corollary 7.3** Let $B$ and $C_d$ be in the forms of (7.57), and partition the matrix $A_0$ as in (7.58). Then, the following conclusions hold

1. (7.64) has a solution if and only if

$$A_{22}^0 > \epsilon_0 I,$$

and in this case, a general solution is given by (7.59), with $K_{d0} \in \mathbb{S}^{\min\{r, m_d\}}$ satisfying

$$K_{d0} < A_{11}^0 - A_{12}^0 \left(A_{22}^0 - \epsilon_0 I\right)^{-1} \left(A_{12}^0\right)^{\mathrm{T}} - \epsilon_0 I. \tag{7.68}$$

2. (7.66) has a solution if and only if

$$A_{22}^0 > \varepsilon_0 \sqrt{l_0} I,$$

and in this case, a general solution is given by (7.59), with $K_{d0} \in \mathbb{S}^{\min\{r, m_d\}}$ satisfying

$$K_{d0} < A_{11}^0 - A_{12}^0 \left(A_{22}^0 - \varepsilon_0 \sqrt{l_0} I\right)^{-1} \left(A_{12}^0\right)^{\mathrm{T}} - \varepsilon_0 \sqrt{l_0} I. \tag{7.69}$$

**Example 7.14**

Consider the control problem of satellite rendezvous maneuver. Under certain conditions, the relative orbital dynamic model of the object satellite with respect to the chasing one can be described by the following well-known C-W's equations:

$$
\begin{cases}
m\ddot{r}_x - 2m\omega_0\dot{r}_y - 3m\omega_0^2 r_x = T_x + d_x \\
m\ddot{r}_y + 2m\omega_0\dot{r}_x = T_y + d_y \\
m\ddot{r}_z + m\omega_0^2 r_z = T_z + d_z,
\end{cases}
\tag{7.70}
$$

where

- $r_x$, $r_y$ and $r_z$ are the components of the relative position
- $\omega_0 = \pi/12\,(\text{rad/h})$ is the orbital angular velocity of the object satellite
- $m$ is the mass of the chaser
- $T_i$ ($i = x, y, z$) is the $i$-th component of the control input force acting on the relative motion dynamics
- $d_i$ ($i = x, y, z$) is the $i$-th components of the external disturbance, which is assumed to be zero here

Choosing $C_d = C_p = I_3$, and denoting $B = I$,

$$
A_2 = mI_3, \quad A_0 = \text{diag}(-3m\omega_0^2, 0, m\omega_0^2),
\tag{7.71}
$$

$$
A_1 = \begin{bmatrix} 0 & -2m\omega_0 & 0 \\ 2m\omega_0 & 0 & 0 \\ 0 & 0 & 0 \end{bmatrix},
\tag{7.72}
$$

then we can write the above model into the form of (7.62).

When $m = 100$, and the perturbation bounds are taken as follows:

$$
\|\Delta A_2\|_2 \le 0.5, \quad \|\Delta A_1\|_2 \le 0.5, \quad \|\Delta A_0\|_2 \le 0.5.
$$

With the Matlab function `feasp` in the Matlab LMI toolbox, the corresponding LMI problem with constraints (7.64) and (7.65) is solved and the following gain matrices are obtained:

$$
K_p = -227132925.2854I_3, \quad K_d = -454265850.5708I_3.
$$

Corresponding to this pair of gain matrices, we have

$$
A_0 - BK_dC_d = 454265850.5708I_3 > 0.5I_3,
$$

and

$$
\langle A_1 - BK_pC_p \rangle_s = 454265850.5708I_3 > 0.5I_3.
$$

Obviously, the magnitudes of the above gain matrices are very large. This will consume too much control energy and also may give difficulty in practical realization. To prevent this we add an additional constraint, that is

$$
A_0 - BK_dC_d < 10I_3.
$$

With this constraints considered, we obtain a pair of feedback matrices as follows:

$$K_p = -3.0419 I_3, \quad K_d = -5.2500 I_3.$$

Correspondingly, we have

$$A_0 - BK_d C_d = 5.2500 I_3 > 0.5 I_3,$$

and

$$\langle A_1 - BK_p C_p \rangle_s = 6.0837 I_3 > 0.5 I_3.$$

## 7.7 Stabilization of Time-Delay Systems

In this section, we consider the stabilization of time-delay systems in the form of

$$\begin{cases} \dot{x}(t) = Ax(t) + A_d x(t - d) + Bu(t) \\ x(t) = \phi(t), \quad t \in [0, d], \quad 0 < d \le \bar{d}, \end{cases} \tag{7.73}$$

where

$A, A_d \in \mathbb{R}^{n \times n}, B \in \mathbb{R}^{n \times r}$ are the system coefficient matrices

$\phi(t)$ is the initial condition

$d$ represents the time-delay

$\bar{d}$ is a known upper bound of $d$

**Problem 7.13** For the time-delay system (7.73), design a memoryless state feedback control law

$$u = Kx,$$

such that the closed-loop system

$$\begin{cases} \dot{x}(t) = (A + BK)x(t) + A_d x(t - d) \\ x(t) = \phi(t), \quad t \in [0, d], \quad 0 < d \le \bar{d} \end{cases} \tag{7.74}$$

is uniformly asymptotically stable.

### 7.7.1 Case of Delay Independence

It follows from Theorem 4.8 that the time-delay linear system (7.74) is asymptotically stable if there exist symmetric positive definite matrices $P$ and $S$, such that

$$\begin{bmatrix} (A + BK)^{\mathrm{T}} P + P(A + BK) + S & PA_d \\ A_d^{\mathrm{T}} P & -S \end{bmatrix} < 0.$$

Pre- and postmultiplying both sides of the aforementioned inequality by $\text{diag}(P^{-1}, P^{-1})$ gives the following equivalent relation:

$$\begin{bmatrix} P^{-1}(A + BK)^{\mathrm{T}} + (A + BK)P^{-1} + P^{-1}SP^{-1} & A_dP^{-1} \\ P^{-1}A_d^{\mathrm{T}} & -P^{-1}SP^{-1} \end{bmatrix} < 0.$$

Denote

$$X = P^{-1}, \quad W = KX, \quad \text{and} \quad Y = XSX,$$

then the aforementioned inequality can be turned into an LMI.

To summarize, we now have the following theorem for solution to Problem 7.13.

**Theorem 7.18** Problem 7.13 has a solution if there exist a matrix $W \in \mathbb{R}^{r \times n}$ and two symmetric positive definite matrices $X$ and $Y$ satisfying

$$\begin{bmatrix} XA^{\mathrm{T}} + AX + BW + W^{\mathrm{T}}B^{\mathrm{T}} + Y & A_dX \\ XA_d^{\mathrm{T}} & -Y \end{bmatrix} < 0. \tag{7.75}$$

In this case, a feedback gain matrix is given by

$$K = WX^{-1}.$$

**Example 7.15**

Consider the following time-delay linear system

$$\dot{x}(t) = \begin{bmatrix} -1 & 0 & 1 \\ 0 & 2 & -1 \\ 2 & 0 & -3 \end{bmatrix} x(t) + \begin{bmatrix} 1 & 0 & 1 \\ 2 & 1 & 1 \\ 0 & 0 & -1 \end{bmatrix} x(t-d) + \begin{bmatrix} 1 & 1 \\ 1 & 2 \\ 0 & 1 \end{bmatrix} u(t).$$

Using the MATLAB function `feasp` in the MATLAB LMI toolbox, the corresponding LMIs problem (7.75) is solved and the following parameter matrices are obtained:

$$X = \begin{bmatrix} 38.1618 & 13.0301 & -50.6435 \\ 13.0301 & 14.9993 & -9.7335 \\ -50.6435 & -9.7335 & 118.6029 \end{bmatrix} > 0,$$

$$Y = \begin{bmatrix} 92.5744 & 37.3352 & -39.7399 \\ 37.3352 & 81.3348 & 32.0054 \\ -39.7399 & 32.0054 & 139.8306 \end{bmatrix} > 0,$$

$$W = 10^4 \times \begin{bmatrix} -1.9427 & -3.8925 & -1.9971 \\ 1.9419 & 1.9402 & 0.0290 \end{bmatrix}.$$

Hence, a gain matrix is given by

$$K = WX^{-1} = \begin{bmatrix} 237.1346 & 2708.3177 & 491.9114 \\ 673.6075 & 947.0163 & 367.7995 \end{bmatrix}.$$

## 7.7.2 Case of Delay Dependence

The previous section has given for Problem 7.13 a delay-independent condition. Corresponding to the delay-dependent LMI condition for stability of time delay systems given in Section 4.5, we have the following delay-dependent condition for stabilization of the time-delay system (7.73).

**Theorem 7.19**    The time-delay system (7.73) is uniformly asymptotically stabilizable by the memoryless feedback controller $u(t) = Kx(t)$ if there exist a scalar $0 < \beta < 1$ and two matrices $X$ and $W$, with $X$ symmetric positive definite, such that

$$\begin{bmatrix} \Phi(X, W) & \bar{d}\left(XA^{\mathrm{T}} + W^{\mathrm{T}}B^{\mathrm{T}}\right) & \bar{d}XA_d^{\mathrm{T}} \\ \bar{d}(AX + BW) & -\bar{d}\beta I & 0 \\ \bar{d}A_d X & 0 & -\bar{d}(1 - \beta)I \end{bmatrix} < 0, \qquad (7.76)$$

where

$$\Phi(X, W) = X(A + A_d)^{\mathrm{T}} + (A + A_d)X + BW + W^{\mathrm{T}}B^{\mathrm{T}} + \bar{d}A_d A_d^{\mathrm{T}}.$$

In this case, a stabilizing control gain matrix is given by

$$K = WX^{-1}.$$

*Proof*    With the given type of memoryless state feedback control, the closed-loop system is given by (7.73), that is,

$$\dot{x}(t) = (A + BK)x(t) + A_d x(t - d). \qquad (7.77)$$

Applying Theorem 4.9 to the aforementioned system gives that the closed-loop system (7.77) is uniformly asymptotically stable if

$$\begin{bmatrix} \Phi(X, K) & \bar{d}X(A + BK)^{\mathrm{T}} & \bar{d}XA_d^{\mathrm{T}} \\ \bar{d}(A + BK)X & -\bar{d}\beta I & 0 \\ \bar{d}A_d X & 0 & -\bar{d}(1 - \beta)I \end{bmatrix} < 0,$$

where

$$\Phi(X, K) = X(A + BK + A_d)^{\mathrm{T}} + (A + BK + A_d)X + \bar{d}A_d A_d^{\mathrm{T}},$$

that is,

$$\begin{bmatrix} \Phi(X, K) & \bar{d}\left(XA^{\mathrm{T}} + X^{\mathrm{T}}K^{\mathrm{T}}B^{\mathrm{T}}\right) & \bar{d}XA_d^{\mathrm{T}} \\ \bar{d}(AX + BKX) & -\bar{d}\beta I & 0 \\ \bar{d}A_d X & 0 & -\bar{d}(1 - \beta)I \end{bmatrix} < 0,$$

with

$$\Phi(X, K) = X (A + A_d)^{\mathrm{T}} + (A + A_d) X + BKX + X^{\mathrm{T}} K^{\mathrm{T}} B^{\mathrm{T}} + \bar{d} A_d A_d^{\mathrm{T}}.$$

Letting $KX = W$ in the aforementioned inequality gives the condition (7.76). The proof is completed. ∎

### Example 7.16

Consider the time-delay linear system in the Example 7.15 with $\bar{d} = 0.1$. Using the MATLAB function feasp in the MATLAB LMI toolbox, the corresponding LMIs problem (7.76) is solved and the following parameter matrices are obtained:

$$X = \begin{bmatrix} 0.4528 & 0.1342 & -0.1700 \\ 0.1342 & 0.2711 & -0.1152 \\ -0.1700 & -0.1152 & 1.4284 \end{bmatrix} > 0,$$

$$W = \begin{bmatrix} 0.4689 & 1.1468 & -7.7048 \\ -1.0662 & -1.7043 & 4.4505 \end{bmatrix},$$

$$\beta = 0.5488.$$

Hence, a gain matrix is given by

$$K = WX^{-1} = \begin{bmatrix} -1.8290 & 2.8495 & -5.3820 \\ 0.2190 & -5.2402 & 2.7194 \end{bmatrix}.$$

## 7.8 Notes and References

This chapter considers several stabilization problems. The LMI conditions and the gain matrix $K$ are summarized in Tables 7.3 through 7.7 in terms of the different choices of region $\mathbb{D}$, and the different types of systems. Particularly, the LMI conditions for insensitive strip region and insensitive disk region design problems by output feedback are shown in Table 7.4. Tables 7.5 and 7.6 respectively summarize the LMI conditions for quadratic stabilization of interval and polytopic systems, while the LMI conditions for delay independence and delay dependence stabilization of time-delay systems are summarized in Table 7.7.

Since stability is the most basic requirement for a practical plant to work normally, stabilization eventually becomes the most important design problem. Within the last decade, with the help of LMI techniques many results on stabilization of various types of systems have been proposed. For instance, LMI approaches for delay-dependent stabilization of time delay systems have been proposed by Zhang et al. (2005b) and Fan et al. (2010). Particularly, Gao et al. (2004) give an LMI approach for delay-dependent output-feedback stabilization of discrete-time systems with time-varying

**Table 7.3  Stabilization of System $(A, B)$**

| Regions | Criteria | Controller |
|---|---|---|
| $\mathbb{H}_0$ | $P > 0$ <br><br> $\langle AP + BW \rangle_s < 0$ | $K = WP^{-1}$ |
| | $P > 0$ <br><br> $AP + PA^T - BB^T < 0$ | $K = -\frac{1}{2}B^T P^{-1}$ |
| $\mathbb{H}_{\alpha,\beta}$ | $P > 0$ <br><br> $\langle AP + BW \rangle_s + 2\alpha P < 0$ <br><br> $\langle AP + BW \rangle_s + 2\beta P > 0$ | $K = WP^{-1}$ |
| $\mathbb{D}_{(0,1)}$ | $\begin{bmatrix} -P & * \\ AP + BW & -P \end{bmatrix} < 0$ | $K = WP^{-1}$ |
| | $\begin{bmatrix} -P & PA^T \\ * & -P - BB^T \end{bmatrix} < 0$ | $K = -X^{-1}B^T P^{-1}A$ <br><br> $X = 2I + B^T P^{-1}B$ |
| $\mathbb{D}_{(q,r)}$ | $\begin{bmatrix} -rP & * \\ qP + PA^T + W^T B^T & -rP \end{bmatrix} < 0$ | $K = WP^{-1}$ |
| $\mathbb{D}_{(L,M)}$ | $P > 0$ <br><br> $L \otimes P + \langle M \otimes (AP) + M \otimes (BW) \rangle_s < 0$ | $K = WP^{-1}$ |

state delay. Besides time-delay systems, results for other types of systems have also been proposed, these include quantized control systems (Zhou et al. 2010a), switched discrete-time systems (Saif et al. 2009), discrete-time periodic linear systems with actuator saturation (Zhou et al. 2011b), and discrete-time descriptor linear systems with input saturation (Zhou et al. 2009).

For robust stabilization using LMI approaches, the results are also numerous and cover many types of uncertain systems. For example, Liu and Duan (2006c) considered robust stabilization of switched systems with time-delay, Guan et al. (2002) investigated decentralized stabilization of discrete time-delay large-scale systems with structured uncertainties, Wu et al. (2004a) and Li and deSouza (1997) both treat stabilization of uncertain time-delay systems using delay-dependent criteria, while Xu et al. (2002) and Cai et al. (2012) studied stabilization of uncertain time-delay singular systems and parameter-dependent interval systems, respectively.

**Table 7.4  Insensitive Region Design**

| Regions | Criteria | Controller |
|---|---|---|
| Strip | $2\gamma_1 I < (A + BKC)^\mathsf{T} + (A + BKC) < 2\gamma_2 I$ | $K$ |
| Disk | $\begin{bmatrix} -\gamma I & (A + BKC + qI) \\ (A + BKC + qI)^\mathsf{T} & -\gamma I \end{bmatrix} < 0$ | $K$ |

**Table 7.5  Quadratic Stabilization of Interval Systems**

| | Criteria | Controller |
|---|---|---|
| Interval | $P > 0$ <br> $\langle M \otimes (A(\delta(t))P) + M \otimes (B(\delta(t))W)\rangle_s$ <br> $+ L \otimes P < 0, \ \forall \delta \in \Delta_E$ | $K = WP^{-1}$ |
| Hurwitz | $P > 0$ <br> $\langle A(\delta(t))P + B(\delta(t))W\rangle_s < 0, \ \forall \delta \in \Delta_E$ | $K = WP^{-1}$ |
| | $P > 0$ <br> $\langle A(\delta(t))P\rangle_s - B_0 B_0^\mathsf{T} < 0, \ \forall \delta \in \Delta_E$ | $K = -\frac{1}{2}B_0^\mathsf{T} P^{-1}$ |
| Schur | $\begin{bmatrix} -P & A(\delta(t))P + B(\delta(t))W \\ * & -P \end{bmatrix} < 0, \ \forall \delta \in \Delta_E$ | $K = WP^{-1}$ |

## Exercises

**7.1** Prove Theorem 7.1.

**7.2** Find a stabilizing state feedback controller for the following continuous-time linear system

$$\dot{x}(t) = \begin{bmatrix} 0 & 1 & 0 \\ 0 & 0 & 1 \\ 250 & 0 & -5 \end{bmatrix} x(t) + \begin{bmatrix} 0 \\ 0 \\ 10 \end{bmatrix} u(t).$$

**7.3** Find a stabilizing state feedback controller for the following discrete-time linear system

$$x(k+1) = \begin{bmatrix} -1 & 2 & 0 \\ 0 & 0.5 & 0 \\ 1 & 0 & -2 \end{bmatrix} x(k) + \begin{bmatrix} 1 & 0 \\ 0 & 1 \\ 1 & 0 \end{bmatrix} u(k).$$

**Table 7.6　Quadratic Stabilization of Polytopic Systems**

| | Criteria | Controller |
|---|---|---|
| Polytopic | $P > 0$<br><br>$\langle M \otimes ((A_0 + A_i)\,P) + M \otimes ((B_0 + B_i)\,W)\rangle_\mathrm{s}$<br><br>$+ L \otimes P < 0,\ i = 1, 2, \ldots, k.$ | $K = WP^{-1}$ |
| Hurwitz | $P > 0$<br><br>$\langle (A_0 + A_i)\,P + (B_0 + B_i)\,W\rangle_\mathrm{s} < 0,$<br><br>$i = 1, 2, \ldots, k$ | $K = WP^{-1}$ |
| Hurwitz | $P > 0$<br><br>$\langle (A_0 + A_i)\,P\rangle_\mathrm{s} - B_0 B_0^\mathsf{T} < 0,$<br><br>$i = 1, 2, \ldots, k$ | $K = -\frac{1}{2} B_0^\mathsf{T} P^{-1}$ |
| Schur | $\begin{bmatrix} -P & (A_0 + A_i)\,P + (B_0 + B_i)\,W \\ * & -P \end{bmatrix} < 0,$<br><br>$i = 1, 2, \ldots, k$ | $K = WP^{-1}$ |

**Table 7.7　Stabilization of Time-Delay Systems**

| Conditions | Criteria | Controller |
|---|---|---|
| Delay independent | $X > 0$<br><br>$\begin{bmatrix} \langle AX + BW\rangle_\mathrm{s} + Y & * \\ X A_d^\mathsf{T} & -Y \end{bmatrix} < 0$ | $K = WX^{-1}$ |
| Delay dependent | $X > 0$<br><br>$\begin{bmatrix} \Phi(X, W) & * & * \\ \bar{d}\,(AX + BW) & -\bar{d}\beta I & * \\ \bar{d} A_d X & 0 & -\bar{d}\,(1 - \beta)\,I \end{bmatrix} < 0$<br><br>$\Phi(X, W) = \langle (A + A_d)\,X + BW\rangle_\mathrm{s} + \bar{d} A_d A_d^\mathsf{T}$ | $K = WX^{-1}$ |

**7.4** Verify that the following linear system

$$\dot{x}(t) = \begin{bmatrix} -1 & -2 \\ 1.5 & -5 \end{bmatrix} x(t) + \begin{bmatrix} 1 \\ 1.5 \end{bmatrix} u(t)$$

is $\mathbb{H}_{(1,3)}$-stable.

**7.5** Find for the following linear system

$$\dot{x}(t) = \begin{bmatrix} -4 & 0 \\ 0 & 1 \end{bmatrix} x(t) + \begin{bmatrix} 1 \\ 2 \end{bmatrix} u(t),$$

a state feedback controller such that the closed-loop system is $\mathbb{D}_{(2,1)}$-stable.

**7.6** Let the matrices associated with the characteristic function of the LMI region $\mathbb{D}$ be

$$L = \begin{bmatrix} -1 & 2 \\ 2 & -1 \end{bmatrix}, \quad M = \begin{bmatrix} 0 & \frac{3}{4} \\ \frac{1}{4} & 0 \end{bmatrix}.$$

Find a $\mathbb{D}$-stabilizing state feedback controller for the following linear system

$$\dot{x}(t) = \begin{bmatrix} -2.5 & 0 \\ 0 & 1 \end{bmatrix} x(t) + \begin{bmatrix} 1 \\ 1 \end{bmatrix} u(t).$$

**7.7** Consider a continuous-time linear system (Wilson et al. 1992, Liu et al. 2000b) in the form of (7.1) with

$$A = \begin{bmatrix} -0.5010 & -0.9850 & 0.1740 & 0 \\ 16.8300 & -0.5750 & 0.0123 & 0 \\ -3227 & 0.3210 & -2.1000 & 0 \\ 0 & 0 & 1 & 0 \end{bmatrix},$$

$$B = \begin{bmatrix} 0.1090 & 0.0070 \\ -132.8000 & 27.1900 \\ -1620 & -1240 \\ 0 & 0 \end{bmatrix}.$$

Design a stabilizing state feedback controller for the system.

**7.8** Consider the following linear system, which was introduced in Duan and Patton (2001):

$$\dot{x}(t) = \begin{bmatrix} 0 & 3 & 4 \\ 1 & 2 & 3 \\ 0 & 2 & 5 \end{bmatrix} x(t) + \begin{bmatrix} 1 & 0 \\ 0 & 0 \\ 0 & 1 \end{bmatrix} u(t).$$

Design a state feedback by LMI method such that the closed-system eigenvalues are located in the region

$$\mathbb{H}_{(1,3)} = \{ x + yj \mid -3 < x < -1 < 0 \}.$$

# Chapter 8

# H$_\infty$/H$_2$ Control

In this chapter, we study some basic H$_\infty$ and H$_2$ problems. These include the H$_\infty$ and H$_2$ state feedback control, and also the robust H$_\infty$ and H$_2$ control problems.

The results in this chapter are closely related to the stability analysis results covered in Chapter 5. In Chapter 5, we have given the analysis conditions for the H$_\infty$ and H$_2$ indices of a linear system to be bounded by certain level. But in this chapter, we will first apply those conditions to the closed-loop system of the state feedback control system and then, through a variable substitution, turn the design conditions for H$_\infty$ and H$_2$ control into LMIs, and meanwhile get the expression of the gain matrix. Such a procedure is shown in Figure 8.1.

## 8.1 H$_\infty$ State Feedback Control

H$_\infty$ control has been a very active field since late 1980s, and by now there have been tremendous results. As an introduction to H$_\infty$ control, here we only deal with a type of very basic H$_\infty$ state feedback control problem.

### 8.1.1 Problem

In this section, we study linear systems of the following form:

$$\begin{cases} \dot{x} = Ax + B_1 u + B_2 w \\ z = Cx + D_1 u + D_2 w, \end{cases} \tag{8.1}$$

where

- $x \in \mathbb{R}^n$ and $z \in \mathbb{R}^m$ are, respectively, the state vector and the output vector
- $w \in \mathbb{R}^p$ is the disturbance vector

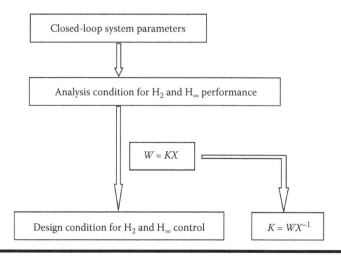

**Figure 8.1  Basic procedure for $H_\infty$ and $H_2$ control.**

- $u \in \mathbb{R}^r$ is the control vector
- $A$, $C$, $D_1$, $D_2$, $B_1$, and $B_2$ are the system coefficient matrices of appropriate dimensions

For the linear system (8.1), designing the state feedback control law

$$u = Kx, \tag{8.2}$$

gives the following closed-loop system

$$\begin{cases} \dot{x} = (A + B_1 K)x + B_2 w \\ z = (C + D_1 K)x + D_2 w. \end{cases} \tag{8.3}$$

Therefore, the influence of the disturbance $w$ to the output $z$ is determined by

$$z(s) = G_{zw}(s)w(s),$$

where

$$G_{zw}(s) = (C + D_1 K)(sI - (A + B_1 K))^{-1} B_2 + D_2. \tag{8.4}$$

Recalling the results in Section 5.1, we have

$$||z(t)||_2 \le ||G_{zw}(s)||_\infty ||w(t)||_2,$$

which states that $||G_{zw}(s)||_\infty$ performs the function of an amplifier of the disturbance to the system output. With this observation, we propose the following problem.

**Problem 8.1** For the linear system (8.1), design a state feedback control law (8.2) such that

$$||G_{zw}(s)||_\infty < \gamma$$

holds for a given positive scalar $\gamma$.

## 8.1.2 Solution

Regarding the solution to Problem 8.1, we have the following theorem.

**Theorem 8.1** The $H_\infty$ problem (Problem 8.1) has a solution if and only if there exist a matrix $W$, and a symmetric positive definite matrix $X$, such that

$$\begin{bmatrix} (AX+B_1W)^{\mathrm{T}}+AX+B_1W & B_2 & (CX+D_1W)^{\mathrm{T}} \\ B_2^{\mathrm{T}} & -\gamma I & D_2^{\mathrm{T}} \\ CX+D_1W & D_2 & -\gamma I \end{bmatrix} < 0. \qquad (8.5)$$

When such a pair of matrices $W$ and $X$ are found, a solution to the problem can be given as

$$K = WX^{-1}. \qquad (8.6)$$

*Proof* It follows from Theorem 5.4 that the second condition in the problem is satisfied if and only if there exists a symmetric positive definite matrix $X$, such that,

$$\begin{bmatrix} X(A+B_1K)^{\mathrm{T}}+(A+B_1K)X & B_2 & X(C+D_1K)^{\mathrm{T}} \\ B_2^{\mathrm{T}} & -\gamma I & D_2^{\mathrm{T}} \\ (C+D_1K)X & D_2 & -\gamma I \end{bmatrix} < 0. \qquad (8.7)$$

That is,

$$\begin{bmatrix} (AX+B_1KX)^{\mathrm{T}}+(AX+B_1KX) & B_2 & (CX+D_1KX)^{\mathrm{T}} \\ B_2^{\mathrm{T}} & -\gamma I & D_2^{\mathrm{T}} \\ CX+D_1KX & D_2 & -\gamma I \end{bmatrix} < 0.$$

By defining

$$W = KX,$$

the aforementioned inequality is turned into the LMI form (8.5). ■

With a prescribed attenuation level, the problem is turned into an LMI feasibility problem in the form of (8.5). The problem with a minimal attenuation level $\gamma$ can be sought via the following optimization problem:

$$
\begin{cases}
\min \quad \gamma \\
\text{s.t.} \quad X > 0 \\
\quad \begin{bmatrix}
(AX + B_1 W)^{\mathrm{T}} + AX + B_1 W & B_2 & (CX + D_1 W)^{\mathrm{T}} \\
B_2^{\mathrm{T}} & -\gamma I & D_2^{\mathrm{T}} \\
CX + D_1 W & D_2 & -\gamma I
\end{bmatrix} < 0.
\end{cases}
$$

(8.8)

The following corollary reveals that the condition $||G_{zw}(s)||_\infty < \gamma$ implies the closed-loop stability.

**Corollary 8.1** The $H_\infty$ problem (Problem 8.1) has a solution only if $(A\ B_1)$ is stabilizable, and with the solution given by (8.6) the closed-loop system is asymptotically stable.

*Proof* Condition (8.5) clearly implies

$$(AX + B_1 W)^{\mathrm{T}} + AX + B_1 W < 0. \tag{8.9}$$

This gives the stabilizability of the matrix pair $(A,\ B_1)$.

Using (8.9) and the solution given by (8.6), we have

$$
\begin{aligned}
(A + B_1 K)X &+ X(A + B_1 K)^{\mathrm{T}} \\
&= AX + B_1 KX + (AX + B_1 KX)^{\mathrm{T}} \\
&= (AX + B_1 W)^{\mathrm{T}} + AX + B_1 W \\
&< 0.
\end{aligned}
$$

Thus, the closed-loop system is asymptotically stable. ■

### 8.1.3 Other Conditions

In the previous section, we have obtained the design condition (8.5) based on the first condition in Theorem 5.4, that is, (5.27). Corresponding to the second and third conditions, here without proof we give the design conditions for $H_\infty$ control respectively as

$$
\begin{bmatrix}
(AX + B_1 W)^{\mathrm{T}} + AX + B_1 W & B_2 & (CX + D_1 W)^{\mathrm{T}} \\
B_2^{\mathrm{T}} & -\gamma^2 I & D_2^{\mathrm{T}} \\
CX + D_1 W & D_2 & -I
\end{bmatrix} < 0,
$$

and

$$\begin{bmatrix} (AX + B_1 W)^{\mathrm{T}} + AX + B_1 W + B_2 B_2^{\mathrm{T}} & (CX + D_1 W)^{\mathrm{T}} + B_2 D_2^{\mathrm{T}} \\ CX + D_1 W + D_2 B_2^{\mathrm{T}} & -\gamma^2 I + D_2 D_2^{\mathrm{T}} \end{bmatrix} < 0.$$

By now we have got all the three design conditions corresponding to the three analysis conditions in Theorem 5.4. One may wonder why we do not use the three conditions in Theorem 5.3 to handle the H$_\infty$ problem, the answer is very simple. As we have seen earlier, applying the conditions in Theorem 5.4 to the problem, the nonlinear term in the inequality we encountered is $KX$. This term can be easily avoided by introducing the simple variable substitution, $W = KX$, since from which we can immediately get the gain matrix as $K = WX^{-1}$. While with the conditions in Theorem 5.3, we have to face the term

$$(A + BK)^{\mathrm{T}} X + X (A + BK)$$

$$= A^{\mathrm{T}} X + XA + K^{\mathrm{T}} B^{\mathrm{T}} X + XBK.$$

If, similarly, we handle the nonlinear term by letting

$$W = XBK,$$

then from this we generally cannot get the gain matrix $K$ except for the special case of $B$ being nonsingular. Nevertheless, as we will find out in Chapter 9, the conditions in Theorem 5.3 will perform important functions in observer design.

To complete this section, we finally give the design conditions corresponding to the deduced condition in Corollary 5.4.

**Theorem 8.2**    The H$_\infty$ problem (Problem 8.1) has a solution if and only if there exist matrices $V$, $W$ and a symmetric matrix $X$ such that

$$\begin{bmatrix} -V - V^{\mathrm{T}} & (AV + B_1 W)^{\mathrm{T}} + X & (CV + D_1 W)^{\mathrm{T}} & V^{\mathrm{T}} & 0 \\ AV + B_1 W + X & -X & 0 & 0 & B_2 \\ CV + D_1 W & 0 & -\gamma I_m & 0 & D_2 \\ V & 0 & 0 & -X & 0 \\ 0 & B_2^{\mathrm{T}} & D_2^{\mathrm{T}} & 0 & -\gamma I_r \end{bmatrix} < 0.$$

When such a triple of matrices $V$, $W$, and $X$ are found, a solution to the problem can be given as

$$K = WV^{-1}.$$

Corresponding to the second deduced condition in Corollary 5.4, the condition turns to be

$$
\begin{bmatrix}
-V - V^{\mathrm{T}} & (AV + B_1 W)^{\mathrm{T}} + X & (CV + D_1 W)^{\mathrm{T}} & V^{\mathrm{T}} & 0 \\
AV + B_1 W + X & -X & 0 & 0 & B_2 \\
CV + D_1 W & 0 & -I_m & 0 & D_2 \\
V & 0 & 0 & -X & 0 \\
0 & B_2^{\mathrm{T}} & D_2^{\mathrm{T}} & 0 & -\gamma^2 I_r
\end{bmatrix} < 0.
$$

### Example 8.1

Consider the linear system with the following parameters:

$$
\begin{cases}
\dot{x} = \begin{bmatrix} 2 & 1 & -2 \\ 1 & -1 & -3 \\ 4 & 0 & -1 \end{bmatrix} x + \begin{bmatrix} 1 & 0 \\ 0 & 3 \\ 3 & 1 \end{bmatrix} u + \begin{bmatrix} 1 \\ 0.2 \\ -0.5 \end{bmatrix} w \\
z = \begin{bmatrix} 2 & 1 & -0.5 \end{bmatrix} x + \begin{bmatrix} 1 & 1 \end{bmatrix} u + 0.05w.
\end{cases}
$$

Using the function `mincx` in LMI toolbox to the optimization problem (8.8), we have

$$
X = 10^7 \times \begin{bmatrix} 8.7047 & -6.5321 & 4.2500 \\ -6.5321 & 8.8601 & -4.2821 \\ 4.2500 & -4.2821 & 5.0227 \end{bmatrix} > 0,
$$

$$
W = 10^7 \times \begin{bmatrix} -5.2786 & 2.9364 & -7.7991 \\ -3.4736 & -0.8733 & 6.0925 \end{bmatrix},
$$

$$
\gamma = 0.0500,
$$

and the state feedback gain

$$
K = WX^{-1} = \begin{bmatrix} -0.2159 & -0.8333 & -2.0805 \\ -1.7841 & -0.1667 & 2.5805 \end{bmatrix}.
$$

Note that the attenuation level $\gamma$ is very small, the disturbance $\omega$ must be effectively attenuated. To verify this, we have plotted the output z in Figure 8.2, for the cases of $\omega(t) = 0$ and $\omega(t) = \sin t$. It is seen from this figure that the disturbance indeed has very small affection to the system output.

Using the function `place` in control system toolbox to the same system, we obtain a feedback gain as

$$
K_1 = \begin{bmatrix} 1.2850 & 0.5978 & -2.0283 \\ -1.4101 & -0.7807 & 1.4861 \end{bmatrix}.
$$

The closed-loop system outputs for the cases of $\omega(t) = 0$ and $\omega(t) = \sin t$ corresponding to the aforementioned feedback gain is shown in Figure 8.3.

Comparing the two Figures 8.2 and 8.3, it is clearly seen that the effect of disturbance is dramatically reduced when $\|G_{zw}(s)\|_\infty$ is minimized. Without this consideration, the disturbance may thoroughly alter the output of the system.

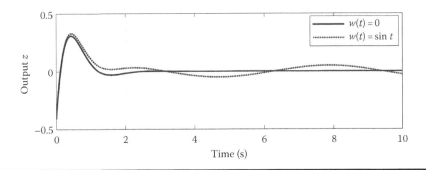

**Figure 8.2   The output of closed-loop system, Example 8.1.**

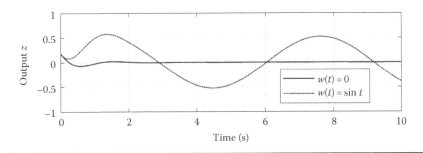

**Figure 8.3   System output corresponding to the feedback $K_1$, Example 8.1.**

## 8.2 H₂ State Feedback Control

First let us look at the H₂ state feedback control problem. For simplicity, here only the very basic problem is treated.

### 8.2.1 Problem

In this section, we study linear systems of the following form:

$$\begin{cases} \dot{x} = Ax + B_1 u + B_2 w \\ z = Cx + Du, \end{cases} \tag{8.10}$$

where
   $x \in \mathbb{R}^n$ and $z \in \mathbb{R}^m$ are respectively the state vector and the output vector
   $w \in \mathbb{R}^p$ is the disturbance vector
   $u \in \mathbb{R}^r$ is the control vector
   $A, C, D, B_1$, and $B_2$ are the system coefficient matrices of appropriate dimensions

In this section, we treat the problem of designing a state feedback controller for system (8.10) such that the closed-loop system is stable and the effect of the disturbance is prohibited to a desired level.

For the linear system (8.10), designing the state feedback control law

$$u = Kx, \tag{8.11}$$

gives the following closed-loop system

$$\begin{cases} \dot{x} = (A + B_1 K) x + B_2 w \\ z = (C + DK) x. \end{cases} \tag{8.12}$$

Therefore, the influence of the disturbance $w$ to the output $z$ is determined by

$$z(s) = G_{zw}(s) w(s),$$

where

$$G_{zw}(s) = (C + DK) (sI - (A + B_1 K))^{-1} B_2. \tag{8.13}$$

Recalling the results in Section 5.1, we have

$$||z(t)||_P \leq ||G_{zw}(s)||_2 ||w(t)||_P,$$

which states that $||G_{zw}(s)||_2$ performs the function of an amplifier of the disturbance to the system output. With this observation, we propose the following problem.

**Problem 8.2** For the linear system (8.10), design a state feedback control law (8.11) such that

$$||G_{zw}(s)||_2 < \gamma$$

holds for a given positive scalar $\gamma$.

## 8.2.2 Solution

It follows from Theorem 5.6 that Problem 8.2 has a solution if and only if there exists a symmetric positive definite matrix $X$ such that

$$\begin{cases} (A + B_1 K) X + X (A + B_1 K)^{\mathrm{T}} + B_2 B_2^{\mathrm{T}} < 0 \\ \mathrm{trace}\left((C + DK) X (C + DK)^{\mathrm{T}}\right) < \gamma^2. \end{cases}$$

Using Lemma 2.13, the aforementioned set of inequalities can be equivalently converted into

$$\begin{cases} (A + B_1 K) X + X (A + B_1 K)^T + B_2 B_2^T < 0 \\ (C + DK) X (C + DK)^T < Z \\ \text{trace}(Z) < \gamma^2, \end{cases}$$

where $Z$ is some symmetric matrix. Further, using Schur complement lemma, gives

$$\begin{cases} (A + B_1 K) X + X (A + B_1 K)^T + B_2 B_2^T < 0 \\ \begin{bmatrix} -Z & (C + DK) X \\ X (C + DK)^T & -X \end{bmatrix} < 0 \\ \text{trace}(Z) < \gamma^2. \end{cases}$$

Finally, by defining

$$W = KX,$$

we can obtain the solution to Problem 8.2.

**Theorem 8.3**   The H₂ problem (Problem 8.2) has a solution if and only if there exist a matrix $W$, two symmetric matrices $Z$ and $X$ such that

$$\begin{cases} AX + B_1 W + (AX + B_1 W)^T + B_2 B_2^T < 0 \\ \begin{bmatrix} -Z & CX + DW \\ (CX + DW)^T & -X \end{bmatrix} < 0 \\ \text{trace}(Z) < \gamma^2. \end{cases} \tag{8.14}$$

When such a triple of matrices are obtained, a gain matrix to the problem can be given by

$$K = WX^{-1}. \tag{8.15}$$

The aforementioned theorem again converts the control system design problem into an LMI feasibility problem, namely, (8.14), which is solvable using the LMI toolbox command `feasp`. In applications, we are often concerned with the problem of finding the minimal attenuation level $\gamma$. This problem can be solved via the following optimization:

$$
\begin{cases}
\min \quad \rho \\
\text{s.t.} \quad AX + B_1 W + (AX + B_1 W)^{\mathrm{T}} + B_2 B_2^{\mathrm{T}} < 0 \\
\qquad \begin{bmatrix} -Z & CX + DW \\ (CX + DW)^{\mathrm{T}} & -X \end{bmatrix} < 0 \\
\qquad \text{trace}\,(Z) < \rho,
\end{cases} \tag{8.16}
$$

where $\rho = \gamma^2$. This minimization problem (8.16) is solvable using the LMI toolbox command mincx.

The following corollary reveals that the condition $||G_{zw}(s)||_2 < \gamma$ implies the closed-loop stability.

**Corollary 8.2**   The $H_2$ Problem 8.2 has a solution only if $(A\ B_1)$ is stabilizable, and with the solution given by (8.15) the closed-loop system is asymptotically stable.

*Proof*   The first inequality in (8.14) clearly implies

$$
(AX + B_1 W)^{\mathrm{T}} + AX + B_1 W < -B_2 B_2^{\mathrm{T}} \le 0. \tag{8.17}
$$

This gives the stabilizability of the matrix pair $(A,\ B_1)$.

Using (8.17) and the solution given by (8.15), we have

$$
(A + B_1 K)X + X(A + B_1 K)^{\mathrm{T}}
$$
$$
= AX + B_1 KX + (AX + B_1 KX)^{\mathrm{T}}
$$
$$
= (AX + B_1 W)^{\mathrm{T}} + AX + B_1 W
$$
$$
< 0.
$$

Thus, the closed-loop system is asymptotically stable.   ■

### 8.2.3 Other Condition

In the previous section, we have obtained the design condition (8.14) based on the first condition in Theorem 5.6, that is, (5.42). Due to the same reason mentioned in Section 8.1.3 about the condition for the $H_\infty$ control problem, we conclude that the second condition in Theorem 5.6 is not directly applicable to the $H_2$ problem.

Again, due to the same reason, the first deduced condition in Theorem 5.7 is also not directly applicable. Now to complete this section, we finally give the design conditions corresponding to the second deduced condition in Theorem 5.7.

**Theorem 8.4** The $H_2$ problem (Problem 8.2) has a solution if and only if there exist matrices $V$, $W$ and two symmetric matrices $P$ and $Z$ such that

$$
\begin{cases}
\text{trace}\,(Z) < \gamma^2 \\[4pt]
\begin{bmatrix} -Z & B_2^{\mathrm{T}} \\ B_2 & -P \end{bmatrix} < 0 \\[16pt]
\begin{bmatrix}
-(V + V^{\mathrm{T}}) & (AV + B_1 W)^{\mathrm{T}} + P & (CV + DW)^{\mathrm{T}} & V^{\mathrm{T}} \\
AV + B_1 W + P & -P & 0 & 0 \\
CV + DW & 0 & -I & 0 \\
V & 0 & 0 & -P
\end{bmatrix} < 0.
\end{cases}
$$

When such a set of matrices $V$, $W$, $P$, and $Z$ are found, a solution to the problem can be given as

$$
K = WV^{-1}.
$$

**Example 8.2**

Consider the following linear system

$$
\begin{cases}
\dot{x} = \begin{bmatrix} -3 & -2 & 1 \\ 1 & 2 & 1 \\ 1 & -1 & -1 \end{bmatrix} x + \begin{bmatrix} 2 & 0 \\ 0 & 2 \\ 0 & 1 \end{bmatrix} u + \begin{bmatrix} 3 \\ 0 \\ 1 \end{bmatrix} \omega \\[16pt]
z = \begin{bmatrix} 1 & 0 & 1 \\ 0 & 1 & 1 \end{bmatrix} x + \begin{bmatrix} 1 & 1 \\ 0 & 1 \end{bmatrix} u.
\end{cases}
$$

Applying the MATLAB® function `mincx` in the LMI toolbox to the optimization problem (8.16), we get

$$
X = 10^3 \times \begin{bmatrix} 0.7799 & -0.0053 & 0.2621 \\ -0.0053 & 0.1400 & 0.4207 \\ 0.2621 & 0.4207 & 1.3635 \end{bmatrix} > 0,
$$

$$
W = 10^3 \times \begin{bmatrix} -0.7851 & 0.1452 & 0.1586 \\ -0.2569 & -0.5607 & -1.7842 \end{bmatrix},
$$

$$
Z = 10^{-3} \times \begin{bmatrix} 0.0089 & 0.0538 \\ 0.0538 & 0.3260 \end{bmatrix},
$$

$$
\rho = 3.3491 \times 10^{-4},
$$

and the corresponding state feedback gain is obtained as

$$
K = WX^{-1} = \begin{bmatrix} -0.9112 & 1.7519 & -0.2491 \\ -0.1063 & -1.9004 & -0.7018 \end{bmatrix},
$$

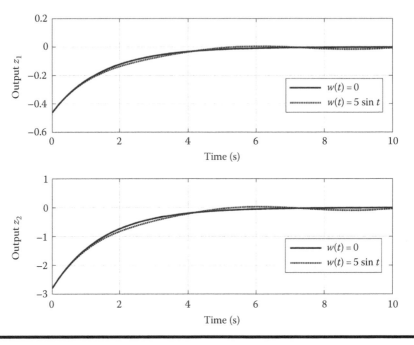

**Figure 8.4** **The output of closed-loop system, Example 8.2.**

the corresponding minimal attenuation level is $\gamma = \sqrt{\rho} = 0.0183$. Note that the attenuation level $\gamma$ is very small, the disturbance $\omega$ must be effectively attenuated. To verify this, we have plotted the output $z$ in Figure 8.4, for the cases of $\omega(t) = 0$ and $\omega(t) = 5\sin t$. It is seen from this figure that the disturbance indeed has very little affection to the system output.

Using the MATLAB function `place` in control system toolbox to the same system, we obtain a feedback gain as

$$K_1 = \begin{bmatrix} 0.3965 & 0.9895 & -0.4499 \\ -0.4185 & -3.0160 & -1.0858 \end{bmatrix},$$

which assigns the same set of closed poles as the optimal gain matrix $K$, that is,

$$\lambda(A + B_1 K_1) = \{-5.4444, -2.2181, -0.6625\}.$$

The closed-loop system output corresponding to the aforementioned feedback gain $K_1$ is shown in Figure 8.5, also for the cases of $\omega(t) = 0$ and $\omega(t) = 5\sin t$.

Comparison of the two Figures 8.4 and 8.5 clearly reveals the effect of disturbance attenuation via $H_2$ control.

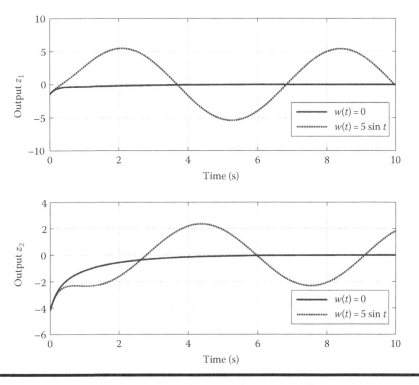

**Figure 8.5**  System output corresponding to gain $K_1$, Example 8.2.

## 8.3 Robust H$_\infty$/H$_2$ State Feedback Control

In this section, we further investigate the problems of robust state feedback control of linear systems with parameter uncertainties.

### 8.3.1 Problems

#### 8.3.1.1 Uncertain System

Consider the following linear system with uncertainty

$$\begin{cases} \dot{x}(t) = (A + \Delta A)x(t) + (B_1 + \Delta B_1)u(t) + B_2 w(t) \\ z(t) = Cx(t) + D_1 u(t) + D_2 w(t), \end{cases} \quad (8.18)$$

where

- $x \in \mathbb{R}^n$ and $z \in \mathbb{R}^m$ are respectively the state vector and the output vector
- $w \in \mathbb{R}^p$ is the disturbance vector

- $u \in \mathbb{R}^r$ is the control vector
- $A, B_1, B_2, C, D_1$, and $D_2$ are known real constant system coefficient matrices of appropriate dimensions

$\Delta A$ and $\Delta B_1$ are real-valued matrix functions that represent the time-varying parameter uncertainties and are of the form

$$\begin{bmatrix} \Delta A & \Delta B_1 \end{bmatrix} = HF \begin{bmatrix} E_1 & E_2 \end{bmatrix}, \tag{8.19}$$

where matrices $E_1$, $E_2$, and $H$ are some known matrices of appropriate dimensions, while $F$ is a matrix which contains the uncertain parameters and satisfies

$$F^T F \le I. \tag{8.20}$$

For the type of perturbation, we obviously have

$$\begin{bmatrix} \Delta A & \Delta B_1 \end{bmatrix} = \begin{cases} \begin{bmatrix} 0 & 0 \end{bmatrix}, & \text{when } F = 0 \\ H \begin{bmatrix} E_1 & E_2 \end{bmatrix}, & \text{when } F = I. \end{cases}$$

### 8.3.1.2 Forms of Perturbation

Remember that in Chapter 4 we handled the problem of robust stability, where the perturbation of the system coefficient matrix takes the following form:

$$\Delta A = \delta_1 A_1 + \delta_2 A_2 + \cdots + \delta_k A_k, \tag{8.21}$$

where
$A_i$, $i = 1, 2, \ldots, k$, are some known matrices
$\delta_i$, $i = 1, 2, \ldots, k$, are the perturbation parameters satisfying

$$|\delta_i| < r_i, \quad i = 1, 2, \ldots, k. \tag{8.22}$$

Note that

$$\Delta A = \begin{bmatrix} A_1 & A_2 & \cdots & A_k \end{bmatrix} \begin{bmatrix} \delta_1 I \\ \delta_2 I \\ \vdots \\ \delta_k I \end{bmatrix},$$

we can thus write $\Delta A$ in the form of

$$\Delta A = HFE \tag{8.23}$$

with

$$H = \begin{bmatrix} A_1 & A_2 & \cdots & A_k \end{bmatrix}, \quad E = \left(\sum_{i=1}^{k} r_i^2\right)^{\frac{1}{2}},$$

and

$$F = \left(\sum_{i=1}^{k} r_i^2\right)^{-\frac{1}{2}} \begin{bmatrix} \delta_1 I \\ \delta_2 I \\ \vdots \\ \delta_k I \end{bmatrix}.$$

On the other hand, suppose that we can separate out the perturbation parameters in the matrix $F$ as follows:

$$F = \delta_1 F_1 + \delta_2 F_2 + \cdots + \delta_k F_k.$$

Then, we have form (8.23),

$$
\begin{aligned}
\Delta A &= HFE \\
&= H\left(\delta_1 F_1 + \delta_2 F_2 + \cdots + \delta_k F_k\right) E \\
&= \delta_1 A_1 + \delta_2 A_2 + \cdots + \delta_k A_k,
\end{aligned}
$$

with

$$A_i = HF_i E, \quad i = 1, 2, \ldots, k.$$

Further, note that

$$
\begin{aligned}
F^{\mathrm{T}} F \\
&= \left(\delta_1 F_1 + \delta_2 F_2 + \cdots + \delta_k F_k\right)^{\mathrm{T}} \left(\delta_1 F_1 + \delta_2 F_2 + \cdots + \delta_k F_k\right) \\
&= \begin{bmatrix} F_1^{\mathrm{T}} & F_2^{\mathrm{T}} & \cdots & F_k^{\mathrm{T}} \end{bmatrix} \operatorname{diag}\left(\delta_1^2 I, \delta_2^2 I, \ldots, \delta_k^2 I\right) \begin{bmatrix} F_1 \\ F_2 \\ \vdots \\ F_k \end{bmatrix},
\end{aligned}
$$

then condition (8.22) becomes

$$\begin{bmatrix} F_1^{\mathrm{T}} & F_2^{\mathrm{T}} & \cdots & F_k^{\mathrm{T}} \end{bmatrix} \operatorname{diag}\left(\delta_1^2 I, \delta_2^2 I, \ldots, \delta_k^2 I\right) \begin{bmatrix} F_1 \\ F_2 \\ \vdots \\ F_k \end{bmatrix} \leq I.$$

Therefore, the two types of perturbations (8.23) and (8.21) are mutually transferable.

### 8.3.1.3 Problem Formulation

With the state feedback control law

$$u(t) = Kx(t), \tag{8.24}$$

the closed-loop system is

$$\begin{cases} \dot{x}(t) = [(A + \Delta A) + (B_1 + \Delta B_1) K] x(t) + B_2 w(t) \\ z(t) = (C + D_1 K) x(t) + D_2 w(t), \end{cases} \tag{8.25}$$

and its transfer function is

$$G_{zw}(s) = (C + D_1 K) (sI - [(A + \Delta A) + (B_1 + \Delta B_1) K])^{-1} B_2 + D_2.$$

**Problem 8.3** (Robust $H_\infty$ control) Given the uncertain linear system (8.18) and a positive scalar $\gamma$, design a state feedback control law $u(t)$ in the form of (8.24) such that the closed-loop system (8.25) is asymptotically stable and satisfies

$$\| G_{zw}(s) \|_\infty < \gamma.$$

**Problem 8.4** (Robust $H_2$ control) Given the uncertain linear system (8.18) with $D_2 = 0$, and a scalar $\gamma > 0$, design a state feedback control law $u(t)$ in the form of (8.24) such that the closed-loop system (8.25) is asymptotically stable and satisfies

$$\| G_{zw}(s) \|_2 < \gamma.$$

## 8.3.2 Solution to the Robust $H_\infty$ Control Problem

For solution to the $H_\infty$ state feedback control problem stated earlier, we have the following result.

**Theorem 8.5** The $H_\infty$ state feedback control problem (Problem 8.3) has a solution if and only if there exist a scalar $\alpha$, a matrix $W$, and a symmetric positive definite matrix $X$ satisfying the following LMI:

$$\begin{bmatrix} \Psi(X, W) & B_2 & (CX + D_1 W)^{\mathrm{T}} & (E_1 X + E_2 W)^{\mathrm{T}} \\ B_2^{\mathrm{T}} & -\gamma I & D_2^{\mathrm{T}} & 0 \\ CX + D_1 W & D_2 & -\gamma I & 0 \\ E_1 X + E_2 W & 0 & 0 & -\alpha I \end{bmatrix} < 0, \tag{8.26}$$

where

$$\Psi(X, W) = \langle AX + B_1 W \rangle_s + \alpha H H^{\mathrm{T}}.$$

In this case, an H$_\infty$ state feedback control law is given by

$$u(t) = W X^{-1} x(t).$$

*Proof*   It follows from Theorem 8.1 that the H$_\infty$ state feedback control problem (Problem 8.3) has a solution if and only if there exist a matrix $W$ and a symmetric positive definite matrix $X$ satisfying

$$\begin{bmatrix} \langle (A + \Delta A) X + (B_1 + \Delta B_1) W \rangle_s & B_2 & (CX + D_1 W)^{\mathrm{T}} \\ B_2^{\mathrm{T}} & -\gamma I & D_2^{\mathrm{T}} \\ CX + D_1 W & D_2 & -\gamma I \end{bmatrix} < 0. \quad (8.27)$$

Consider

$$\begin{bmatrix} \langle (A + \Delta A) X + (B_1 + \Delta B_1) W \rangle_s & B_2 & (CX + D_1 W)^{\mathrm{T}} \\ B_2^{\mathrm{T}} & -\gamma I & D_2^{\mathrm{T}} \\ CX + D_1 W & D_2 & -\gamma I \end{bmatrix}$$

$$= \begin{bmatrix} \langle AX + B_1 W \rangle_s & B_2 & (CX + D_1 W)^{\mathrm{T}} \\ B_2^{\mathrm{T}} & -\gamma I & D_2^{\mathrm{T}} \\ CX + D_1 W & D_2 & -\gamma I \end{bmatrix}$$

$$+ \begin{bmatrix} \langle \Delta AX + \Delta B_1 W \rangle_s & 0 & 0 \\ 0 & 0 & 0 \\ 0 & 0 & 0 \end{bmatrix}.$$

In view of the relations in (8.19), we further have

$$\begin{bmatrix} \langle \Delta AX + \Delta B_1 W \rangle_s & 0 & 0 \\ 0 & 0 & 0 \\ 0 & 0 & 0 \end{bmatrix}$$

$$= \begin{bmatrix} \langle HF(E_1 X + E_2 W) \rangle_s & 0 & 0 \\ 0 & 0 & 0 \\ 0 & 0 & 0 \end{bmatrix}$$

$$= \left\langle \begin{bmatrix} H \\ 0 \\ 0 \end{bmatrix} F \begin{bmatrix} E_1 X + E_2 W & 0 & 0 \end{bmatrix} \right\rangle_s.$$

Thus, the inequality (8.27) can be equivalently written as

$$
\begin{bmatrix} \langle AX + B_1 W \rangle_s & B_2 & (CX + D_1 W)^{\mathrm{T}} \\ B_2^{\mathrm{T}} & -\gamma I & D_2^{\mathrm{T}} \\ CX + D_1 W & D_2 & -\gamma I \end{bmatrix}
$$

$$
+ \left\langle \begin{bmatrix} H \\ 0 \\ 0 \end{bmatrix} F \begin{bmatrix} E_1 X + E_2 W & 0 & 0 \end{bmatrix} \right\rangle_s < 0.
$$

Applying Lemma 2.5, we can equivalently convert the earlier inequality into

$$
\begin{bmatrix} \langle AX + B_1 W \rangle_s + \alpha HH^{\mathrm{T}} & B_2 & (CX + D_1 W)^{\mathrm{T}} \\ B_2^{\mathrm{T}} & -\gamma I & D_2^{\mathrm{T}} \\ CX + D_1 W & D_2 & -\gamma I \end{bmatrix}
$$

$$
+ \alpha^{-1} \begin{bmatrix} E_1 X + E_2 W & 0 & 0 \end{bmatrix}^{\mathrm{T}} \begin{bmatrix} E_1 X + E_2 W & 0 & 0 \end{bmatrix} < 0. \qquad (8.28)
$$

Finally, by using Schur complement lemma, the inequality (8.28) can be equivalently arranged into the form of inequality (8.26). The proof is then completed. ■

In applications, we are often concerned with the problem of finding the minimal attenuation level $\gamma$. This problem can be solved via the following optimization:

$$
\begin{cases} \min & \gamma \\ \text{s.t.} & X > 0 \\ & \begin{bmatrix} \Psi(X, W) & B_2 & (CX + D_1 W)^{\mathrm{T}} & (E_1 X + E_2 W)^{\mathrm{T}} \\ B_2^{\mathrm{T}} & -\gamma I & D_2^{\mathrm{T}} & 0 \\ CX + D_1 W & D_2 & -\gamma I & 0 \\ E_1 X + E_2 W & 0 & 0 & -\alpha I \end{bmatrix} < 0. \end{cases}
$$

$$(8.29)$$

This minimization problem can be solved using the LMI toolbox command `mincx`.

### Example 8.3

Consider a linear system in the form of (8.18) with the following parameters:

$$
A = \begin{bmatrix} -0.7467 & -3.8410 & -9.9319 \\ -2.9689 & -8.8444 & 4.9235 \\ 1.7507 & 0.1229 & -8.7683 \end{bmatrix},
$$

$$
B_1 = \begin{bmatrix} 1.0877 & 0.0169 \\ -0.7231 & 1.2568 \\ 1.9908 & -0.4266 \end{bmatrix}, \quad B_2 = \begin{bmatrix} -0.0193 \\ 0.0220 \\ 0.0312 \end{bmatrix},
$$

$$C = \begin{bmatrix} -3.2704 & -2.4767 & -3.6348 \\ 4.7975 & 3.7574 & -4.8824 \\ -2.2855 & 2.3731 & 3.9390 \end{bmatrix},$$

$$D_1 = \begin{bmatrix} -0.2034 & -0.1624 \\ -0.8051 & -0.1231 \\ 0.6458 & -1.7409 \end{bmatrix}, \quad D_2 = \begin{bmatrix} 0.0313 \\ -0.0137 \\ 0.0385 \end{bmatrix},$$

and $H = I_3$,

$$E_1 = \begin{bmatrix} -0.1134 & -0.1707 & -0.0203 \\ 0.0953 & 0.0675 & -0.0397 \\ 0.0920 & -0.0075 & 0.1172 \end{bmatrix}, \quad E_2 = \begin{bmatrix} 0.1342 & 0.1092 \\ 0.0488 & 0.1759 \\ -0.0954 & -0.1677 \end{bmatrix}.$$

Using the function mincx in the LMI toolbox to the $H_\infty$ optimization problem (8.29), we obtain the following optimal parameters

$$X = \begin{bmatrix} 0.0338 & -0.0236 & 0.0358 \\ -0.0236 & 0.0645 & -0.0446 \\ 0.0358 & -0.0446 & 0.0504 \end{bmatrix} > 0,$$

$$W = \begin{bmatrix} -0.0644 & 0.1373 & -0.1618 \\ -0.0016 & -0.0149 & 0.0207 \end{bmatrix},$$

$$\alpha = 0.0021, \quad \gamma = 0.0949,$$

and thus the state feedback gain is given by

$$K = WX^{-1} = \begin{bmatrix} 9.1630 & -3.2219 & -12.5748 \\ -3.0468 & 1.1270 & 3.5736 \end{bmatrix}.$$

Corresponding to this gain matrix, we have plotted the closed-loop system output $z$ in Figure 8.6, with $\omega(t) = \sin 2t$. The solid lines represent the case of $F = 0_{3 \times 3}$, while the dash lines represent the case of $F = I_3$. It is seen from this figure that the uncertainty indeed has very little affection to the system output.

Using the function place in MATLAB control system toolbox to the same system, we can obtain a feedback gain as

$$K_1 = \begin{bmatrix} 6.8814 & -3.3725 & -10.2539 \\ 9.5152 & -6.4425 & -12.9601 \end{bmatrix},$$

which assigns the same set of closed-loop poles as the gain matrix $K$, that is,

$$\lambda(A + BK_1) = \lambda(A + BK) = \{-9.0896 \pm 4.1847i, -13.0774\}.$$

The closed-loop system output corresponding to this feedback gain $K_1$ is shown in Figure 8.7, for the same choices of $\omega(t)$ and $F$ as in Figure 8.6. Comparison of the two figures clearly reveals the advantage of the $H_2$ control technique in the aspect of disturbance attenuation.

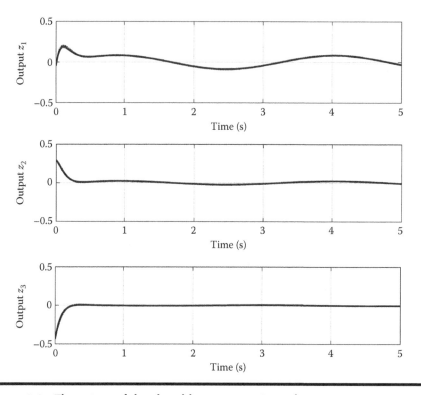

**Figure 8.6** The output of the closed-loop system, Example 8.3.

### 8.3.3 Solution to the Robust $H_2$ Control Problem

For solution to the $H_2$ state feedback control problem stated earlier, we have the following result.

**Theorem 8.6** The $H_2$ state feedback control problem (Problem 8.4) has a solution if and only if there exist a scalar $\beta$, a matrix $W$, two symmetric matrices $Z$ and $X$ satisfying the following LMIs:

$$
\begin{cases}
\begin{bmatrix} \langle AX + B_1 W \rangle_s + B_2 B_2^{\mathrm{T}} + \beta HH^{\mathrm{T}} & (E_1 X + E_2 W)^{\mathrm{T}} \\ E_1 X + E_2 W & -\beta I \end{bmatrix} < 0 \\[2ex]
\begin{bmatrix} -Z & CX + D_1 W \\ (CX + D_1 W)^{\mathrm{T}} & -X \end{bmatrix} < 0 \\[2ex]
\mathrm{trace}\,(Z) < \gamma^2.
\end{cases}
\tag{8.30}
$$

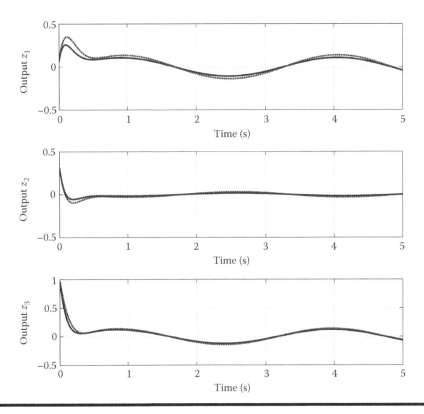

**Figure 8.7** **System output corresponding to gain $K_1$, Example 8.3.**

In this case, an $H_2$ state feedback control law is given by

$$u(t) = WX^{-1}x(t).$$

*Proof* It follows from Theorem 8.3 that the $H_2$ state feedback control problem (Problem 8.4) has a solution if and only if there exist a matrix $W$, two symmetric matrices $Z$ and $X$ satisfying

$$\begin{cases} \langle (A + \Delta A) X + (B_1 + \Delta B_1) W \rangle_s + B_2 B_2^\mathsf{T} < 0 \\ \begin{bmatrix} -Z & CX + D_1 W \\ (CX + D_1 W)^\mathsf{T} & -X \end{bmatrix} < 0 \\ \operatorname{trace}(Z) < \gamma^2. \end{cases} \tag{8.31}$$

Thus, we need only to show the equivalence between the first inequality in (8.30) and that in (8.31).

Since

$$\langle (A + \Delta A) X + (B_1 + \Delta B_1) W \rangle_s + B_2 B_2^{\mathrm{T}}$$
$$= \langle AX + B_1 W \rangle_s + B_2 B_2^{\mathrm{T}} + \langle \Delta AX + \Delta B_1 W \rangle_s,$$

and, in view of the relations in (8.19),

$$\langle \Delta AX + \Delta B_1 W \rangle_s = \langle HF (E_1 X + E_2 W) \rangle_s,$$

the first inequality in (8.31) can be equivalently written as

$$\langle AX + B_1 W \rangle_s + B_2 B_2^{\mathrm{T}} + \langle HF (E_1 X + E_2 W) \rangle_s < 0.$$

Applying Lemma 2.5, we can equivalently convert the aforementioned inequality into

$$\langle AX + B_1 W \rangle_s + B_2 B_2^{\mathrm{T}} + \beta HH^{\mathrm{T}}$$
$$+ \beta^{-1} (E_1 X + E_2 W)^{\mathrm{T}} (E_1 X + E_2 W) < 0. \tag{8.32}$$

Finally, by using Schur complement lemma, the inequality (8.32) can be equivalently arranged into the form of the first inequality in (8.30). The proof is then completed. ∎

In applications, we are often concerned with the problem of finding the minimal attenuation level $\gamma$. This problem can be solved via the following optimization:

$$\begin{cases} \min \quad \rho \\ \text{s.t.} \quad \begin{bmatrix} \langle AX + B_1 W \rangle_s + B_2 B_2^{\mathrm{T}} + \beta HH^{\mathrm{T}} & (E_1 X + E_2 W)^{\mathrm{T}} \\ E_1 X + E_2 W & -\beta I \end{bmatrix} < 0 \\ \begin{bmatrix} -Z & CX + D_1 W \\ (CX + D_1 W)^{\mathrm{T}} & -X \end{bmatrix} < 0 \\ \text{trace} (Z) < \rho, \end{cases} \tag{8.33}$$

where $\rho$ is a positive optimizing variable. This minimization problem can be solved using the LMI toolbox command `mincx`. When this $\rho$ is obtained, the minimal attenuation level is given by $\sqrt{\rho}$.

**Example 8.4**

Consider a linear system in the form of (8.18) with the following parameters:

$$A = \begin{bmatrix} -1.8764 & 2.2925 & 8.2951 \\ -2.6505 & -1.8239 & -9.5310 \\ 6.0722 & 8.8747 & -6.3834 \end{bmatrix},$$

$$B_1 = \begin{bmatrix} 1.0877 & -0.9443 \\ 2.8711 & -0.1364 \\ 1.9588 & -0.1735 \end{bmatrix}, \quad B_2 = \begin{bmatrix} 0.1103 \\ -0.0536 \\ 0.1214 \end{bmatrix},$$

$$C = \begin{bmatrix} 0.5588 & 2.6838 & 1.7343 \\ -2.7756 & -4.5748 & 1.6906 \end{bmatrix},$$

$$D_1 = \begin{bmatrix} 0.2699 & 0.0016 \\ 0.1486 & 0.7459 \end{bmatrix}, \quad D_2 = 0,$$

and $H = I_3$,

$$E_1 = \begin{bmatrix} 0.0873 & 0.1531 & -0.1706 \\ 0.1777 & 0.0006 & -0.1456 \\ 0.1330 & 0.0811 & -0.1166 \end{bmatrix}, \quad E_2 = \begin{bmatrix} -0.0078 & -0.1237 \\ -0.0297 & -0.1814 \\ -0.0764 & -0.1313 \end{bmatrix}.$$

Applying the MATLAB function `mincx` in the LMI toolbox to the $H_2$ optimization problem (8.33), we get

$$X = \begin{bmatrix} 0.0429 & 0.0080 & -0.0154 \\ 0.0080 & 0.0707 & 0.0373 \\ -0.0154 & 0.0373 & 0.0365 \end{bmatrix} > 0,$$

$$W = \begin{bmatrix} -0.1070 & -0.9185 & -0.5773 \\ 0.2715 & 0.5540 & 0.2040 \end{bmatrix},$$

$$Z = \begin{bmatrix} 0.0245 & -0.0121 \\ -0.0121 & 0.0060 \end{bmatrix},$$

and

$$\rho = 0.0305, \quad \beta = 0.0029.$$

The corresponding state feedback gain is thus obtained as

$$K = WX^{-1} = \begin{bmatrix} -5.8671 & -5.7982 & -12.3917 \\ 5.5622 & 6.5411 & 1.2663 \end{bmatrix},$$

while the optimal attenuation level is $\gamma = \sqrt{\rho} = 0.1746$.
    We have plotted the output $z$ in Figure 8.8, for the cases of $F = 0_{3\times3}$ and $F = I_3$, where $w(t) = 0.5 \sin 1.5t$. It is seen from this figure that the uncertainty indeed has very little affection to the system output.

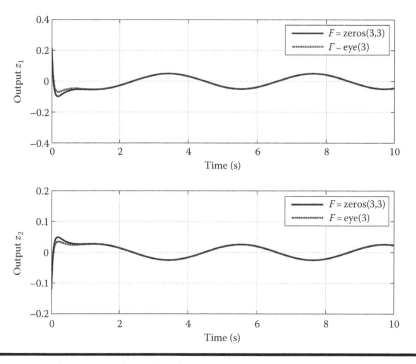

**Figure 8.8** The output of the closed-loop system, Example 8.4.

Using the function `place` in MATLAB control system toolbox to the same system, we obtain a feedback gain as

$$K_1 = \begin{bmatrix} -1.9046 & -12.0954 & -2.7428 \\ 11.7365 & -4.0700 & 5.5739 \end{bmatrix},$$

which set the same set of closed-loop poles, that is, $-45.8913$, $-3.1314$, $-14.7269$, as the gain $K$. The closed-loop system output corresponding to the feedback gain $K_1$ is shown in Figure 8.9, for the same choices of $w(t)$ and $F$ as in Figure 8.8.

## 8.4 LQ Regulation via H₂ Control

In this section, the well-known quadratic optimal regulation problem is treated. We will see that, besides the traditional Riccati equation approach, the problem can be equivalently converted into a standard $H_2$ problem, and hence can be solved via LMI techniques.

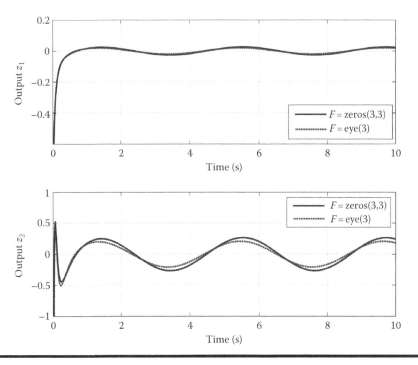

**Figure 8.9** System output corresponding to gain $K_1$, Example 8.4.

## 8.4.1 Problem Description

Consider the constant linear multi-variable system

$$\dot{x} = Ax + Bu, \quad x(0) = x_0, \tag{8.34}$$

where
  $x \in \mathbb{R}^n$ and $u \in \mathbb{R}^r$ are the state vector and input vector, respectively
  $A$ and $B$ are the system coefficient matrices of appropriate dimensions

The well-known linear quadratic optimal regulation problem for the aforementioned system is stated as follows.

**Problem 8.5** Design, for system (8.34), an optimal state feedback controller $u = Kx$ such that the following quadratic performance index

$$J(x, u) = \int_0^\infty (x^T Q x + u^T R u)\, dt \tag{8.35}$$

is minimized, where

$$Q = Q^T \geq 0, \quad R = R^T > 0. \tag{8.36}$$

For a traditional solution to this problem we have a basic result which is based on the following two typical assumptions:

**A1.** $(A, B)$ is stabilizable.
**A2.** $(A, L)$ is observable, with $L = Q^{1/2}$.

**Theorem 8.7** Let Assumptions A1 and A2 hold, then the following algebraic Riccati equation

$$A^T P + PA - PBR^{-1}B^T P + Q = 0$$

has a unique symmetric positive definite solution $P$, and in this case the optimal solution to the aforementioned LQR problem is given, in terms of this positive definite matrix $P$, as follows:

$$u(t) = -R^{-1}B^T Px(t). \tag{8.37}$$

The corresponding closed-loop system is given by

$$\dot{x}(t) = \left(A - BR^{-1}B^T P\right)x(t), \tag{8.38}$$

and the minimum value of the performance index is

$$\gamma = \min_u J(x, u) = x_0^T Px_0. \tag{8.39}$$

## 8.4.2 Relation to $H_2$ Performance

Introduce the following auxiliary system

$$\begin{cases} \dot{x} = Ax + Bu + x_0\omega \\ y = Cx + Du, \end{cases} \tag{8.40}$$

where

$$C = \begin{bmatrix} Q^{\frac{1}{2}} \\ 0 \end{bmatrix}, \quad D = \begin{bmatrix} 0 \\ R^{\frac{1}{2}} \end{bmatrix}, \tag{8.41}$$

and $\omega$ represents an impulse disturbance. With the state feedback controller $u = Kx$ applied to the auxiliary system (8.40), the closed-loop system is obtained as

$$\begin{cases} \dot{x} = (A + BK)x + x_0\omega \\ y = (C + DK)x. \end{cases} \tag{8.42}$$

Thus, the transfer function of the aforementioned system (8.42) from the disturbance $\omega$ to the output $y$ is

$$G_{y\omega}(s) = (C + DK)[sI - (A + BK)]^{-1}x_0. \tag{8.43}$$

The following theorem tells us that the LQR performance for system (8.34) can be reformulated into an $H_2$ performance for the auxiliary system (8.40) and (8.41).

**Theorem 8.8**    Given the linear system (8.34) and the quadratic performance index (8.35) satisfying (8.36) and Assumptions A1 and A2, then

$$J(x, u) = \left\| G_{y\omega}(s) \right\|_2^2, \tag{8.44}$$

where $G_{y\omega}(s)$ is the transfer function given by (8.43).

*Proof*    Define

$$S = \begin{bmatrix} Q^{\frac{1}{2}} \\ R^{\frac{1}{2}}K \end{bmatrix},$$

then it is easy to verify

$$S = C + DK, \tag{8.45}$$

and from (8.35) and $u = Kx$, we obtain

$$J = \int_0^\infty (x^T Qx + u^T Ru)\, dt$$

$$= \int_0^\infty (x^T Qx + x^T K^T RKx)\, dt$$

$$= \int_0^\infty x^T \left(Q + K^T RK\right)x\, dt$$

$$= \int_0^\infty x^T S^T Sx\, dt.$$

Let $X(s)$ denote the Laplace transform of the state vector $x(t)$ of the system (8.42), that is, $X(s) = \mathcal{L}[x(t)]$, then,

$$X(s) = (sI - A - BK)^{-1}x_0. \tag{8.46}$$

According to the definition of $H_2$ norm, we have

$$\|SX(s)\|_2 = \text{trace}\left(\frac{1}{2\pi}\int_{-\infty}^{\infty} SX(j\omega)X^H(j\omega)S^T d\omega\right)^{1/2},$$

meanwhile, according to the definition of $L_2$ norm, we have

$$\|Sx(t)\|_2 = \left(\int_{-\infty}^{\infty} x^T(t)S^T Sx(t)\, dt\right)^{1/2}.$$

Recalling the Parseval lemma yields

$$\|Sx(t)\|_2 = \|SX(s)\|_2.$$

Thus, further using (8.45) and (8.46), gives

$$J = \int_0^{\infty} x^T(t)S^T Sx(t)\, dt$$
$$= \|Sx(t)\|_2^2$$
$$= \|SX(s)\|_2^2$$
$$= \left\|S(sI - A - BK)^{-1}x_0\right\|_2^2$$
$$= \left\|G_{y\omega}(s)\right\|_2^2.$$

Thus, the conclusion holds true. ∎

## 8.4.3 Solution

The importance of the earlier result is that it reveals the fact that the LQR performance index of a system equals to the $H_2$ index of an associated auxiliary system, and with such a fact, an LQR design can be solved via an $H_2$ problem, and hence can be solved with good numerical reliability.

Based on Theorems 5.6 and 8.8, we can obtain the following result.

**Theorem 8.9** Let Assumptions A1 and A2 hold, then a state feedback control in the form of $u = Kx$ exists such that $J(x, u) < \gamma$ if and only if there exist $X \in \mathbb{S}^n$, $Y \in \mathbb{S}^r$ and $W \in \mathbb{R}^{r \times n}$ satisfying

$$(AX + BW) + (AX + BW)^{\mathrm{T}} + x_0 x_0^{\mathrm{T}} < 0, \tag{8.47}$$

$$\mathrm{trace}\left( Q^{\frac{1}{2}} X \left( Q^{\frac{1}{2}} \right)^{\mathrm{T}} \right) + \mathrm{trace}\,(Y) < \gamma, \tag{8.48}$$

$$\begin{bmatrix} -Y & R^{\frac{1}{2}} W \\ \left( R^{\frac{1}{2}} W \right)^{\mathrm{T}} & -X \end{bmatrix} < 0. \tag{8.49}$$

In this case, a feedback gain is given by

$$K = WX^{-1}. \tag{8.50}$$

*Proof* According to Theorem 8.8, $J(x, u) < \gamma$ is equivalent to

$$\left\| (C + DK)\,[sI - (A + BK)]^{-1}\,x_0 \right\|_2^2 < \gamma.$$

Further using Theorem 5.6, we know that the earlier relation is equivalent to the existence of an $X > 0$ satisfying the following two LMIs:

$$(A + BK)\,X + X\,(A + BK)^{\mathrm{T}} + x_0 x_0^{\mathrm{T}} < 0, \tag{8.51}$$

$$\mathrm{trace}\left( \begin{bmatrix} Q^{\frac{1}{2}} \\ R^{\frac{1}{2}} K \end{bmatrix} X \begin{bmatrix} Q^{\frac{1}{2}} \\ R^{\frac{1}{2}} K \end{bmatrix}^{\mathrm{T}} \right) < \gamma. \tag{8.52}$$

Through some deductions, (8.52) can be equivalently converted into

$$\mathrm{trace}\left( Q^{\frac{1}{2}} X \left( Q^{\frac{1}{2}} \right)^{\mathrm{T}} \right) + \mathrm{trace}\left( R^{\frac{1}{2}} KX \left( R^{\frac{1}{2}} K \right)^{\mathrm{T}} \right) < \gamma,$$

which is equivalent to (8.48) and

$$R^{\frac{1}{2}} KX \left( R^{\frac{1}{2}} K \right)^{\mathrm{T}} < Y, \tag{8.53}$$

with $Y$ being a symmetric positive definite matrix.
Letting

$$KX = W,$$

from which we can get the expression (8.50) for the feedback gain, and (8.51) can be converted into (8.47), and (8.53) becomes

$$R^{\frac{1}{2}} WX^{-1} \left( R^{\frac{1}{2}} W \right)^{\mathrm{T}} < Y,$$

which turns to be equivalent to (8.49) according to the Schur complement lemma. ■

It clearly follows from the earlier theorem that Problem 8.5 can be solved via the following optimization problem:

$$\begin{cases} \min & \gamma \\ \text{s.t.} & (8.47) - (8.49), \end{cases} \tag{8.54}$$

the produced optimal $\gamma$ is then the minimum value of the performance index.

**Example 8.5**

Consider the following system

$$\dot{x}(t) = Ax(t) + Bu(t), \quad x(0) = \begin{bmatrix} 1 & 0 \end{bmatrix}^{\mathrm{T}},$$

where

$$A = \begin{bmatrix} 0 & 1 \\ -1 & 1 \end{bmatrix}, \quad B = \begin{bmatrix} 0 \\ 1 \end{bmatrix}.$$

The performance index is in the form of (8.35), with

$$Q = I_2, \quad R = 1.$$

By solving the corresponding optimization problem (8.54), we obtain the optimal parameters as

$$X = \begin{bmatrix} 0.8919 & -0.5000 \\ -0.5000 & 0.4204 \end{bmatrix} > 0,$$

$$W = \begin{bmatrix} 0.9715 & -0.9204 \end{bmatrix},$$

$$Y = 2.0661 > 0,$$

$$\gamma = 3.3784.$$

Then, the optimal state feedback guaranteed cost control law can be computed as

$$K = WX^{-1} = \begin{bmatrix} -0.4142 & -2.6818 \end{bmatrix}. \tag{8.55}$$

On the other hand, applying the MATLAB function `lqr` in the MATLAB optimal toolbox to the system, we obtain the solution of the corresponding Riccati equation as

$$P = \begin{bmatrix} 3.3784 & 0.4142 \\ 0.4142 & 2.6818 \end{bmatrix} > 0,$$

and the state feedback given by (8.37) is computed exactly the same as in (8.55). The corresponding optimal quadratic performance index is then given by

$$J^* = x(0)^{\mathsf{T}} P x(0) = 3.3784,$$

which is indeed equal to the optimal $\gamma$.

## 8.5 Notes and References

### 8.5.1 Summary and References

This chapter studies LMI approaches for $H_\infty/H_2$ state feedback control and also the robust $H_\infty/H_2$ control problems. The LMI conditions for the basic $H_\infty$ and $H_2$ state feedback control problems are given in Table 8.1, while those for the robust $H_\infty$ and $H_2$ state feedback control problems are given in Table 8.2.

In this chapter, the quadratic performance index is also treated in Section 8.4. It is shown that the quadratic optimal regulation problem can be equivalently converted into a standard $H_2$ problem, and hence can be solved via LMI techniques.

**Table 8.1  $H_\infty$ and $H_2$ Control**

| Problems | Criteria | Controller |
|---|---|---|
| $H_\infty$ | $X > 0$ <br><br> $\begin{bmatrix} (AX + B_1 W)_s & * & * \\ B_2^{\mathsf{T}} & -\gamma I & * \\ CX + D_1 W & D_2 & -\gamma I \end{bmatrix} < 0$ | $K = WX^{-1}$ |
| $H_2$ | $\begin{bmatrix} -Z & CX + DW \\ (CX + DW)^{\mathsf{T}} & -X \end{bmatrix} < 0$ <br><br> $AX + B_1 W + (AX + B_1 W)^{\mathsf{T}} + B_2 B_2^{\mathsf{T}} < 0$ <br><br> trace $(Z) < \gamma^2$ | $K = WX^{-1}$ |

**Table 8.2  Robust H$_\infty$ and H$_2$ Control**

| Problems | Criteria | Controller |
|---|---|---|
| H$_\infty$ | $X > 0$<br><br>$$\begin{bmatrix} \Psi(X, W) & * & * & * \\ B_2^T & -\gamma I & * & * \\ CX + D_1 W & D_2 & -\gamma I & * \\ E_1 X + E_2 W & 0 & 0 & -\beta I \end{bmatrix} < 0$$<br><br>$\Psi(X, W) = (AX + B_1 W)_s + \beta HH^T$ | $K = WX^{-1}$ |
| H$_2$ | trace $(Z) < \gamma^2$<br><br>$$\begin{bmatrix} -Z & * \\ (CX + D_1 W)^T & -X \end{bmatrix} < 0$$<br><br>$$\begin{bmatrix} (AX + B_1 W)_s + B_2 B_2^T + \beta HH^T & * \\ E_1 X + E_2 W & -\beta I \end{bmatrix} < 0$$ | $K = WX^{-1}$ |

H$_\infty$/H$_2$ control became popular by the end of the 1990s. In 1989, a celebrated result was proposed by Doyle et al. (1989) for the standard H$_2$ and H$_\infty$ control problems based on algebraic Riccati equations. Later, this algebraic Riccati equation approach to the solution of mixed H$_\infty$/H$_2$ control was further strengthened (Doyle et al. 1994, Zhou et al. 1994). These contributions really laid a milestone for the development of H$_\infty$/H$_2$ control theory and have inspired a great atmosphere in the investigation of H$_\infty$/H$_2$ control theories and applications. Essential advances in the LMI approaches for H$_\infty$/H$_2$ control really occurred after the algebraic Riccati equation approach.

Duan and his coauthors have given LMI approaches for generalized H$_2$ control of linear time-delay systems (Liu and Duan 2006a, Liu et al. 2006). In the literature, H$_\infty$ control has attracted much more attention than H$_2$ control. Earlier results on H$_\infty$ control include Peterson et al. (1991), Gahinet and Apkarian (1993, 1994), Choi and Chung (1997), Bouhtouri et al. (1999), and Kim and Park (1999). In the 2000's, Fridman and Shaked (2002) gave an LMI method for *H*$_\infty$ control of linear time-delay systems, and Zhang and Duan (2005b), Liu et al. (2007), and Zhang et al. (2007, 2008) all studied H$_\infty$ control of discrete-time systems, while Gao et al. (2008) and Seiler and Sengupta (2005) both proposed LMI-based H$_\infty$ approach to networked control. Different from these, very recently Zhou et al.

(2012b) considered a Lyapunov inequality characterization and a Riccati inequality approach to H$_\infty$ and L$_2$ low-gain feedback.

Robust H$_\infty$ control based on LMIs have also been considered by many authors for many types of uncertain systems, for example, mechanical systems (Yang et al. 2010), systems with time-varying delays (Cao et al. 1998), Markovian jump systems (Cao and Lam 2000), stochastic systems (Xu and Chen 2002, Zhang et al. 2005a), T-S fuzzy systems (He and Duan 2006).

## 8.5.2 Dissipative, Passive, and Nonexpansive Control

In Chapter 6 we have discussed in Sections 6.3 through 6.5 the following three concepts concerning the properties of a linear system in the form of (8.1):

1. Dissipativity, which is associated with a supply function

$$s(u, y) = \begin{bmatrix} y \\ u \end{bmatrix}^{\mathrm{H}} Q \begin{bmatrix} y \\ u \end{bmatrix}.$$

2. Passivity, which is equivalent to positive-realness.
3. Nonexpansivity, which is equivalent to bounded-realness.

Corresponding to these three concepts, we can also formulate three types of design problems.

**Problem 8.6**    For the linear system (8.1), design a state feedback controller (8.2) such that the closed-loop system (8.3) is dissipative (passive or nonexpansive) with respect to the input $w(t)$ and output $z(t)$.

It follows from Theorem 6.15 that nonexpansivity is equivalent to the bounded-realness of the system, and hence is, following from Proposition 6.3, equivalent to $\|G(s)\|_\infty < 1$. Therefore, the problem of nonexpansive control is a special H$_\infty$ control problem, which does not need further investigation.

Recall from Theorems 6.12 and 6.14 that a linear system in the form of

$$\begin{cases} \dot{x}(t) = Ax(t) + Bu(t) \\ y(t) = Cx(t) + Du(t), \end{cases} \tag{8.56}$$

is dissipative or passive if and only if there exists a $P > 0$ such that

$$\begin{bmatrix} A^{\mathrm{T}}P + PA & PB \\ B^{\mathrm{T}}P & 0 \end{bmatrix} - \begin{bmatrix} C & D \\ 0 & I \end{bmatrix}^{\mathrm{T}} Q \begin{bmatrix} C & D \\ 0 & I \end{bmatrix} \leq 0, \tag{8.57}$$

or

$$\begin{bmatrix} A^T P + PA & PB - C^T \\ B^T P - C & -D^T - D \end{bmatrix} \leq 0. \tag{8.58}$$

Therefore, following the same treatment as in Section 8.1 for $H_\infty$ control, or the treatment in Sections 7.1, 7.2, and 7.7 for state feedback stabilization, solutions to the problems of dissipative control and passive control can easily be obtained via the LMIs in (8.57) and (8.58), respectively. Readers are strongly recommended to establish the full formulations as well as the solutions for these problems.

For references regarding the problems of dissipative control, one can refer to Qin and Duan (2006), Li et al. (2006a), Li and Duan (2008); while for passive control, Fu and Duan (2004), Li et al. (2006b), Duan and Li (2009), and Liu et al. (2010).

## Exercises

**8.1** Consider the linear system in the form of (8.1) with

$$A = \begin{bmatrix} 0 & 1 & 0 \\ 0 & 0 & 1 \\ 0 & -2 & -3 \end{bmatrix}, \quad B_1 = \begin{bmatrix} 0 \\ 0 \\ 1 \end{bmatrix}, \quad B_2 = \begin{bmatrix} 1 \\ 0 \\ 0 \end{bmatrix},$$

$$C = \begin{bmatrix} 1 & 0 & 0 \end{bmatrix}, \quad D_1 = 1, \quad D_2 = 0.05.$$

Design a state feedback control law $u = Kx$ such that the closed-loop system is stable and the transfer function matrix

$$G_{zw}(s) = (C + D_1 K)(sI - (A + B_1 K))^{-1} B_2 + D_2$$

satisfies

$$\|G_{zw}(s)\|_\infty < \gamma$$

for a minimal $\gamma$.

**8.2** Consider the linear system in the form of (8.10) with

$$A_1 = \begin{bmatrix} -5 & 1 & 0 \\ 0 & 1 & 1 \\ 1 & 1 & 1 \end{bmatrix}, \quad B_1 = \begin{bmatrix} 0 & 0 \\ 0 & 1 \\ 1 & 0 \end{bmatrix}, \quad B_2 = \begin{bmatrix} 0.05 \\ 0 \\ 0.03 \end{bmatrix},$$

$$C = \begin{bmatrix} 1 & 0 & 0 \\ 0 & 2 & 1 \end{bmatrix}, \quad D = \begin{bmatrix} 1 & 3 \\ 1 & 0 \end{bmatrix}.$$

Design a state feedback control law $u = Kx$ such that the closed-loop system is stable and the transfer function matrix

$$G_{zw}(s) = (C + DK)(sI - (A + B_1 K))^{-1} B_2$$

satisfies

$$\|G_{zw}(s)\|_2 < \gamma$$

for a minimal $\gamma$.

**8.3** Consider a linear system in the form of (8.18) with the following parameters

$$A = \begin{bmatrix} 1.95 & 0.78 \\ 0.76 & 1.87 \end{bmatrix}, \quad B_1 = \begin{bmatrix} 1.50 \\ 0.45 \end{bmatrix}, \quad B_2 = \begin{bmatrix} 0.01 \\ 0.03 \end{bmatrix},$$

$$C = \begin{bmatrix} 0.95 & 0.78 \end{bmatrix}, \quad D_1 = 0.50, \quad D_2 = 0.02,$$

$$\begin{bmatrix} \Delta A & \Delta B_1 \end{bmatrix} = HF \begin{bmatrix} E_1 & E_2 \end{bmatrix}, \quad F^T F \leq I,$$

$$H = \begin{bmatrix} 1 & 0 \\ 0 & 1 \end{bmatrix}, \quad E_1 = \begin{bmatrix} 0.04 & 0.02 \\ -0.01 & 0.01 \end{bmatrix}, \quad E_2 = \begin{bmatrix} 0.03 \\ 0.01 \end{bmatrix}.$$

Design a state feedback control law $u = Kx$ such that the closed-loop system is asymptotically stable and the transfer function matrix

$$G_{zw}(s) = (C + D_1 K)(sI - [(A + \Delta A) + (B_1 + \Delta B_1) K])^{-1} B_2 + D_2$$

satisfies

$$\|G_{zw}(s)\|_\infty < \gamma$$

for a minimal $\gamma$.

**8.4** Consider a linear system in the form of (8.18) with the following parameters

$$A = \begin{bmatrix} 1.7465 & -3.2684 & 6.2368 \\ 3.0268 & 9.4581 & 2.3659 \\ 4.3658 & 3.2478 & -2.3724 \end{bmatrix},$$

$$B_1 = \begin{bmatrix} 0.2369 & 1.4533 \\ 0.2361 & -3.4874 \\ 1.3687 & -5.2365 \end{bmatrix}, \quad B_2 = \begin{bmatrix} 0.0003 \\ 0.0036 \\ 0.1578 \end{bmatrix},$$

$$C = \begin{bmatrix} 0.7635 & 0.3547 & 5.2654 \\ 1.7566 & -0.5348 & 1.6366 \end{bmatrix},$$

$$D_1 = \begin{bmatrix} 1.0024 & 0.7853 \\ 0.1365 & 1.3959 \end{bmatrix}, \quad D_2 = 0,$$

$$\begin{bmatrix} \Delta A & \Delta B_1 \end{bmatrix} = HF \begin{bmatrix} E_1 & E_2 \end{bmatrix}, \quad F^\mathsf{T} F \le I,$$

$$E_1 = \begin{bmatrix} 0.0473 & -0.1401 & 0.4006 \\ 0.1227 & -0.0006 & 0.1456 \\ 0.1440 & -0.0811 & 0.2366 \end{bmatrix},$$

$$E_2 = \begin{bmatrix} 0.5478 & 0.4537 \\ 0.0237 & 0.1014 \\ 0.0124 & 0.3613 \end{bmatrix}, \quad H = \begin{bmatrix} 1 & 0 & 0 \\ 0 & 1 & 0 \\ 0 & 0 & 1 \end{bmatrix}.$$

Design a state feedback control law $u = Kx$ such that the closed-loop system is asymptotically stable and the transfer function matrix

$$G_{zw}(s) = (C + D_1 K)(sI - [(A + \Delta A) + (B_1 + \Delta B_1)K])^{-1} B_2$$

satisfies

$$\| G_{zw}(s) \|_2 < \gamma$$

for a minimal $\gamma$.

**8.5** Consider a linear system in the form of

$$\begin{cases} \dot{x} = Ax + B_1 u + B_2 w \\ y = Cx + D_1 u + D_2 w, \end{cases}$$

where $w$ is a measurable disturbance modeled by

$$\dot{w} = Fw + Dv,$$

with $v$ being a driving disturbance. Give the solution to the design of a feedback controller in the form of

$$u = K_1 x + K_2 w,$$

such that $\| G_{yv}(s) \|_\infty$ and $\| G_{yv}(s) \|_2$ are minimized.

**8.6** Solve the problem in Example 8.1 using Theorem 8.2 and compare the obtained results with those given in the example.

**8.7** Solve the problem in Example 8.2 using Theorem 8.4 and compare the obtained results with those given in the example.

# Chapter 9

# State Observation and Filtering

State observation in a control system is a very important problem and has applications in many problems, such as state feedback control, fault detection and isolation, etc. In this chapter, basic LMI approaches for the design of state observers of linear systems are presented. The full- and reduced-order state observers, $H_\infty/H_2$ full-order state observers, and $H_\infty/H_2$ state filtering are covered.

## 9.1 Full- and Reduced-Order State Observers

In this section, we examine the problem of designing state observers for the following system

$$\begin{cases} \dot{x} = Ax + Bu \\ y = Cx, \end{cases} \tag{9.1}$$

where $x \in \mathbb{R}^n$, $u \in \mathbb{R}^r$, and $y \in \mathbb{R}^m$ are the state vector, the input vector, and the output vector, respectively. Without loss of generality, it is assumed that $\mathrm{rank}(C) = m \leq n$.

## 9.1.1 Full-Order State Observers

For the system (9.1), a full-order state observer takes the following form

$$\dot{\hat{x}} = A\hat{x} + Bu + L(C\hat{x} - y), \tag{9.2}$$

where
$\hat{x}$ is the state observation vector
$L \in \mathbb{R}^{n \times m}$ is the observer gain

Denote

$$e = x - \hat{x},$$

then we obtain from (9.1) and (9.2) the following observation error equation:

$$\dot{e} = (A + LC) e. \tag{9.3}$$

The problem of full-order state observers design for a continuous-time linear system in the form of (9.1) can be stated as follows.

**Problem 9.1**    For the continuous-time linear system (9.1), design a full-order state observer in the form of (9.2) such that the error system (9.3) is asymptotically stable. As a consequence, for arbitrary control input $u(t)$ and arbitrary system initial values $x(0)$ and $\hat{x}(0)$ the observation error approaches to zero, that is,

$$\lim_{t \to \infty} \left( x(t) - \hat{x}(t) \right) = 0.$$

The aforementioned full-order observer design problem is obviously a dual problem of the state feedback stabilization problem formulated in Section 7.1.1. Thus for solution to this problem, we immediately have the following result.

**Theorem 9.1**    For the full-order state observers problem (Problem 9.1) the following conclusions hold.

1. It has a solution if and only if there exist a symmetric positive definite matrix $P$ and a matrix $W$ satisfying

$$PA + A^{\mathrm{T}}P + WC + C^{\mathrm{T}}W^{\mathrm{T}} < 0. \tag{9.4}$$

   In this case, a solution to the problem is given by

$$L = P^{-1}W. \tag{9.5}$$

2. It has a solution if and only if there exists a symmetric positive definite matrix $P$ satisfying

$$PA + A^\mathrm{T}P - C^\mathrm{T}C < 0. \tag{9.6}$$

In this case, a solution to the problem is given by

$$L = -\frac{1}{2}P^{-1}C^\mathrm{T}. \tag{9.7}$$

*Proof*  Note that

$$(A + LC)^\mathrm{T} = \tilde{A} + \tilde{B}K,$$

where

$$\tilde{A} = A^\mathrm{T}, \quad \tilde{B} = C^\mathrm{T}, \quad K = L^\mathrm{T}.$$

With this dual relationship between state feedback stabilization and full-order state observer design, the conclusions directly follow from Theorem 6.5.  ■

### Example 9.1

Consider the benchmark system in Example 7.1 again. Applying the MATLAB® function `feasp` in the MATLAB LMI toolbox to the full-order observer design condition (9.4), we get the parameters $P$ ($> 0$) and $W$. Based on the obtained parameter matrices $W$ and $P$, the observer gain is obtained by

$$L = \begin{bmatrix} l_1 & l_2 & \cdots & l_{20} \end{bmatrix}^\mathrm{T} = WP^{-1}.$$

Entries in the gain matrix $L$ are given in Table 9.1. By using the MATLAB function `eig`, the set of observer eigenvalues are obtained as

$$\begin{aligned}
\lambda(A + LC) = \{ &-16.4953 \pm 67.5403i, -6.3405 \pm 39.2256i, \\
&-3.4178 \pm 27.6323i, -2.1414 \pm 21.3719i, \\
&-1.7870 \pm 17.3286i, -2.0263 \pm 13.5890i, \\
&-2.4872 \pm 9.5010i, -3.0999 \pm 5.1472i, \\
&-157.7079 \pm 275.6389i, -3.8856, -6.0092 \}.
\end{aligned}$$

To verify the result, we have plotted

$$\epsilon(k) = \|x(k) - \hat{x}(k)\|_2$$

in Figure 9.1. It is seen from this figure that the observation gives a very high precision in about 2 s.

Table 9.1  Entries of Gain Matrix *L*

| *j* | $l_{3j+1}$ | $l_{3j+2}$ | $l_{3j+3}$ |
|---|---|---|---|
| 0 | −36096254.4923 | −278355214.5243 | −1045326768.8413 |
| 1 | −2542948598.5469 | −4496268739.0632 | −6144778948.3535 |
| 2 | −6737706111.3431 | −6072176176.1721 | −4569771871.4278 |
| 3 | −2901039846.7673 | −1562328508.0585 | −715017855.2874 |
| 4 | −277529403.6193 | −90794608.7237 | −24753951.0927 |
| 5 | −5521345.2195 | −978540.6858 | −131136.0375 |
| 6 | −12303.7153 | −610.9012 | |

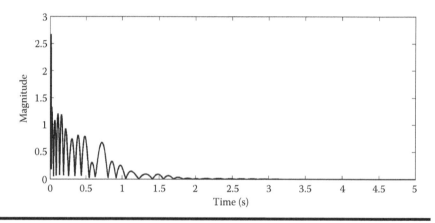

**Figure 9.1  Magnitude of the error vector, Example 9.1.**

## 9.1.2 Reduced-Order State Observer Design

The previous section gives an LMI approach for the design of full-order state observers for linear systems. In this section, we further investigate the problem of reduced-order observer design for the linear system (9.1).

### 9.1.2.1 Problem Formulation

In the design of reduced-order state observers for linear systems in the form of (9.1), the following lemma performs a fundamental role. For a proof, readers may refer to Duan (2004a).

**Lemma 9.1**  Given the linear system (9.1), and let $R \in \mathbb{R}^{(n-m) \times n}$ be an arbitrarily chosen matrix which makes the matrix

$$T = \begin{bmatrix} C \\ R \end{bmatrix}$$

nonsingular, then

$$CT^{-1} = \begin{bmatrix} I_m & 0 \end{bmatrix}. \tag{9.8}$$

Furthermore, let

$$TAT^{-1} = \begin{bmatrix} A_{11} & A_{12} \\ A_{21} & A_{22} \end{bmatrix}, \quad A_{11} \in \mathbb{R}^{m \times m}, \tag{9.9}$$

then the matrix pair $(A_{22}, A_{12})$ is detectable if and only if $(A, C)$ is detectable.

Let

$$Tx = \begin{bmatrix} x_1 \\ x_2 \end{bmatrix}, \quad TB = \begin{bmatrix} B_1 \\ B_2 \end{bmatrix}, \tag{9.10}$$

then it follows from the relations in (9.8) through (9.10) that system (9.1) is equivalent to

$$\begin{cases} \begin{bmatrix} \dot{x}_1 \\ \dot{x}_2 \end{bmatrix} = \begin{bmatrix} A_{11} & A_{12} \\ A_{21} & A_{22} \end{bmatrix} \begin{bmatrix} x_1 \\ x_2 \end{bmatrix} + \begin{bmatrix} B_1 \\ B_2 \end{bmatrix} u \\ y = x_1. \end{cases} \tag{9.11}$$

In the equivalent system (9.11), the substate vector $x_1$ is directly equal to the output $y$ of the original system (9.1). Thus to reconstruct the state of the system (9.1), we suffice to get an estimate of the substate vector $x_2$, namely, $\hat{x}_2$, from the earlier equivalent system (9.11). Once an estimate $\hat{x}_2$ is obtained, an estimate of $x(t)$, that is, the state vector of system (9.1), can be obtained as

$$\hat{x}(t) = T^{-1} \begin{bmatrix} y(t) \\ \hat{x}_2(t) \end{bmatrix}.$$

This brings about the problem of reduced-order state observer design.

**Problem 9.2**  For the continuous-time linear system (9.11), design a reduced-order state observer in the form of

$$\begin{cases} \dot{z} = Fz + Gy + Hu \\ \hat{x}_2 = Mz + Ny, \end{cases} \tag{9.12}$$

such that for arbitrary control input $u(t)$, and arbitrary initial system values $x_1(0)$, $x_2(0)$, and $z(0)$, there holds

$$\lim_{t \to \infty} \left( x_2(t) - \hat{x}_2(t) \right) = 0.$$

### 9.1.2.2 Solution

For solution to the aforementioned reduced-order state observer design problem, we have the following result.

**Theorem 9.2**    Problem 9.2 has a solution if and only if one of the following two conditions holds:

1. There exist a symmetric positive definite matrix $P$ and a matrix $W$ satisfying

$$PA_{22} + A_{22}^{\mathrm{T}}P + WA_{12} + A_{12}^{\mathrm{T}}W^{\mathrm{T}} < 0. \qquad (9.13)$$

2. There exists a symmetric positive definite matrix $P$ satisfying

$$PA_{22} + A_{22}^{\mathrm{T}}P - A_{12}^{\mathrm{T}}A_{12} < 0. \qquad (9.14)$$

In this case, a reduced-order state observer can be obtained as in (9.12) with

$$F = A_{22} + LA_{12}, \quad G = (A_{21} + LA_{11}) - (A_{22} + LA_{12})\,L, \qquad (9.15)$$

$$H = B_2 + LB_1, \quad M = I, \quad N = -L, \qquad (9.16)$$

where

$$L = P^{-1}W,$$

with $W$ and $P > 0$ being a pair of feasible solutions to the inequality (9.13) or

$$L = -\frac{1}{2}P^{-1}A_{12}^{\mathrm{T}},$$

with $P > 0$ being a solution to the inequality (9.14).

*Proof*    From the equivalent system (9.11), we obtain the state and output equations of the subsystem associated with substate $x_2$ as

$$\begin{cases} \dot{x}_2 = A_{22}x_2 + (A_{21}y + B_2u) \\ \dot{y} - A_{11}y - B_1u = A_{12}x_2. \end{cases} \qquad (9.17)$$

Let

$$v = A_{21}y + B_2u, \quad w = \dot{y} - A_{11}y - B_1u, \tag{9.18}$$

then (9.17) can be rewritten as

$$\begin{cases} \dot{x}_2 = A_{22}x_2 + v \\ w = A_{12}x_2. \end{cases} \tag{9.19}$$

It follows from Theorem 9.1 that the system (9.19) has a full-order state observer of the form

$$\dot{\hat{x}}_2 = (A_{22} + LA_{12})\,\hat{x}_2 - Lw + v \tag{9.20}$$

if and only if one of the two conditions in the theorem holds, and the observer gain matrix $L$ can be given by $L = P^{-1}W$, with $W$ and $P > 0$ being a pair of feasible solutions to the inequality (9.13), or given by $L = -\frac{1}{2}P^{-1}A_{12}^{\mathrm{T}}$, with $P > 0$ being a solution to the inequality (9.14). Therefore, in the following we need only to show that (9.20) can be turned into the form of (9.12).

Substituting (9.18) into (9.20) yields

$$\dot{\hat{x}}_2 = (A_{22} + LA_{12})\,\hat{x}_2 - L\left(\dot{y} - A_{11}y - B_1u\right) + A_{21}y + B_2u. \tag{9.21}$$

To eliminate $\dot{y}$ in the observer, we define a new observer state vector as

$$z = \hat{x}_2 + Ly.$$

Thus, using (9.21), we have

$$\begin{aligned}
\dot{z} &= \dot{\hat{x}}_2 + L\dot{y} \\
&= (A_{22} + LA_{12})\,\hat{x}_2 + L\left(A_{11}y + B_1u\right) + A_{21}y + B_2u \\
&= (A_{22} + LA_{12})\left(z - Ly\right) + L\left(A_{11}y + B_1u\right) + A_{21}y + B_2u \\
&= (A_{22} + LA_{12})\,z + [(A_{21} + LA_{11}) - (A_{22} + LA_{12})\,L]\,y + (B_2 + LB_1)\,u.
\end{aligned}$$

Therefore, a reduced-order state observer in the new state vector $z$ is obtained as

$$\begin{cases} \dot{z} = (A_{22} + LA_{12})\,z + [(A_{21} + LA_{11}) - (A_{22} + LA_{12})\,L]\,y + (B_2 + LB_1)\,u \\ \hat{x}_2 = z - Ly. \end{cases}$$

This is in the form of (9.12) with the coefficients given by (9.15) and (9.16). The proof is then completed. ∎

**Example 9.2**

Consider the following linear system, which has been studied by Duan (2003):

$$
\begin{cases}
\dot{x}(t) = \begin{bmatrix} -0.5 & 0 & 0 \\ 0 & -2 & 10 \\ 0 & 1 & -2 \end{bmatrix} x(t) + \begin{bmatrix} 1 & 0 \\ -2 & 2 \\ 0 & 1 \end{bmatrix} u(t) \\
y = \begin{bmatrix} 1 & 0 & 0 \\ 0 & 0 & 1 \end{bmatrix} x.
\end{cases}
$$

Choosing

$$
T = \begin{bmatrix} 1 & 0 & 0 \\ 0 & 0 & 1 \\ 0 & 1 & 0 \end{bmatrix},
$$

we can easily obtain, according to (9.9) and (9.10),

$$
A_{11} = \begin{bmatrix} -0.5 & 0 \\ 0 & -2 \end{bmatrix}, \quad A_{12} = \begin{bmatrix} 0 \\ 1 \end{bmatrix},
$$

$$
A_{21} = \begin{bmatrix} 0 & 10 \end{bmatrix}, \quad A_{22} = -2,
$$

and

$$
B_1 = I_2, \quad B_2 = \begin{bmatrix} -2 & 2 \end{bmatrix}.
$$

Applying the MATLAB function feasp in the MATLAB LMI toolbox to the reduced-order observer design condition (9.13), we get the parameters

$$
P = 4.8024 \times 10^8 > 0, \quad W = 10^8 \times \begin{bmatrix} 0 & 7.2037 \end{bmatrix},
$$

hence

$$
L = P^{-1}W = \begin{bmatrix} 0 & 1.5000 \end{bmatrix}.
$$

Thus, the coefficient matrices of the reduced-order observer can be obtained according to (9.15) and (9.16) as

$$
F = -0.5000, \quad G = \begin{bmatrix} 0 & 7.7500 \end{bmatrix}, \quad M = 1,
$$
$$
H = \begin{bmatrix} -2.0000 & 3.5000 \end{bmatrix}, \quad N = \begin{bmatrix} 0 & -1.5000 \end{bmatrix}.
$$

To demonstrate the state observation effect, the observation error $e = x_2 - \hat{x}_2$ is shown in Figure 9.2.

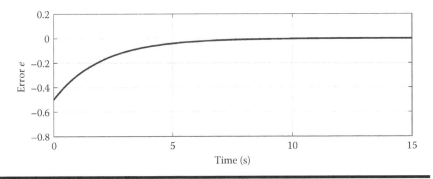

**Figure 9.2  The observation error, Example 9.2.**

## 9.2 Full-Order $H_\infty/H_2$ State Observers

In this section, we study linear systems of the following form:

$$\begin{cases} \dot{x}(t) = Ax(t) + B_1 u(t) + B_2 w(t), \ x(0) = x_0 \\ y(t) = C_1 x(t) + D_1 u(t) + D_2 w(t) \\ z(t) = C_2 x(t), \end{cases} \tag{9.22}$$

where

- $x \in \mathbb{R}^n$, $y \in \mathbb{R}^l$, $z \in \mathbb{R}^m$ are respectively the state vector, the measured output vector, and the output vector of interests
- $w \in \mathbb{R}^p$ and $u \in \mathbb{R}^r$ are the disturbance vector and the control vector, respectively
- $A$, $B_1$, $B_2$, $C_1$, $C_2$, $D_1$, and $D_2$ are the system coefficient matrices of appropriate dimensions

In this section, we treat the problem of designing a full-order state observer for system (9.22) such that the effect of the disturbance $w(t)$ to the estimate error is prohibited to a desired level.

### 9.2.1 Problem Formulation

For the system (9.22), we introduce a full-order state observer in the following form:

$$\dot{\hat{x}} = (A + LC_1)\hat{x} - Ly + (B_1 + LD_1)u \tag{9.23}$$

where $\hat{x}$ is the state observation vector and $L \in \mathbb{R}^{n \times m}$ is the observer gain. Obviously, the estimate of the interested output is given by

$$\hat{z}(t) = C_2 \hat{x}(t),$$

which is desired to have as little affection as possible from the disturbance $w(t)$.

Using (9.22), we have

$$\dot{x}(t) = Ax(t) + B_1 u(t) + B_2 w(t)$$
$$= Ax(t) + Ly - Ly + B_1 u(t) + B_2 w(t)$$
$$= (A + LC_1) x(t) - Ly$$
$$+ (B_1 + LD_1) u(t) + (B_2 + LD_2) w(t). \tag{9.24}$$

Denoting

$$\begin{cases} e(t) = x(t) - \hat{x}(t) \\ \tilde{z}(t) = z(t) - \hat{z}(t), \end{cases}$$

we obtain from (9.23) and (9.24) the following observation error equation

$$\begin{cases} \dot{e} = (A + LC_1) e + (B_2 + LD_2) w \\ \tilde{z}(t) = C_2 e. \end{cases} \tag{9.25}$$

The transfer function of the system is clearly given by

$$G_{\tilde{z}w}(s) = C_2 (sI - A - LC_1)^{-1} (B_2 + LD_2).$$

With the aforementioned preparation, the problems of $H_\infty$ and $H_2$ state observer designs can be stated as follows:

**Problem 9.3** ($H_\infty$ state observers) Given system (9.22) and a positive scalar $\gamma$, find a matrix $L$ such that

$$||G_{\tilde{z}w}(s)||_\infty < \gamma. \tag{9.26}$$

**Problem 9.4** ($H_2$ state observers) Given system (9.22) and a positive scalar $\gamma$, find a matrix $L$ such that

$$||G_{\tilde{z}w}(s)||_2 < \gamma. \tag{9.27}$$

As a consequence of the requirements (9.26) and (9.27), the error system (9.25) is asymptotically stable, and hence we have

$$e(t) = x(t) - \hat{x}(t) \longrightarrow 0, \text{ as } t \longrightarrow \infty.$$

This states that $\hat{x}(t)$ is an asymptotic estimate of $x(t)$.

## 9.2.2 Solutions to Problems

### 9.2.2.1 $H_\infty$ State Observer Design

Regarding the solution to the problem of $H_\infty$ state observers design, we have the following theorem.

**Theorem 9.3** The $H_\infty$ state observers problem (Problem 9.3) has a solution if and only if there exist a matrix $W$ and a symmetric positive definite matrix $P$ such that

$$\begin{bmatrix} A^\mathrm{T} P + C_1^\mathrm{T} W^\mathrm{T} + PA + WC_1 & PB_2 + WD_2 & C_2^\mathrm{T} \\ (PB_2 + WD_2)^\mathrm{T} & -\gamma I & 0 \\ C_2 & 0 & -\gamma I \end{bmatrix} < 0. \tag{9.28}$$

When such a pair of matrices $W$ and $P$ are found, a solution to the problem is given as

$$L = P^{-1} W. \tag{9.29}$$

*Proof* It follows from Theorem 5.3 that the problem has a solution if and only if there exists a symmetric positive definite matrix $P$ such that

$$\begin{bmatrix} (A + LC_1)^\mathrm{T} P + P(A + LC_1) & P(B_2 + LD_2) & C_2^\mathrm{T} \\ (B_2 + LD_2)^\mathrm{T} P & -\gamma I & 0 \\ C_2 & 0 & -\gamma I \end{bmatrix} < 0. \tag{9.30}$$

By defining

$$W = PL,$$

the inequality (9.30) is equivalently turned into the LMI form (9.28), and the gain matrix $L$ is obviously solved from the aforementioned relation as in (9.29). ▪

With a prescribed attenuation level, the problem of $H_\infty$ state observers design is turned into an LMI feasibility problem in the form of (9.28). The problem with a minimal attenuation level $\gamma$ can be sought via the following optimization problem:

$$\begin{cases} \min \quad \gamma \\ \text{s.t.} \quad P > 0 \\ \qquad \begin{bmatrix} A^\mathrm{T} P + C_1^\mathrm{T} W^\mathrm{T} + PA + WC_1 & PB_2 + WD_2 & C_2^\mathrm{T} \\ (PB_2 + WD_2)^\mathrm{T} & -\gamma I & 0 \\ C_2 & 0 & -\gamma I \end{bmatrix} < 0. \end{cases} \tag{9.31}$$

### 9.2.2.2 $H_2$ State Observer Design

Regarding the solution to the problem of $H_2$ state observers design, we have the following theorem.

**Theorem 9.4**  For the $H_2$ state observers problem (Problem 9.4) the following two conclusions hold.

1. It has a solution if and only if there exist a matrix $W$, a symmetric matrix $Q$, and a symmetric matrix $X$ such that

$$\begin{cases} \begin{bmatrix} XA + WC_1 + (XA + WC_1)^{\mathrm{T}} & XB_2 + WD_2 \\ (XB_2 + WD_2)^{\mathrm{T}} & -I \end{bmatrix} < 0 \\ \begin{bmatrix} -Q & C_2 \\ C_2^{\mathrm{T}} & -X \end{bmatrix} < 0 \\ \operatorname{trace}(Q) < \gamma^2. \end{cases} \tag{9.32}$$

When such a triple of matrices are obtained, a solution to the problem is given as

$$L = X^{-1}W.$$

2. It has a solution if and only if there exist a matrix $V$, a symmetric matrix $Z$, and a symmetric matrix $Y$ such that

$$\begin{cases} A^{\mathrm{T}}Y + C_1^{\mathrm{T}}V^{\mathrm{T}} + YA + VC_1 + C_2^{\mathrm{T}}C_2 < 0 \\ \begin{bmatrix} -Z & (YB_2 + VD_2)^{\mathrm{T}} \\ YB_2 + VD_2 & -Y \end{bmatrix} < 0 \\ \operatorname{trace}(Z) < \gamma^2. \end{cases} \tag{9.33}$$

When such a triple of matrices are obtained, a solution to the problem can be given as

$$L = Y^{-1}V.$$

*Proof*  It follows from Theorem 5.6 that Problem 9.4 has a solution if and only if there exists a symmetric positive definite matrix $P$ such that

$$\begin{cases} (A + LC_1)P + P(A + LC_1)^{\mathrm{T}} + (B_2 + LD_2)(B_2 + LD_2)^{\mathrm{T}} < 0 \\ \operatorname{trace}(C_2PC_2^{\mathrm{T}}) < \gamma^2. \end{cases} \tag{9.34}$$

Pre- and postmultiplying both sides of the first inequality in (9.34) by $P^{-1}$ yields

$$P^{-1}(A + LC_1) + (A + LC_1)^{\mathrm{T}} P^{-1} + P^{-1}(B_2 + LD_2)(B_2 + LD_2)^{\mathrm{T}} P^{-1} < 0, \tag{9.35}$$

which can be converted into the following form according to the Schur complement lemma:

$$\begin{bmatrix} P^{-1}(A + LC_1) + (A + LC_1)^{\mathrm{T}} P^{-1} & P^{-1}(B_2 + LD_2) \\ (B_2 + LD_2)^{\mathrm{T}} P^{-1} & -I \end{bmatrix} < 0. \tag{9.36}$$

Let

$$X = P^{-1}, \quad W = XL,$$

then we can rewrite (9.36) as

$$\begin{bmatrix} XA + WC_1 + (XA + WC_1)^{\mathrm{T}} & XB_2 + WD_2 \\ (XB_2 + WD_2)^{\mathrm{T}} & -I \end{bmatrix} < 0. \tag{9.37}$$

Therefore, by now we have shown that the first inequality in (9.34) holds for some $P > 0$ if and only if (9.37) holds for some $W$ and $X > 0$.

Now let us treat the second equation in (9.34). Using Lemma 2.13, the second inequality in (9.34) can be equivalently converted into

$$C_2 P C_2^{\mathrm{T}} - Q < 0, \ \operatorname{trace}(Q) < \gamma^2,$$

which is clearly equivalent to

$$\begin{bmatrix} -Q & C_2 \\ C_2^{\mathrm{T}} & -P^{-1} \end{bmatrix} < 0, \ \operatorname{trace}(Q) < \gamma^2. \tag{9.38}$$

Substituting the relation $X = P^{-1}$ into the first inequality in (9.38) gives

$$\begin{bmatrix} -Q & C_2 \\ C_2^{\mathrm{T}} & -X \end{bmatrix} < 0. \tag{9.39}$$

Thus by now we have shown that the second condition in (9.34) holds for some $P > 0$ if and only if $\operatorname{trace}(Q) < \gamma^2$ and (9.39) holds for some symmetric matrix $Q$ and an $X > 0$.

With the aforementioned two aspects, the proof of first conclusion is completed. The second one can be shown similarly. ■

The aforementioned theorem again converts the problem of $H_2$ state observer design into an LMI feasibility problem, namely, (9.32) or (9.33), which are solvable using the LMI toolbox command `feasp`. In applications, we are often concerned with the problem of finding the minimal attenuation level $\gamma$. This problem can be solved via the optimization

$$
\begin{cases}
\min \quad \rho \\
\text{s.t.} \quad \begin{bmatrix} XA + WC_1 + (XA + WC_1)^{\mathrm{T}} & XB_2 + WD_2 \\ (XB_2 + WD_2)^{\mathrm{T}} & -I \end{bmatrix} < 0 \\
\qquad \begin{bmatrix} -Q & C_2 \\ C_2^{\mathrm{T}} & -X \end{bmatrix} < 0 \\
\qquad \text{trace}\,(Q) < \rho,
\end{cases}
\tag{9.40}
$$

or

$$
\begin{cases}
\min \quad \rho \\
\text{s.t.} \quad A^{\mathrm{T}}Y + C_1^{\mathrm{T}}V^{\mathrm{T}} + YA + VC_1 + C_2^{\mathrm{T}}C_2 < 0 \\
\qquad \begin{bmatrix} -Z & (YB_2 + VD_2)^{\mathrm{T}} \\ YB_2 + VD_2 & -Y \end{bmatrix} < 0 \\
\qquad \text{trace}\,(Z) < \rho.
\end{cases}
\tag{9.41}
$$

When a minimal $\rho$ is obtained, the minimal attenuation level is $\gamma = \sqrt{\rho}$.

## 9.2.3 Examples

### Example 9.3

Consider a linear system in the form of (9.22) with the following parameters:

$$
A = \begin{bmatrix} 2.8982 & -3.1606 & 0.6816 \\ 6.3595 & -4.2055 & 4.5423 \\ 3.2046 & -3.1761 & -3.8142 \end{bmatrix},
$$

$$
B_1 = \begin{bmatrix} 2.0310 & 1.2164 \\ 0.4084 & 0.2794 \\ -0.7775 & -0.3307 \end{bmatrix}, \quad B_2 = \begin{bmatrix} -0.0785 \\ -0.0853 \\ -0.0986 \end{bmatrix},
$$

$$
C_1 = \begin{bmatrix} -0.8778 & -4.9442 & -4.5084 \\ 4.0161 & -2.0259 & 1.9318 \end{bmatrix},
$$

$$
C_2 = \begin{bmatrix} 0.9607 & 1.5600 & 2.8558 \\ -2.4371 & 1.3634 & 0.0095 \end{bmatrix},
$$

$$
D_1 = \begin{bmatrix} 0.6004 & 0.2107 \\ 1.9320 & -0.3997 \end{bmatrix}, \quad D_2 = \begin{bmatrix} 0.0330 \\ -0.0414 \end{bmatrix}.
$$

Applying the function `mincx` in the MATLAB LMI toolbox to the full-order $H_\infty$ optimal state observer condition (9.31), we get the optimal parameters as

$$P = 10^8 \times \begin{bmatrix} 4.5460 & -1.3537 & 0.3240 \\ -1.3537 & 2.0404 & 0.0781 \\ 0.3240 & 0.0781 & 1.4478 \end{bmatrix} > 0,$$

$$W = 10^8 \times \begin{bmatrix} -0.2425 & -6.7957 \\ 2.0709 & -0.1726 \\ 4.3510 & -0.7553 \end{bmatrix},$$

$$\gamma = 1.6072 \times 10^{-9},$$

and the corresponding observer gain is computed as

$$L = P^{-1}W = \begin{bmatrix} 0.0057 & -1.8916 \\ 0.9056 & -1.3385 \\ 2.9550 & -0.0262 \end{bmatrix}.$$

Note that the attenuation level $\gamma$ is very small, the disturbance $w$ must be effectively attenuated. To verify this, we have plotted the observation error $\xi$ in Figure 9.3, for the case of $w(t) =$wgn(1, 101, 100', real'), which generates a 1-by-101 matrix of real white noise with power of 100 dBW. It is seen from this figure that the disturbance indeed has very little affection to the observation error.

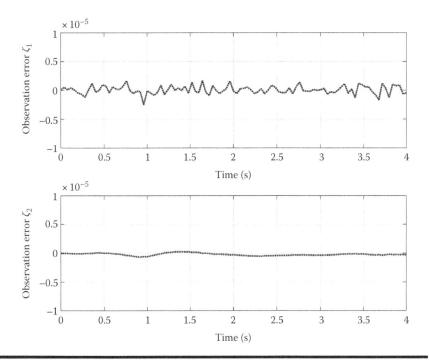

**Figure 9.3   The observation error, Example 9.3.**

**Example 9.4**

Consider the linear system in the form of (9.22) with the following parameters:

$$A = \begin{bmatrix} -3.2216 & 2.8762 & 0.6450 \\ -7.0385 & -3.0859 & -2.1610 \\ 9.1159 & 4.5239 & -6.3258 \end{bmatrix},$$

$$B_1 = \begin{bmatrix} 2.6106 & 2.1274 \\ 0.8399 & 2.0541 \\ 0.2641 & -1.8041 \end{bmatrix}, \; B_2 = \begin{bmatrix} 0.1198 \\ -0.0304 \\ -0.0101 \end{bmatrix},$$

$$C_1 = \begin{bmatrix} 3.0453 & 2.1022 & 1.9242 \\ 4.4915 & 1.7021 & -0.6568 \end{bmatrix},$$

$$C_2 = \begin{bmatrix} 1.9683 & -1.4475 & -2.2344 \\ -0.1943 & 0.6912 & -1.3730 \end{bmatrix},$$

$$D_1 = \begin{bmatrix} -1.6956 & 1.7291 \\ 0.5102 & 1.1855 \end{bmatrix}, \; D_2 = \begin{bmatrix} 0.0677 \\ -0.0154 \end{bmatrix}.$$

Applying the function $\mathtt{mincx}$ in the MATLAB LMI toolbox to the full-order $H_2$ optimal state observer condition (9.40) gives the following optimal parameters:

$$X = 10^8 \times \begin{bmatrix} 2.6341 & -1.2013 & -1.2783 \\ -1.2013 & 3.1159 & 0.2705 \\ -1.2783 & 0.2705 & 5.3733 \end{bmatrix} > 0,$$

$$W = 10^8 \times \begin{bmatrix} -5.3168 & 0.3068 \\ 3.2888 & -1.2212 \\ 2.8040 & -1.6540 \end{bmatrix},$$

$$Q = 10^{-7} \times \begin{bmatrix} 0.1973 & 0.0124 \\ 0.0124 & 0.0569 \end{bmatrix} > 0,$$

$$\rho = 2.5416 \times 10^{-8},$$

and the corresponding observer gain is computed as

$$L = X^{-1}W = \begin{bmatrix} -1.8272 & -0.2633 \\ 0.3450 & -0.4633 \\ 0.0698 & -0.3471 \end{bmatrix},$$

while the optimal attenuation level is $\gamma = \sqrt{\rho} = 1.5942 \times 10^{-4}$. Note that the attenuation level $\gamma$ is very small, the disturbance $w$ must be effectively attenuated. To verify this, we have plotted the observation error $\xi$ in Figure 9.4, for the case of

$$w(t) = \mathrm{wgn}(1, 101, 20', \mathrm{real}'),$$

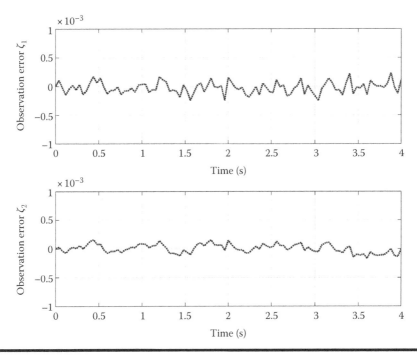

**Figure 9.4    The observation error, Example 9.4.**

which generates a 1-by-101 matrix of real white noise with power of 20dBW. It is seen from this figure that the disturbance indeed has very little affection to the observation error.

## 9.3 H∞ Filtering

In this section, we consider the H∞ filtering of a linear system in the following form:

$$\begin{cases} \dot{x} = Ax + Bw, \ x(0) = x_0 \\ y = Cx + Dw \\ z = Lx, \end{cases} \tag{9.42}$$

where

- $x \in \mathbb{R}^n$, $y \in \mathbb{R}^l$, $z \in \mathbb{R}^m$ are respectively the state vector, the measured output vector, and the output vector of interests
- $w \in \mathbb{R}^p$ is the disturbance vector
- $A$, $B$, $C$, $D$, and $L$ are the system matrices of appropriate dimensions

The word filtering is normally used for state estimation in stochastic systems, where the effect of the stochastic noise is attenuated to a minimum level determined by the variance of the noise. But in this section, the disturbance $w$ is not necessarily stochastic noise. However, the design purpose is the same: to eliminate the effect of the disturbance as completely as possible. From this point of view, such a problem is in essence a problem of robust estimation, and yet it has been commonly termed as filtering.

We remark that (9.43) is usually already the closed-loop system resulted in by certain feedback control, and therefore, the matrix $A$ is assumed to be stable in this section.

## 9.3.1 Problem Formulation

The filter to be designed for the system (9.42) is chosen to possess the following form:

$$\begin{cases} \dot{\varsigma} = A_f \varsigma + B_f y, \ \varsigma(0) = \varsigma_0 \\ \hat{z} = C_f \varsigma + D_f y, \end{cases} \qquad (9.43)$$

where

$\varsigma \in \mathbb{R}^n$ is the state vector
$\hat{z} \in \mathbb{R}^m$ is the estimation vector of $z$
$A_f, B_f, C_f$, and $D_f$ are the coefficient matrices of appropriate dimensions

With the given system (9.42) and the observer (9.43), the combined complete system can be easily shown to be

$$\begin{cases} \dot{x}_e = \tilde{A} x_e + \tilde{B} w, \ x_e(0) = x_{e0} \\ \tilde{z} = \tilde{C} x_e + \tilde{D} w, \end{cases}$$

where

$$\tilde{z} = z - \hat{z} \in \mathbb{R}^m$$

is the estimation error;

$$x_e = \begin{bmatrix} x \\ \varsigma \end{bmatrix}$$

is the state vector of the system; and $\tilde{A}, \tilde{B}, \tilde{C}$, and $\tilde{D}$ are the coefficient matrices which are given by

$$\tilde{A} = \begin{bmatrix} A & 0 \\ B_f C & A_f \end{bmatrix}, \quad \tilde{B} = \begin{bmatrix} B \\ B_f D \end{bmatrix}, \qquad (9.44)$$

$$\tilde{C} = \begin{bmatrix} L - D_f C & -C_f \end{bmatrix}, \quad \tilde{D} = -D_f D. \qquad (9.45)$$

With the aforementioned preparation, the problems to be solved in this section can be stated as follows.

**Problem 9.5** ($H_\infty$ filtering) For the given system (9.42) and a given positive scalar $\gamma$, find the matrices $A_f$, $B_f$, $C_f$, and $D_f$ such that

$$||G_{\tilde{z}w}(s)||_\infty < \gamma, \tag{9.46}$$

where

$$G_{\tilde{z}w}(s) = \tilde{C}\left(sI - \tilde{A}\right)^{-1}\tilde{B} + \tilde{D}.$$

In the absence of the disturbance, the error system becomes

$$\begin{cases} \dot{x}_e = \tilde{A}x_e, \ x_e(0) = x_{e0} \\ \tilde{z} = \tilde{C}x_e. \end{cases}$$

As a consequence of the $H_\infty$ requirement in the aforementioned problem, the error system is stable, and thus

$$\tilde{z} = z - \hat{z} = \tilde{C}x_e \to 0, \text{ as } t \to \infty.$$

This states that $\hat{z}$ is an asymptotic estimate of the system output $z$. In the presence of the disturbance, our purpose is obviously to let the effect of the disturbance to the estimation error be restricted within a given level.

### 9.3.2 Solution to $H_\infty$ Filtering

The following theorem gives the solution to the $H_\infty$ filtering problem.

**Theorem 9.5** The $H_\infty$ filtering problem has a solution if and only if there exist matrices $R$, $X$, $M$, $N$, $Z$, and $D_f$ such that

$$\begin{cases} X > 0 \\ R - X > 0 \\ \begin{bmatrix} RA + A^\mathsf{T}R + ZC + C^\mathsf{T}Z^\mathsf{T} & * & * & * \\ M^\mathsf{T} + ZC + XA & M^\mathsf{T} + M & * & * \\ B^\mathsf{T}R + D^\mathsf{T}Z^\mathsf{T} & B^\mathsf{T}X + D^\mathsf{T}Z^\mathsf{T} & -\gamma I & * \\ L - D_f C & -N & -D_f D & -\gamma I \end{bmatrix} < 0, \end{cases} \tag{9.47}$$

where "$*$" denotes a corresponding symmetric term in the matrix. When such a set of matrices are obtained, the filter coefficient matrices are given by

$$A_f = X^{-1}M, \quad B_f = X^{-1}Z, \quad C_f = N. \tag{9.48}$$

*Proof*  The proof is composed of three steps.

*Step 1. Derivation of the third condition in (9.47)*
It follows from Theorem 5.3 that condition (9.46) holds if and only if there exists a positive definite symmetric matrix $P$ such that

$$\begin{bmatrix} \tilde{A}^T P + P\tilde{A} & P\tilde{B} & \tilde{C}^T \\ \tilde{B}^T P & -\gamma I & \tilde{D}^T \\ \tilde{C} & \tilde{D} & -\gamma I \end{bmatrix} < 0, \tag{9.49}$$

where the matrices $\tilde{A}$, $\tilde{B}$, $\tilde{C}$, and $\tilde{D}$ are given by (9.44) and (9.45), hence (9.49) is actually an inequality in the variables $A_f$, $B_f$, $C_f$, $D_f$, and $P$.
Let

$$P = \begin{bmatrix} X_{11} & X_{12} \\ X_{12}^T & X_{22} \end{bmatrix}, \quad X_{11}, X_{22} \in \mathbb{R}^n,$$

be a feasible solution to (9.49). Without loss of generality we assume $\det(X_{12}) \neq 0$, since otherwise we can choose

$$P = \begin{bmatrix} X_{11} & X_{12} + \alpha I \\ X_{12}^T + \alpha I & X_{22} \end{bmatrix},$$

with $|\alpha|$ sufficiently small so that the $P$ matrix is still positive definite and also satisfies (9.49), and furthermore, $\det(X_{12} + \alpha I) \neq 0$ holds.
Using the definitions of $\tilde{A}$, $\tilde{B}$, $\tilde{C}$ and $\tilde{D}$ given in (9.44) and (9.45) and the partitioned form of $P$, we can rewrite (9.49) as

$$\begin{bmatrix} (X_{11}A + X_{12}B_fC)_s & * & * & * \\ A_f^TX_{12}^T + X_{22}B_fC + X_{12}^TA & X_{22}A_f + A_f^TX_{22} & * & * \\ B^TX_{11} + D^TB_f^TX_{12}^T & B^TX_{12} + D^TB_f^TX_{22} & -\gamma I & * \\ L - D_fC & -C_f & -D_fD & -\gamma I \end{bmatrix} < 0. \tag{9.50}$$

Denoting

$$J = \text{diag}(I, X_{12}X_{22}^{-1}, I, I),$$

and pre- and postmultiplying (9.50) by $J$ and $J^\mathrm{T}$, respectively, yields

$$\begin{bmatrix} RA + A^\mathrm{T}R + ZC + C^\mathrm{T}Z^\mathrm{T} & * & * & * \\ M^\mathrm{T} + ZC + XA & M^\mathrm{T} + M & * & * \\ B^\mathrm{T}R + D^\mathrm{T}Z^\mathrm{T} & B^\mathrm{T}X + D^\mathrm{T}Z^\mathrm{T} & -\gamma I & * \\ L - D_f C & -N & -D_f D & -\gamma I \end{bmatrix} < 0, \quad (9.51)$$

where

$$R = X_{11}, \quad X = X_{12}X_{22}^{-1}X_{12}^\mathrm{T}, \quad Z = X_{12}B_f, \quad (9.52)$$

$$M = X_{12}A_f X_{22}^{-1}X_{12}^\mathrm{T}, \quad N = C_f X_{22}^{-1}X_{12}^\mathrm{T}. \quad (9.53)$$

Therefore, by now we have shown that (9.49) holds for some $P$ if and only if (9.51) holds for some $R$, $X$, $M$, $N$, $Z$, and $D_f$.

*Step 2. Derivation of the first two conditions in (9.47)*
In this step, we need to show

$$P > 0 \Longleftrightarrow X > 0, \quad R - X > 0. \quad (9.54)$$

Using Schur complement lemma, we know that

$$P > 0 \Longleftrightarrow \begin{cases} X_{22} > 0 \\ S_{ch}(X_{22}) = X_{11} - X_{12}X_{22}^{-1}X_{12}^\mathrm{T} > 0. \end{cases}$$

Since $X_{22} > 0$ is equivalent to $X = X_{12}X_{22}^{-1}X_{12}^\mathrm{T} > 0$, and $R - X = S_{ch}(X_{22})$, the relation in (9.54) clearly holds.

*Step 3. Derivation of the coefficients*
From relations in (9.52) and (9.53), we can obtain the coefficient matrices of the filter as

$$A_f = X_{12}^{-1}MX_{12}^{-\mathrm{T}}X_{22}, \quad B_f = X_{12}^{-1}Z, \quad C_f = NX_{12}^{-\mathrm{T}}X_{22}.$$

For an expression directly given by the decision variables $R$, $X$, $M$, $N$, $Z$, and $D_f$, we suffice to find an equivalent substitution $\left(A_f', B_f', C_f'\right)$ for $\left(A_f, B_f, C_f\right)$. To do this, we select the following transformation matrix

$$T = X_{12}^{-\mathrm{T}}X_{22},$$

and under this transformation, we have

$$A'_f = TA_f T^{-1} = X^{-1}M,$$

$$B'_f = TB_f = X^{-1}Z,$$

$$C'_f = C_f T^{-1} = N.$$

Therefore, a set of coefficient matrices of the filter can be given as in (9.48). ▪

The aforementioned theorem turns the $H_\infty$ filtering problem into an LMI feasibility problem, while the optimal solution with the minimal attenuation level can be solved via the following optimization problem:

$$\begin{cases} \min & \gamma \\ \text{s.t.} & \text{the inequalities in (9.47).} \end{cases} \tag{9.55}$$

**Example 9.5**

Consider a linear system in the form of (9.42) with the following parameters:

$$A = \begin{bmatrix} -4.4697 & 9.3324 & 8.7645 \\ -9.6408 & 2.0236 & 2.2076 \\ -6.6937 & 1.3595 & -3.2241 \end{bmatrix}, \quad B = \begin{bmatrix} 0.2309 \\ -0.4584 \\ -0.8390 \end{bmatrix},$$

$$C = \begin{bmatrix} 0.7813 & -4.9301 & -4.3341 \\ 4.9421 & -3.4866 & -4.1502 \end{bmatrix}, \quad D = \begin{bmatrix} 0.0074 \\ -0.0088 \end{bmatrix},$$

$$L = \begin{bmatrix} -1.1604 & 1.7916 & -1.4033 \\ -2.0351 & 1.5369 & -0.3777 \end{bmatrix}.$$

Applying the function `mincx` in the MATLAB LMI toolbox to the full-order $H_\infty$ filtering condition (9.55) yields an optimal $H_\infty$ filter in the form of (9.43) with the following parameters

$$A_f = \begin{bmatrix} 24.5876 & 78.0725 & 59.3187 \\ -54.3299 & -155.1420 & -118.7542 \\ -95.6715 & -274.8571 & -213.2328 \end{bmatrix},$$

$$B_f = \begin{bmatrix} -20.3795 & 9.1014 \\ 43.0915 & -15.8549 \\ 77.4143 & -30.2425 \end{bmatrix},$$

$$C_f = \begin{bmatrix} 0.9307 & -1.4258 & 1.7673 \\ 1.3605 & -0.4626 & 1.4468 \end{bmatrix}.$$

The corresponding attenuation level is $\gamma = 1.3280 \times 10^{-7}$. Note that the attenuation level $\gamma$ is very small, the disturbance $w$ must be effectively attenuated.

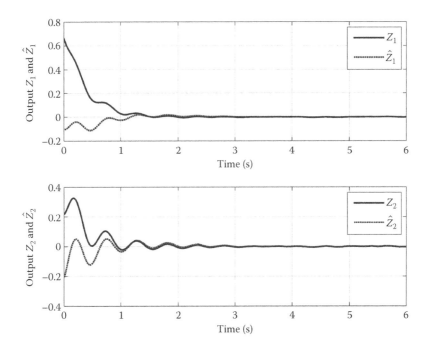

**Figure 9.5   The system output *z* and estimation output *ẑ*, Example 9.5.**

To verify this, we have plotted the system output $z$ and the estimation output $\hat{z}$ in Figure 9.5, for the case $w(t) = \text{wgn}(1, 4001, 10,' \text{real}')$, which generates a 1-by-4001 matrix of real white noise with power of 10 dBW. It is seen from this figure that the disturbance indeed has very little affection to the observation error.

## 9.4  H₂ Filtering

In this section, we consider the $H_2$ filtering of a linear system in the following form:

$$\begin{cases} \dot{x} = Ax + Bw, \ x(0) = x_0 \\ y = Cx + Dw \\ z = Lx, \end{cases} \tag{9.56}$$

where

- $x \in \mathbb{R}^n$, $y \in \mathbb{R}^l$, $z \in \mathbb{R}^m$ are respectively the state vector, the measured output vector, and the output vector of interests
- $w \in \mathbb{R}^p$ is the disturbance vector
- $A$, $B$, $C$, and $L$ are the system matrices of appropriate dimensions

Again, as in Section 9.3, the matrix $A$ is assumed to be stable.

## 9.4.1 Problem Formulation

The filter to be designed for the system (9.56) is chosen to possess the following form:

$$\begin{cases} \dot{\varsigma} = A_f\varsigma + B_f y, \ \varsigma(0) = \varsigma_0 \\ \hat{z} = C_f\varsigma, \end{cases} \tag{9.57}$$

where

$\varsigma \in \mathbb{R}^n$ is the state vector
$\hat{z} \in \mathbb{R}^m$ is the estimation vector
$A_f$, $B_f$ and $C_f$ are the coefficient matrices of appropriate dimensions

With the given system (9.56) and the observer (9.57), the combined complete system can be easily shown to be

$$\begin{cases} \dot{x}_e = \tilde{A}x_e + \tilde{B}w, \ x_e(0) = x_{e0} \\ \tilde{z} = \tilde{C}x_e, \end{cases}$$

where

$$x_e = \begin{bmatrix} x \\ \varsigma \end{bmatrix}, \quad \tilde{z} = z - \hat{z} \in \mathbb{R}^m,$$

and $\tilde{A}$, $\tilde{B}$, $\tilde{C}$, and $\tilde{D}$ are the coefficient matrices that are given by

$$\tilde{A} = \begin{bmatrix} A & 0 \\ B_f C & A_f \end{bmatrix}, \quad \tilde{B} = \begin{bmatrix} B \\ B_f D \end{bmatrix}, \tag{9.58}$$

$$\tilde{C} = \begin{bmatrix} L & -C_f \end{bmatrix}. \tag{9.59}$$

The transfer function matrix of this system is obviously

$$G_{\tilde{z}w}(s) = \tilde{C} \left( sI - \tilde{A} \right)^{-1} \tilde{B}.$$

With the aforementioned preparation, the problem to be solved in this section can be stated as follows.

**Problem 9.6** (H₂ filtering) For the given system (9.42) and a given positive scalar $\gamma$, find the matrices $A_f$, $B_f$, and $C_f$ such that

$$\|G_{\tilde{z}w}(s)\|_2 < \gamma. \tag{9.60}$$

As a consequence of the H$_2$ requirement in the aforementioned problem, the closed-loop error system is stable. In the absence of the disturbance, we have

$$\begin{cases} \dot{x}_e = \tilde{A}x_e, \ x_e(0) = x_{e0} \\ \tilde{z} = \tilde{C}x_e, \end{cases}$$

and thus

$$\tilde{z} = z - \hat{z} = \tilde{C}x_e \longrightarrow 0, \text{ as } t \longrightarrow \infty.$$

This states that $\hat{z}$ is an asymptotic estimate of the system output $z$. In the presence of the disturbance, our purpose is to let the effect of the disturbance to the estimation error be restricted within a given level.

## 9.4.2 Solution to H$_2$ Filtering

The following theorem gives the solution to the H$_2$ filtering problem.

**Theorem 9.6** The H$_2$ filtering problem has a solution if and only if one of the following conditions holds:

1. There exist matrices $R$, $X$, $M$, $N$, $Z$, and $Q$ such that

$$\begin{cases} R - X > 0 \\ \text{trace}(Q) < \gamma^2 \\ \begin{bmatrix} -Q & * & * \\ L^T & -R & * \\ -N^T & -X & -X \end{bmatrix} < 0 \\ \begin{bmatrix} RA + A^T R + ZC + C^T Z^T & * & * \\ M^T + ZC + XA & M^T + M & * \\ B^T R + D^T Z^T & B^T X + D^T Z^T & -I \end{bmatrix} < 0. \end{cases} \tag{9.61}$$

When such a set of matrices are obtained, the filter coefficient matrices are given by

$$A_f = X^{-1}M, \quad B_f = X^{-1}Z, \quad C_f = N. \tag{9.62}$$

2. There exist matrices $\bar{R}$, $\bar{X}$, $\bar{M}$, $\bar{N}$, $\bar{Z}$, and $\bar{Q}$ such that

$$
\begin{cases}
\bar{R} - \bar{X} > 0 \\
\mathrm{trace}(\bar{Q}) < \gamma^2 \\
\begin{bmatrix} -\bar{Q} & * & * \\ \bar{R}B + \bar{Z}D & -\bar{R} & * \\ \bar{X}B + \bar{Z}D & -\bar{X} & -\bar{X} \end{bmatrix} < 0 \\
\begin{bmatrix} \bar{R}A + A^{\mathrm{T}}\bar{R} + \bar{Z}C + C^{\mathrm{T}}\bar{Z}^{\mathrm{T}} & * & * \\ \bar{M}^{\mathrm{T}} + \bar{Z}C + \bar{X}A & \bar{M}^{\mathrm{T}} + \bar{M} & * \\ L & -\bar{N} & -I \end{bmatrix} < 0.
\end{cases}
\tag{9.63}
$$

When such a set of matrices are obtained, the filter coefficient matrices are given by

$$
A_f = \bar{X}^{-1}\bar{M}, \quad B_f = \bar{X}^{-1}\bar{Z}, \quad C_f = \bar{N}.
\tag{9.64}
$$

*Proof*  Since the two conclusions can be proven similarly, here we only prove the first one.

*Step 1. Derivation of the fourth condition in (9.63)*
It follows from Theorem 5.6 that condition (9.60) holds if and only if there exists a positive definite matrix $P$ satisfying

$$
\tilde{A}P + P\tilde{A}^{\mathrm{T}} + \tilde{B}\tilde{B}^{\mathrm{T}} < 0,
\tag{9.65}
$$

and

$$
\mathrm{trace}\left(\tilde{C}P\tilde{C}^{\mathrm{T}}\right) < \gamma^2,
\tag{9.66}
$$

where the matrices $\tilde{A}$, $\tilde{B}$, $\tilde{C}$ are given by (9.58) and (9.59), hence (9.65) and (9.66) are inequalities in the variables $A_f$, $B_f$, $C_f$, and $P$.

Pre- and postmultiplying both sides of (9.65) by $P^{-1}$ yields

$$
P^{-1}\tilde{A} + \tilde{A}^{\mathrm{T}}P^{-1} + P^{-1}\tilde{B}\tilde{B}^{\mathrm{T}}P^{-1} < 0,
$$

which can be converted into the following form according to the Schur complement lemma:

$$
\begin{bmatrix} P^{-1}\tilde{A} + \tilde{A}^{\mathrm{T}}P^{-1} & P^{-1}\tilde{B} \\ \tilde{B}^{\mathrm{T}}P^{-1} & -I \end{bmatrix} < 0.
\tag{9.67}
$$

Let

$$
P^{-1} = \begin{bmatrix} X_{11} & X_{12} \\ X_{12}^{\mathrm{T}} & X_{22} \end{bmatrix} > 0.
$$

Without loss of generality, we assume $\det(X_{12}) \neq 0$ since otherwise a desired $P$ can be obtained by slightly perturbing $X_{12}$. Using the definitions of $\tilde{A}$, $\tilde{B}$ given in (9.58) and the aforementioned partitioned form of $P$, we can rewrite (9.67) as

$$
\begin{bmatrix} X_{11}A + A^{\mathrm{T}}X_{11} + X_{12}B_fC + C^{\mathrm{T}}B_f^{\mathrm{T}}X_{12}^{\mathrm{T}} & * & * \\ A_f^{\mathrm{T}}X_{12}^{\mathrm{T}} + X_{22}B_fC + X_{12}^{\mathrm{T}}A & X_{22}A_f + A_f^{\mathrm{T}}X_{22} & * \\ B^{\mathrm{T}}X_{11} + D^{\mathrm{T}}B_f^{\mathrm{T}}X_{12}^{\mathrm{T}} & B^{\mathrm{T}}X_{12} + D^{\mathrm{T}}B_f^{\mathrm{T}}X_{22} & -I \end{bmatrix} < 0.
$$

(9.68)

Denoting

$$
J = \mathrm{diag}(I, X_{12}X_{22}^{-1}, I),
$$

and pre- and postmultiplying (9.68) by $J$ and $J^{\mathrm{T}}$, respectively, gives

$$
\begin{bmatrix} RA + A^{\mathrm{T}}R + ZC + C^{\mathrm{T}}Z^{\mathrm{T}} & * & * \\ M^{\mathrm{T}} + ZC + XA & M^{\mathrm{T}} + M & * \\ B^{\mathrm{T}}R + D^{\mathrm{T}}Z^{\mathrm{T}} & B^{\mathrm{T}}X + D^{\mathrm{T}}Z^{\mathrm{T}} & -I \end{bmatrix} < 0,
\qquad (9.69)
$$

where

$$
R = X_{11}, \quad X = X_{12}X_{22}^{-1}X_{12}^{\mathrm{T}}, \quad Z = X_{12}B_f, \quad M = X_{12}A_fX_{22}^{-1}X_{12}^{\mathrm{T}}. \quad (9.70)
$$

Therefore, by now we have shown that (9.65) holds for some $P$ if and only if (9.69) holds for some $R$, $X$, $M$, $N$, and $Z$.

*Step 2. Derivation of the second and the third conditions in (9.63)*
By Lemma 2.13, the inequality (9.66) holds if and only if there exists $Q > 0$, satisfying

$$
\tilde{C}P\tilde{C}^{\mathrm{T}} - Q < 0, \ \mathrm{trace}\,(Q) < \gamma^2,
$$

where the first one is clearly equivalent to

$$
\begin{bmatrix} -Q & \tilde{C} \\ \tilde{C}^{\mathrm{T}} & -P^{-1} \end{bmatrix} < 0.
$$

Substituting the expressions of $\tilde{C}$ and $P^{-1}$ into the earlier expression yields

$$\begin{bmatrix} -Q & * & * \\ L^{\mathrm{T}} & -X_{11} & * \\ -C_f^{\mathrm{T}} & -X_{12}^{\mathrm{T}} & -X_{22} \end{bmatrix} < 0. \tag{9.71}$$

Further, letting $J_1 = \mathrm{diag}(I, I, X_{12}X_{22}^{-1})$ and pre- and postmultiplying (9.71) by $J_1$ and $J_1^{\mathrm{T}}$, respectively, produces

$$\begin{bmatrix} -Q & * & * \\ L^{\mathrm{T}} & -R & * \\ -N^{\mathrm{T}} & -X & -X \end{bmatrix} < 0, \tag{9.72}$$

where

$$N = C_f X_{22}^{-1} X_{12}^{\mathrm{T}}. \tag{9.73}$$

By now we have shown that (9.66) holds for some $P$ if and only if $\mathrm{trace}(Q) < \gamma^2$ and (9.72) holds for some $R$, $X$, $M$, $N$, $Z$, and $Q$.

*Step 3. Derivation of the first condition in (9.63)*
Noting that condition (9.72) implies $X > 0$, to complete the proof of the existence condition (9.61), we now need to show, when $X > 0$, there holds

$$P > 0 \iff R - X > 0. \tag{9.74}$$

Using the partitioned form of $P^{-1}$, the Schur complement lemma and the nonsingularity of $X_{12}$, we have

$$P > 0 \iff P^{-1} > 0$$

$$\iff X_{22} > 0, \; S_{ch}(X_{22}) > 0$$

$$\iff X = X_{12}X_{22}^{-1}X_{12}^{\mathrm{T}} > 0, \quad S_{ch}(X_{22}) > 0.$$

Further note that

$$R - X = X_{11} - X_{12}X_{22}^{-1}X_{12}^{\mathrm{T}} = S_{ch}(X_{22}),$$

the relation (9.74) clearly follows.

*Step 4. Derivation of coefficients*
From (9.70) and (9.73), the coefficient matrices of the filter are solved as

$$A_f = X_{12}^{-1}MX_{12}^{-\mathrm{T}}X_{22}, \quad B_f = X_{12}^{-1}Z, \quad C_f = NX_{12}^{-\mathrm{T}}X_{22}.$$

For an expression directly given by the decision variables $R$, $X$, $M$, $N$, $Z$, and $Q$, we suffice to find an equivalent substitution $\left(A'_f, \quad B'_f, \quad C'_f\right)$ for $\left(A_f, B_f, C_f\right)$. To do this, we select the following transformation matrix

$$T = X_{12}^{-T} X_{22},$$

and under this transformation, we have

$$A'_f = TA_f T^{-1} = X^{-1}M,$$
$$B'_f = TB_f = X^{-1}Z,$$
$$C'_f = C_f T^{-1} = N.$$

Therefore, a set of coefficient matrices of the filter can be given as in (9.62).　■

The aforementioned theorem turns the $H_2$ filtering problem into an LMI feasibility problem, while the optimal solution with the minimal attenuation level $\rho = \gamma^2$ can be solved via the following optimization problems

$$\begin{cases} \min & \rho \\ \text{s.t.} & \text{the inequalities in (9.61).} \end{cases} \tag{9.75}$$

or

$$\begin{cases} \min & \rho \\ \text{s.t.} & \text{the inequalities in (9.63).} \end{cases} \tag{9.76}$$

**Example 9.6**

Consider a linear system in the form of (9.42) with the following parameters:

$$A = \begin{bmatrix} -6.1153 & -7.0691 & -5.6947 \\ -3.3389 & -8.4648 & -5.1474 \\ 7.2212 & 2.8381 & -1.3190 \end{bmatrix}, \quad B = \begin{bmatrix} -0.7857 \\ 0.3499 \\ 0.0274 \end{bmatrix},$$

$$C = \begin{bmatrix} 4.0933 & -2.3327 & -3.9925 \\ 2.5846 & 2.6715 & 4.5528 \end{bmatrix}, \quad D = \begin{bmatrix} 0.0015 \\ 0.0811 \end{bmatrix},$$

$$L = \begin{bmatrix} -1.9456 & -2.6203 & -0.6425 \\ 2.1679 & -2.3872 & -1.4399 \end{bmatrix}.$$

Applying the function `mincx` in the MATLAB LMI toolbox to the full-order $H_2$ filtering problem (9.75) gives

$$\rho = 1.1172 \times 10^{-6},$$

and the following observer parameter matrices

$$A_f = \begin{bmatrix} 3.8778 & 27.5487 & 53.4339 \\ -5.9253 & -24.8891 & -33.2556 \\ -2.2540 & 6.9799 & 5.7877 \end{bmatrix},$$

$$B_f = \begin{bmatrix} -3.7109 & 9.7477 \\ 2.1073 & -4.3431 \\ -2.1249 & -0.3000 \end{bmatrix},$$

$$C_f = \begin{bmatrix} 1.9456 & 2.6203 & 0.6425 \\ -2.1679 & 2.3871 & 1.4399 \end{bmatrix},$$

while the attenuation level is $\gamma = \sqrt{\rho} = 0.001057$. Note that the attenuation level $\gamma$ is very small, the disturbance $w$ must be effectively attenuated. To verify this, we have plotted the system output $z$ and the estimation output $\hat{z}$ in Figure 9.6, for the case of

$$w(t) = \text{wgn}(1, 401, 10', \text{real}'),$$

which generates a 1-by-401 matrix of real white noise with power of 10 dBW. It is seen from this figure that the disturbance indeed has very little affection to the observation error.

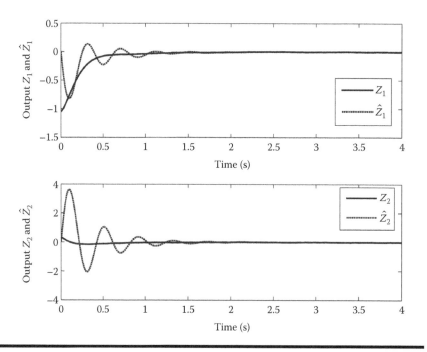

**Figure 9.6** The system output $z$ and estimation output $\hat{z}$, Example 9.6.

## 9.5 Notes and References

This chapter considers state observation and filtering based on LMI approaches, and the results are summarized as follows.

*Full-order state observer*:

$$\dot{\hat{x}} = A\hat{x} + Bu + L(C\hat{x} - y),$$

*Reduced-order state observer*:

$$\begin{cases} \dot{z} = Fz + Gy + Hu \\ \hat{x}_2 = Mz + Ny, \end{cases}$$

where

$$G = (A_{21} + LA_{11}) - (A_{22} + LA_{12})L, \quad M = I,$$

$$F = A_{22} + LA_{12}, \quad H = B_2 + LB_1, \quad N = -L,$$

where the matrices $A_{ij}$ and $B_i$, $i, j = 1, 2$, are obtained via Lemma 9.1.

The LMI conditions for these two types of observers are summarized in Table 9.2. $H_\infty/H_2$ *full-order state observer*:

$$\begin{cases} \dot{\hat{x}} = (A + LC_1)\hat{x} - Ly + (B_1 + LD_1)u \\ \hat{z}(t) = C_2\hat{x}(t) \\ \tilde{z}(t) = z(t) - \hat{z}(t). \end{cases}$$

**Table 9.2  Full- and Reduced-Order Observers**

|  | Criteria | Observer |
|---|---|---|
| **Full-order** | $P > 0$ <br> $PA + A^{\mathsf{T}}P + WC + C^{\mathsf{T}}W^{\mathsf{T}} < 0$ | $L = P^{-1}W$ |
|  | $P > 0$ <br> $PA + A^{\mathsf{T}}P - C^{\mathsf{T}}C < 0$ | $L = -\frac{1}{2}P^{-1}C^{\mathsf{T}}$ |
| **Reduced-order** | $P > 0$ <br> $PA_{22} + A_{22}^{\mathsf{T}}P + WA_{12} + A_{12}^{\mathsf{T}}W^{\mathsf{T}} < 0$ | $L = P^{-1}W$ |
|  | $P > 0$ <br> $PA_{22} + A_{22}^{\mathsf{T}}P - A_{12}^{\mathsf{T}}A_{12} < 0$ | $L = -\frac{1}{2}P^{-1}A_{12}^{\mathsf{T}}$ |

**Table 9.3   $H_\infty/H_2$ Full-Order Observers**

| | Criteria | Observer |
|---|---|---|
| $\|G_{\tilde{z}w}(s)\|_\infty < \gamma$ | $P > 0$ <br><br> $\begin{bmatrix} (PA + WC_1)_s & * & * \\ (PB_2 + WD_2)^T & -\gamma I & * \\ C_2 & 0 & -\gamma I \end{bmatrix} < 0$ | $L = P^{-1}W$ |
| $\|G_{\tilde{z}w}(s)\|_2 < \gamma$ | $\begin{bmatrix} (XA + WC_1)_s & * \\ (XB_2 + WD_2)^T & -I \end{bmatrix} < 0$ <br><br> $\begin{bmatrix} -Q & C_2 \\ C_2^T & -X \end{bmatrix} < 0$ <br><br> trace $(Q) < \gamma^2$ | $L = X^{-1}W$ |

LMI conditions for this type of $H_\infty/H_2$ full-order state observers are given in Table 9.3.

*$H_\infty/H_2$ state filter*:

$$\begin{cases} \dot{\varsigma} = A_f\varsigma + B_f y, \ \varsigma(0) = 0 \\ \hat{z} = C_f\varsigma + D_f y \\ \tilde{z} = z - \hat{z}. \end{cases}$$

LMI conditions for this type of $H_\infty/H_2$ full-order state observers are given in Tables 9.4 and 9.5.

In many circumstances, state observation or filtering is a dual problem with the feedback control. It is not only needed to retrieve state information for state feedback control implementation but is also often used for many other purposes, such as signal analysis, fault detection and isolation (Duan and Patton 2001, Duan et al. 2002, Casavola et al. 2008), etc. Due to its wide scope of applications, state observation and filtering have been investigated and used by many control scientists and engineers (Geromel 1999, Stilwell and Rugh 1999, Gao et al. 2005, Zhang et al. 2006a, Duan et al. 2007, Wu et al. 2007).

As for LMI approaches for state observer designs, the basic ideas illustrated in this chapter have been extended to many different types of systems, such as time-delay systems (Fu et al. 2004, 2006a, Chen 2007a,b, Wu et al. 2009), uncertain discrete-time switched systems (Zhang and Duan (2005a)), and uncertain descriptor time-delay systems with Markovian jumping parameters (Fu and Duan 2005). Meanwhile, $H_\infty$ filtering has also been studied by many researchers,

**Table 9.4  H$_\infty$ State Filtering Design**

| Solutions | Observer |
|---|---|
| $X > 0$ | |
| $R - X > 0$ | |
| $\begin{bmatrix} \langle RA + ZC \rangle_s & * & * & * \\ M^T + ZC + XA & M^T + M & * & * \\ B^T R + D^T Z^T & B^T X + D^T Z^T & -\gamma I & * \\ L - D_f C & -N & -D_f D & -\gamma I \end{bmatrix} < 0$ | $A_f = X^{-1}M$ $B_f = X^{-1}Z$ $C_f = N$ |
| Decision variables are $R$, $X$, $M$, $N$, $Z$, and $D_f$ | |

for various types of systems as well, for example, linear continuous-time systems subject to sensor nonlinearity (Zhou et al. 2011c), continuous-time state-delayed systems (Zhang et al. 2009), and polytopic discrete-time systems (Gao et al. 2005, Kim et al. 2005). For many types of uncertain systems, robust $H_\infty$ filtering has also been investigated with the LMI approaches (Li and Fu 1997, Xu and Van Dooren 2002, Gao and Wang 2004, Xu and Lam 2006, Liu and Duan (2006b), and He et al. 2006a,c).

Parallel to $H_\infty$ and $H_2$ filtering, some other types of filtering designs have also been studied with the LMI techniques, such as the $L_2$-$L_\infty$ filtering (Grigoriadis and Watson 1997, Palhares and Peres 1999, Gao and Wang 2003a,b, Zhang et al. 2006b, 2010), robust passive filtering (Zhao et al. 2007), and robust dissipative filtering (Fu et al. 2006b, Duan et al. 2009a).

## Exercises

**9.1** Prove Lemma 9.1.

**9.2** Consider the following linear system, which has been studied by Duan (2003):

$$\begin{cases} \dot{x}(t) = \begin{bmatrix} -0.5 & 0 & 0 \\ 0 & -2 & 10 \\ 0 & 1 & -2 \end{bmatrix} x(t) + \begin{bmatrix} 1 & 0 \\ -2 & 2 \\ 0 & 1 \end{bmatrix} u(t) \\ y = \begin{bmatrix} 1 & 0 & 0 \\ 0 & 0 & 1 \end{bmatrix} x. \end{cases}$$

Design a full-order observer for the system using LMI technique.

**9.3** Design a reduced-order observer for the benchmark system in Example 7.1 using LMI technique.

**Table 9.5   $H_2$ State Filtering Design**

| Solutions | Observer |
|---|---|
| $R - X > 0$<br><br>$\text{trace}(Q) < \gamma^2$<br><br>$\begin{bmatrix} -Q & * & * \\ L^T & -R & * \\ -N^T & -X & -X \end{bmatrix} < 0$<br><br>$\begin{bmatrix} \langle RA + ZC \rangle_s & * & * \\ M^T + ZC + XA & M^T + M & * \\ B^T R + D^T Z^T & B^T X + D^T Z^T & -I \end{bmatrix} < 0$<br><br>Decision variables are $R,\ X,\ M,\ N,\ Z,$ and $Q$ | $A_f = X^{-1} M$<br><br>$B_f = X^{-1} Z$<br><br>$C_f = N$ |
| $\bar{R} - \bar{X} > 0$<br><br>$\text{trace}(\bar{Q}) < \gamma^2$<br><br>$\begin{bmatrix} -\bar{Q} & * & * \\ \bar{R}B + \bar{Z}D & -\bar{R} & * \\ \bar{X}B + \bar{Z}D & -\bar{X} & -\bar{X} \end{bmatrix} < 0$<br><br>$\begin{bmatrix} \langle RA + ZC \rangle_s & * & * \\ M^T + ZC + XA & M^T + M & * \\ B^T R + D^T Z^T & B^T X + D^T Z^T & -I \end{bmatrix} < 0$<br><br>$\begin{bmatrix} \langle \bar{R}A + \bar{Z}C \rangle_s & * & * \\ \bar{M}^T + \bar{Z}C + \bar{X}A & \bar{M}^T + \bar{M} & * \\ L & -\bar{N} & -I \end{bmatrix} < 0$<br><br>Decision variables are $\bar{R}, \bar{X}, \bar{M}, \bar{N}, \bar{Z},$ and $\bar{Q}$ | $A_f = \bar{X}^{-1} \bar{M}$<br><br>$B_f = \bar{X}^{-1} \bar{Z}$<br><br>$C_f = \bar{N}$ |

**9.4** Consider a linear system in the form of (9.22) with the following parameters:

$$A = \begin{bmatrix} 0 & 1 & 0 \\ 1 & 1 & 0 \\ -1 & 0 & 0 \end{bmatrix}, \quad B_1 = \begin{bmatrix} 0 \\ 1 \\ 0 \end{bmatrix}, \quad B_2 = \begin{bmatrix} 0 \\ 0 \\ 1 \end{bmatrix},$$

$$C_1 = \begin{bmatrix} 1 & 0 & 0 \\ 0 & 0 & 1 \end{bmatrix}, \quad C_2 = \begin{bmatrix} 0 & 1 & 0 \\ 0 & 0 & 1 \end{bmatrix},$$

$$D_1 = \begin{bmatrix} 1 \\ 1 \end{bmatrix}, \quad D_2 = \begin{bmatrix} 1 \\ 0 \end{bmatrix}.$$

Find an observer gain matrix $L$, such that the transfer function of the observation error system (9.25), that is,

$$G_{\tilde{z}w}(s) = C_2 (sI - A - LC_1)^{-1} (B_2 + LD_2),$$

satisfies $||G_{\tilde{z}w}(s)||_\infty < 0.5$.
**9.5** Consider the linear system in Exercise 9.4 again. Find a matrix $L$ and a minimal attenuation level $\gamma$, such that the transfer function of the observation error system (9.25) satisfies $||G_{\tilde{z}w}(s)||_2 < \gamma$
**9.6** Consider a linear system in the form of (9.42) with the following parameters:

$$A = \begin{bmatrix} -1 & 1 & 0 \\ 0 & -1 & 0 \\ 1 & 0 & -5 \end{bmatrix}, \quad B = \begin{bmatrix} 3 \\ 1 \\ 2 \end{bmatrix},$$

$$C = \begin{bmatrix} 1 & 0 & 1 \\ 0 & 1 & 1 \end{bmatrix}, \quad D = \begin{bmatrix} 1 \\ 0 \end{bmatrix}, \quad L = \begin{bmatrix} 1 & 1 & 0 \\ 0 & 0 & 1 \end{bmatrix}.$$

Find an $H_\infty$ filter for the system using the LMI technique.
**9.7** Consider the linear system in Exercise 9.6 again. Find an $H_2$ filter for the system.

# Chapter 10

# Multiple Objective Designs

This chapter deals with multiple objective designs in control linear systems by LMI techniques. As a matter of fact, in previous sections we treated some multiple objective design problems, for example, the $\mathbb{D}$-stabilization problem and the quadratic $\mathbb{D}$-stabilization problem. These problems may be classified into the category of multiple objective designs because the chosen LMI region $\mathbb{D}$ ensures both the closed-loop stability and the system dynamical performance requirements. This category of problems has the feature that multiple design requirements are met by a set of LMI conditions with common decision variables. The problems of insensitive region designs with minimum gain treated in the first section of this chapter belong to this category.

In Sections 10.2 and 10.3, we treat another category of multiple objective design problems in which each design requirement has its own set of LMI conditions with its own decision variables, and the solution to the problem involves a multiple sets of LMI conditions with uncommon decision variables. In order to make all the LMI conditions satisfied, as is demonstrated in these two sections, sometimes certain decision variables need to be set common.

## 10.1 Insensitive Region Designs with Minimum Gain

In Section 7.5, we have solved the insensitive strip region design problem and the insensitive disk region design problem. The nature of these designs is to place the closed-loop eigenvalues in a particular region on the complex plane, and this region has an inner boundary that is insensitive to perturbations to the system parameter matrices. In this section, we deal with a combination of the insensitive region requirement and the minimum gain requirement.

The gain magnitude requirement can be presented as

$$||K||_2 < \gamma, \tag{10.1}$$

with $\gamma$ being some given positive scalar. It is easy to see that this is equivalent to

$$K^T K < \gamma^2 I,$$

which can be equivalently converted into the following LMI in $K$:

$$\begin{bmatrix} -\gamma I & K \\ K^T & -\gamma I \end{bmatrix} < 0. \tag{10.2}$$

## 10.1.1 Insensitive Strip Region Designs with Minimum Gain

In this section, we consider the control of the following continuous-time linear system

$$\begin{cases} \dot{x} = Ax + Bu \\ y = Cx, \end{cases} \tag{10.3}$$

where $x \in \mathbb{R}^n$, $y \in \mathbb{R}^m$, and $u \in \mathbb{R}^r$ are the state vector, the output vector, and the input vector, respectively.

The problem of insensitive strip region designs with minimum gain is stated as follows.

**Problem 10.1** Consider the continuous-time linear system (10.3). For given real scalars $\gamma_1$ and $\gamma_2$, design an output feedback control law $u = Ky$ such that (10.1) holds for a minimal positive scalar $\gamma$, and

$$\gamma_1 < \lambda_i \left( A_c^s \right) < \gamma_2, \quad i = 1, 2, \ldots, n, \tag{10.4}$$

where

$$A_c^s \triangleq \frac{1}{2} \langle A_c \rangle_s = \frac{(A + BKC)^T + (A + BKC)}{2}.$$

Recall from Section 7.5, condition (10.4) implies that the closed-loop eigenvalues are first bounded by $\mathbb{H}_{(-\alpha_2, -\alpha_1)} \subset \mathbb{H}_{(-\gamma_2, -\gamma_1)}$, with

$$\alpha_1 = \lambda_{\min} \left( A_c^s \right), \quad \alpha_2 = \lambda_{\max} \left( A_c^s \right).$$

But the boundaries of $\mathbb{H}_{(-\alpha_2,-\alpha_1)}$ are insensitive to perturbations in the system matrices as well as in the gain matrix $K$. Therefore, this design is robust in the sense that the system has a strip eigenvalue region that is insensitive to small parameter perturbations in the system matrices.

The condition (10.4) is clearly equivalent to

$$2\gamma_1 I < (A + BKC)^{\mathrm{T}} + (A + BKC) < 2\gamma_2 I. \qquad (10.5)$$

Combining the aforementioned LMIs with (10.2), we immediately have the following result.

**Theorem 10.1**  Problem 10.1 has a solution if and only if the following optimization problem

$$\begin{cases} \min & \gamma \\ \text{s.t.} & \begin{bmatrix} -\gamma I & K \\ K^{\mathrm{T}} & -\gamma I \end{bmatrix} < 0 \\ & 2\gamma_1 I < (A + BKC)^{\mathrm{T}} + (A + BKC) < 2\gamma_2 I \end{cases} \qquad (10.6)$$

has a solution with respect to $K$ and $\gamma$.

### Example 10.1

Consider the continuous-time linear system in Example 7.12, with real scalars $\gamma_1 = -5$, $\gamma_2 = -1$. Using the MATLAB® function mincx in the MATLAB LMI toolbox, the corresponding LMI problem (10.6) is solved and a solution is found to be

$$\gamma = 0.3495, \quad K = \begin{bmatrix} -0.2240 \\ 0.2678 \end{bmatrix}.$$

Compared with the gain matrix obtained in Example 7.12, with magnitude 1.0895, this one has a much smaller magnitude. Further, with this solution, we can find

$$\alpha_1 = \lambda_{\min}\left(A_c^s\right) = -4.9748, \quad \alpha_2 = \lambda_{\max}\left(A_c^s\right) = -1.1251.$$

With the help of the MATLAB function eig, the closed-loop eigenvalues are obtained as $-4.9748$, $-1.9843$, and $-1.1251$, which are indeed located within $\mathbb{H}_{(-\alpha_2,-\alpha_1)} \subset \mathbb{H}_{(-\gamma_2,-\gamma_1)}$.

## 10.1.2 Insensitive Disk Region Designs with Minimum Gain

In this section, we consider the control of the following linear system

$$\begin{cases} \rho x = Ax + Bu \\ y = Cx, \end{cases} \qquad (10.7)$$

where $x \in \mathbb{R}^n$, $y \in \mathbb{R}^m$, and $u \in \mathbb{R}^r$ are the state vector, the output vector, and the input vector, respectively. Again, the symbol $\rho$ represents the differential operator (in the continuous-time case), or the one-step shift forward operator, that is, $\rho x(k) = x(k+1)$ (in the discrete-time case).

The problem of insensitive disk region designs with minimum gain is stated as follows.

**Problem 10.2** Consider the linear system (10.7). For given positive scalars $\gamma_0$ and $q$, design an output feedback control law $u = Ky$ such that

$$||K||_2 < \gamma \tag{10.8}$$

holds for a minimal positive scalar $\gamma$, and

$$\eta = ||A + BKC + qI||_2 < \gamma_0. \tag{10.9}$$

It follows from Section 7.5 that with condition (10.9) satisfied, the closed-loop eigenvalues are located in $\mathbb{D}_{(q,\eta)} \subset \mathbb{D}_{(q,\gamma_0)}$, and the boundary of $\mathbb{D}_{(q,\eta)}$ is insensitive to perturbations in the system matrices as well as in the gain matrix $K$. Therefore, this design is robust in the sense that the system has a disk eigenvalue region that is insensitive to small parameter perturbations in the system matrices.

It has been shown in Section 7.5 that condition (10.9) is equivalent to

$$\begin{bmatrix} -\gamma_0 I & (A + BKC + qI) \\ (A + BKC + qI)^{\mathrm{T}} & -\gamma_0 I \end{bmatrix} < 0.$$

Combining the earlier LMI with (10.2), we immediately have the following result.

**Theorem 10.2** Problem 10.2 has a solution if and only if the following optimization problem

$$\begin{cases} \min & \gamma \\ \text{s.t.} & \begin{bmatrix} -\gamma I & K \\ K^{\mathrm{T}} & -\gamma I \end{bmatrix} < 0 \\ & \begin{bmatrix} -\gamma_0 I & A + BKC + qI \\ (A + BKC + qI)^{\mathrm{T}} & -\gamma_0 I \end{bmatrix} < 0 \end{cases} \tag{10.10}$$

has a solution with respect to $K$ and $\gamma$.

Particularly, for the case of Schur stabilization, we choose to solve the earlier optimization problem with $q = 0$ and $\gamma_0 \leq 1$. In this case, the problem gives a stabilizing output feedback controller in the Schur stability sense.

**Example 10.2**

Consider the continuous-time linear system in Example 7.13 with $q = \gamma_0 = 4$, using the MATLAB function mincx in the MATLAB LMI toolbox we solve the minimization problem (10.10) and obtain the solution

$$K = \begin{bmatrix} -1.2246 \\ -0.2592 \end{bmatrix}, \quad \gamma = 1.2528.$$

Compared with the gain matrix obtained in Example 7.13, with magnitude 2.9411, this one has a much smaller magnitude. Further, with this solution, it is computed that

$$\eta = ||A + BKC + qI||_2 = 4.0000.$$

With the MATLAB function eig we find the closed-loop eigenvalues to be $-1.9046 \pm 2.4937i$ and $-1.6421$, which are indeed located within the desired region $\mathbb{D}_{(q,\eta)}$.

# 10.2 Mixed $H_\infty/H_2$ Designs with Desired Pole Regions

In this section, we study the multiobjective control of linear systems of the following form:

$$\begin{cases} \dot{x} = Ax + B_1 u + B_2 w \\ z_\infty = C_\infty x + D_{\infty 1} u + D_{\infty 2} w \\ z_2 = C_2 x + D_{21} u, \end{cases} \tag{10.11}$$

where

- $x \in \mathbb{R}^n$ and $z_2, z_\infty \in \mathbb{R}^m$ are the state vector and the output vectors, respectively
- $w \in \mathbb{R}^p$ and $u \in \mathbb{R}^r$ are the disturbance vector and the control vector
- $A, B_1, B_2, C_\infty, C_2, D_{\infty 1}, D_{\infty 2}$ and $D_{21}$ are the system coefficient matrices of appropriate dimensions

## 10.2.1 Problem

For the linear system (10.11), with the following state feedback control law

$$u = Kx, \tag{10.12}$$

the closed-loop system can be obtained as

$$\begin{cases} \dot{x} = (A + B_1 K) x + B_2 w \\ z_\infty = (C_\infty + D_{\infty 1} K) x + D_{\infty 2} w \\ z_2 = (C_2 + D_{21} K) x, \end{cases} \quad (10.13)$$

and the transfer function matrices $G_{z_\infty w}(s)$ and $G_{z_2 w}(s)$ are given, respectively, by

$$G_{z_\infty w}(s) = (C_\infty + D_{\infty 1} K)(sI - (A + B_1 K))^{-1} B_2 + D_{\infty 2},$$

and

$$G_{z_2 w}(s) = (C_2 + D_{21} K)(sI - (A + B_1 K))^{-1} B_2.$$

Thus, the $H_\infty$ performance and the $H_2$ performance requirements for the system are, respectively,

$$\|G_{z_\infty w}(s)\|_\infty < \gamma_\infty, \quad (10.14)$$

and

$$\|G_{z_2 w}(s)\|_2 < \gamma_2. \quad (10.15)$$

For performance of the system response, we introduce the closed-loop eigenvalue location requirement. Let

$$\mathbb{D} = \{s \mid s \in \mathbb{C}, \ L + sM + \bar{s}M^{\mathrm{T}} < 0\}, \quad (10.16)$$

where $L \in \mathbb{S}^q$, $M \in \mathbb{R}^{q \times q}$. Obviously, $\mathbb{D}$ is an LMI region with the following characteristic function

$$F_{\mathbb{D}} = L + sM + \bar{s}M^{\mathrm{T}}.$$

It is a region on the complex plane, which can be used to restrain the closed-loop eigenvalue locations.

In this section, we treat the following problem of mixed $H_\infty/H_2$ designs with LMI eigenvalue regions for the linear system (10.11).

**Problem 10.3** For the linear system (10.11), design a state feedback control law (10.12), such that

1. The $H_\infty$ performance (10.14) and the $H_2$ performance (10.15) are satisfied
2. The closed-loop eigenvalues are all located in $\mathbb{D}$, that is,

$$\lambda(A + B_1 K) \subset \mathbb{D},$$

with $\mathbb{D}$ being given by (10.16)

## 10.2.2 Solution to the Problem

### 10.2.2.1 Condition for the $H_\infty$ Performance

It follows from Theorem 8.1 that there exists a $K$ such that $\|G_{z_\infty w}(s)\|_\infty < \gamma_\infty$ holds if and only if there exist a matrix $W_\infty$, and an $X_\infty > 0$, such that

$$
\begin{bmatrix}
(AX_\infty + B_1 W_\infty)^{\mathrm{T}} + AX_\infty + B_1 W_\infty & B_2 & (C_\infty X_\infty + D_{\infty 1} W_\infty)^{\mathrm{T}} \\
B_2^{\mathrm{T}} & -\gamma_\infty I & D_{\infty 2}^{\mathrm{T}} \\
C_\infty X_\infty + D_{\infty 1} W_\infty & D_{\infty 2} & -\gamma_\infty I
\end{bmatrix} < 0.
$$

$$(10.17)$$

When this condition is met, such a feedback gain matrix can be given by

$$
K = K_\infty = W_\infty X_\infty^{-1}.
$$

### 10.2.2.2 Condition for the $H_2$ Performance

It follows from Theorem 8.3 that there exists a $K$ such that $\|G_{z_2 w}(s)\|_2 < \gamma_2$ holds if and only if there exist a matrix $W_2$, two symmetric matrices $Z$ and $X_2$, such that

$$
\begin{cases}
AX_2 + B_1 W_2 + (AX_2 + B_1 W_2)^{\mathrm{T}} + B_2 B_2^{\mathrm{T}} < 0 \\
\begin{bmatrix}
-Z & C_2 X_2 + D_{21} W_2 \\
(C_2 X_2 + D_{21} W_2)^{\mathrm{T}} & -X_2
\end{bmatrix} < 0 \\
\mathrm{trace}\,(Z) < \gamma_2^2.
\end{cases}
$$

$$(10.18)$$

When this condition is met, such a feedback gain can be taken as

$$
K = K_2 = W_2 X_2^{-1}.
$$

### 10.2.2.3 Condition for the Closed-Loop Pole Location

It follows from Theorem 7.5 that the pole placement constraint is satisfied if and only if there exist a matrix $W_D$, and an $X_D > 0$, such that

$$
L \otimes X_D + M \otimes (AX_D + B_1 W_D) + M^{\mathrm{T}} \otimes (AX_D + B_1 W_D)^{\mathrm{T}} < 0. \qquad (10.19)
$$

When this condition is met, such a feedback gain matrix can be taken as

$$
K = K_D = W_D X_D^{-1}.
$$

### 10.2.2.4 Solution

Let us define three sets based on the aforementioned three conditions (10.17) through (10.19):

$$\Omega_\infty = \{(X_\infty, W_\infty) \mid X_\infty > 0, \ W_\infty \in \mathbb{R}^{r \times n}, \text{ s.t. } (10.17) \text{ holds}\},$$

$$\Omega_2 = \{(X_2, W_2) \mid X_2 > 0, \ W_2 \in \mathbb{R}^{r \times n}, \exists Z \in \mathbb{S}^m \text{ s.t. } (10.18) \text{ holds}\},$$

and

$$\Omega_D = \{(X_D, W_D) \mid X_D > 0, \ W_D \in \mathbb{R}^{r \times n}, \text{ s.t. } (10.19) \text{ holds}\}.$$

Also, we denote the intersection of the three sets by $\Omega_0$, that is,

$$\Omega_0 = \Omega_\infty \cap \Omega_2 \cap \Omega_D.$$

Since our problem is to seek a single feedback gain $K$ such that all the three requirements (10.17) through (10.19) are simultaneously met, in view of the three expressions for the gain matrix $K$, it is clear that the three requirements (10.17) through (10.19) are linked by

$$W_\infty X_\infty^{-1} = W_2 X_2^{-1} = W_D X_D^{-1}. \tag{10.20}$$

Based on this reasoning, the following result is clearly true.

**Proposition 10.1** Problem 10.3 has a solution if and only if the following parameter set

$$\mathbb{P} = \left\{ (X_\infty, W_\infty, X_2, W_2, X_D, W_D) \; \middle| \; \begin{aligned} &(X_\infty, W_\infty) \in \Omega_\infty \\ &(X_2, W_2) \in \Omega_2 \\ &(X_D, W_D) \in \Omega_D \end{aligned} \right., \\ W_\infty X_\infty^{-1} = W_2 X_2^{-1} = W_D X_D^{-1} \right\}$$

is not null, and in this case, a feedback gain is given by

$$K = W_\infty X_\infty^{-1} = W_2 X_2^{-1} = W_D X_D^{-1}.$$

Since (10.20) is involved in the definition of the parameter set $\mathbb{P}$, finding a $(X_\infty, W_\infty, X_2, W_2, X_D, W_D) \in \mathbb{P}$ is obviously not an LMI problem, and hence can be very difficult. For convenience, let us set

$$X_D = X_\infty = X_2 \triangleq X, \tag{10.21}$$

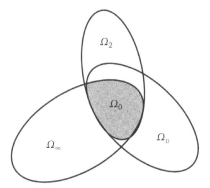

**Figure 10.1**   **Constrained parameter field.**

and

$$W_D = W_\infty = W_2 \triangleq W, \qquad (10.22)$$

then the parameter set $\mathbb{P}$ reduces to

$$
\begin{aligned}
\mathbb{P}_0 &= \{(X, W, X, W, X, W) \mid (X, W) \in \Omega_\infty, (X, W) \in \Omega_2, (X, W) \in \Omega_D\} \\
&= \{(X, W, X, W, X, W) \mid (X, W) \in \Omega_\infty \cap \Omega_2 \cap \Omega_D\} \\
&= \{(X, W, X, W, X, W) \mid (X, W) \in \Omega_0\}.
\end{aligned}
$$

Therefore, to avoid the difficulty, we now may find a $(X, W) \in \Omega_0$ instead of a $(X_\infty, W_\infty, X_2, W_2, X_D, W_D) \in \mathbb{P}$ (see Figure 10.1), and compute the gain matrix by

$$K = WX^{-1}. \qquad (10.23)$$

This leads to the following result for solution to Problem 10.3.

**Theorem 10.3**   Problem 10.3 has a solution if there exist two symmetric matrices $X, Z$ and a matrix $W$, satisfying

$$
\begin{bmatrix}
(AX + B_1 W)^{\mathrm{T}} + AX + B_1 W & B_2 & (C_\infty X + D_{\infty 1} W)^{\mathrm{T}} \\
B_2^{\mathrm{T}} & -\gamma_\infty I & D_{\infty 2}^{\mathrm{T}} \\
C_\infty X + D_{\infty 1} W & D_{\infty 2} & -\gamma_\infty I
\end{bmatrix} < 0, \quad (10.24)
$$

$$
\begin{cases}
AX + B_1 W + (AX + B_1 W)^{\mathrm{T}} + B_2 B_2^{\mathrm{T}} < 0 \\
\begin{bmatrix}
-Z & C_2 X + D_{21} W \\
(C_2 X + D_{21} W)^{\mathrm{T}} & -X
\end{bmatrix} < 0 \\
\mathrm{trace}\,(Z) < \gamma_2^2,
\end{cases}
\qquad (10.25)
$$

and

$$L \otimes X + M \otimes (AX + B_1 W) + M^T \otimes (AX + B_1 W)^T < 0. \qquad (10.26)$$

In this case, a feedback gain can be taken as in (10.23).

It is seen from the aforementioned equation that the parameters $X_D$, $X_\infty$, $X_2$ and $W_D$, $W_\infty$, $W_2$ are specially chosen as in (10.21) and (10.22), or the parameters are sought within the subset $\mathbb{P}_o$ instead of the entire set $\mathbb{P}$, the condition in this theorem is thus only sufficient.

Based on this theorem, the following optimization problem can be easily formulated, which has very important application significance:

$$\begin{cases} \min & c_2 \gamma_2^2 + c_\infty \gamma_\infty \\ \text{s.t.} & (10.24) \text{ through } (10.26), \end{cases} \qquad (10.27)$$

where $c_2 > 0$ and $c_\infty > 0$ are the weighting factors. To avoid nonlinearity in the aforementioned problem we may simply substitute $\gamma_2^2$ with a single variable $\rho$, as treated before.

**Example 10.3**

Consider the linear system

$$\begin{cases} \dot{x} = \begin{bmatrix} 2 & 1 & -2 \\ 1 & -1 & -3 \\ 4 & 0 & -1 \end{bmatrix} x + \begin{bmatrix} 1 & 0 \\ 0 & -1 \\ -2 & 1 \end{bmatrix} u + \begin{bmatrix} 0.1 \\ 0.2 \\ 0 \end{bmatrix} w \\ z_\infty = \begin{bmatrix} 0 & 0.1 & -0.1 \end{bmatrix} x + \begin{bmatrix} 0.1 & -0.1 \end{bmatrix} u + 0.01w \\ z_2 = \begin{bmatrix} 1 & -1 & 5 \end{bmatrix} x + \begin{bmatrix} 1 & 1 \end{bmatrix} u. \end{cases}$$

Design a state feedback control law, such that

1. The closed-loop system is stable, and the transfer function matrix satisfies

$$\left\| G_{z_\infty w}(s) \right\|_\infty < \gamma_\infty,$$

and

$$\left\| G_{z_2 w}(s) \right\|_2 < \gamma_2,$$

for possibly small positive scalars $\gamma_\infty$ and $\gamma_2$.
2. The closed-loop poles locate in the LMI region

$$\mathbb{D} = \left\{ x + yj \,\middle|\, x, y \in R, \ -5 < x < -0.5 \right\}.$$

Choosing

$$c_2 = 1, \quad c_\infty = 50,$$

and using the function `mincx` in the MATLAB LMI toolbox to the optimization problem (10.27), we obtain

$$X = 10^7 \times \begin{bmatrix} 0.1428 & -0.6082 & -0.0146 \\ -0.6082 & 2.5912 & 0.0620 \\ -0.0146 & 0.0620 & 0.0015 \end{bmatrix} > 0,$$

$$W = 10^7 \times \begin{bmatrix} -0.0423 & 0.1800 & 0.0043 \\ -0.6359 & 2.7092 & 0.0649 \end{bmatrix}',$$

$$Z = 0.0018,$$

and

$$\gamma_\infty = 0.0122, \quad \gamma_2^2 = 0.0018.$$

The state feedback gain can be computed as

$$K = WX^{-1} = \begin{bmatrix} -4.2265 & -0.8970 & -1.0699 \\ 3.1769 & 1.8843 & -3.8835 \end{bmatrix}',$$

and the set of the closed-loop poles is

$$\lambda(A + BK) = \{-2.7512 \pm 6.7000i, \ -2.3521\}.$$

To verify the effect of disturbance attenuation, we have plotted the output $z_\infty$ and $z_2$ in Figure 10.2, for the cases of $w(t) = 0$, and $w(t) =$ wgn(101, 1, 10,' real'), which generate a 101-by-1 real matrix of white Gaussian noise with power of 10dBW. It is seen from this figure that the disturbance indeed has very small affection on the system output.

For comparison, using the function `place` in MATLAB control system toolbox to the same system, we obtain a feedback gain as

$$K_1 = \begin{bmatrix} -6.4901 & -2.4961 & -0.5944 \\ -2.3319 & 0.3851 & -2.1699 \end{bmatrix}',$$

which assigns the same set of closed-loop poles, that is, $-2.7521 \pm 6.6925i$, $-2.3521$, as the gain $K$. The closed-loop system output corresponding to the aforementioned feedback gain $K_1$ is also shown in Figure 10.3. Comparing the two figures, it is easy to see that the disturbance effect in the closed-loop system resulted in by $K_1$ is more than 10 times of that in the closed-loop system resulted in by $K$. Therefore, the mixed $H_\infty/H_2$ controller is better since it gives responses with much better disturbance attenuation effect.

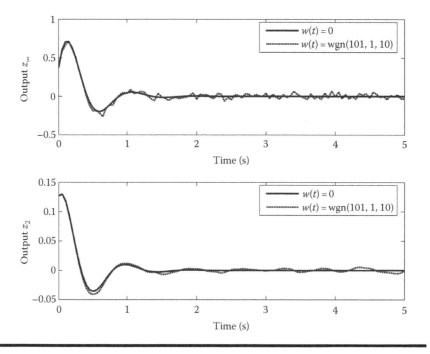

**Figure 10.2** The output $z_\infty$ and $z_2$, Example 10.3.

## 10.3 Mixed Robust $H_\infty/H_2$ Designs with Desired Pole Regions

In the last section, we treated the problem of mixed $H_2/H_\infty$ designs with a desired LMI pole region. In this section, we further investigate the perturbed system case.

### 10.3.1 Problem

Different from Section 10.2, in this section we consider a type of uncertain linear systems in the following form:

$$\begin{cases} \dot{x} = (A + \Delta A)x + (B_1 + \Delta B_1)u + B_2 w \\ z_\infty = C_\infty x + D_{\infty 1}u + D_{\infty 2}w \\ z_2 = C_2 x + D_{21}u, \end{cases} \tag{10.28}$$

where

- $x \in \mathbb{R}^n$ and $z_2 \in \mathbb{R}^m$, $z_\infty \in \mathbb{R}^l$ are, respectively, the state vector and the output vectors
- $w \in \mathbb{R}^p$ and $u \in \mathbb{R}^r$ are the disturbance vector and the control vector

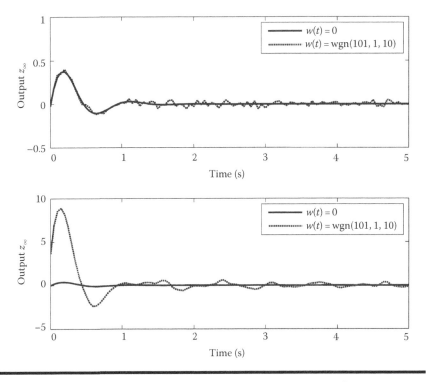

**Figure 10.3   The system output corresponding to gain $K_1$, Example 10.3.**

- $A$, $B_1$, $B_2$, $C_\infty$, $C_2$, $D_{\infty 1}$, $D_{\infty 2}$ and $D_{21}$ are the system coefficient matrices of appropriate dimensions
- $\Delta A$ and $\Delta B_1$ are real-valued matrix functions which represent the time-varying parameter uncertainties

Furthermore, the parameter uncertainties $\Delta A$ and $\Delta B_1$ are in the form of

$$\begin{bmatrix} \Delta A & \Delta B_1 \end{bmatrix} = HF \begin{bmatrix} E_1 & E_2 \end{bmatrix}, \tag{10.29}$$

where

- $H$, $E_1$ and $E_2$ are known matrices of appropriate dimensions
- $F$ is a matrix containing the uncertainty, which satisfies

$$F^\mathrm{T} F \leq I.$$

For the linear system (10.28), design a state feedback control law

$$u = Kx, \tag{10.30}$$

the closed-loop system can be obtained as

$$\begin{cases} \dot{x} = [(A + \Delta A) + (B_1 + \Delta B_1) K] x + B_2 w \\ z_\infty = (C_\infty + D_{\infty 1} K) x + D_{\infty 2} w \\ z_2 = (C_2 + D_{21} K) x, \end{cases} \tag{10.31}$$

and, corresponding to the two output vectors, the transfer function matrices are given, respectively, by

$$\tilde{G}_{z_\infty w}(s) = (C_\infty + D_{\infty 1} K) (sI - [(A + \Delta A) + (B_1 + \Delta B_1) K])^{-1} B_2 + D_{\infty 2},$$

and

$$\tilde{G}_{z_2 w}(s) = (C_2 + D_{21} K) (sI - [(A + \Delta A) + (B_1 + \Delta B_1) K])^{-1} B_2.$$

Thus, the $H_\infty$ performance and the $H_2$ performance requirements are, respectively,

$$||\tilde{G}_{z_\infty w}(s)||_\infty < \gamma_\infty, \tag{10.32}$$

and

$$||\tilde{G}_{z_2 w}(s)||_2 < \gamma_2. \tag{10.33}$$

In this section, we treat the following problem of mixed robust $H_\infty/H_2$ design with an LMI eigenvalues region for the linear system (10.28).

**Problem 10.4** For the linear system (10.28), design a state feedback control law (10.30) such that

1. The $H_\infty$ performance (10.32) and the $H_2$ performance (10.33) are satisfied
2. The nonperturbed closed-loop eigenvalues are all located in $\mathbb{D}$, that is,

$$\lambda (A + B_1 K) \subset \mathbb{D},$$

with $\mathbb{D}$ being given by (10.16).

## 10.3.2 Solution to the Problem

### 10.3.2.1 Condition for the Robust $H_\infty$ Performance

It follows from Theorem 8.5 that $H_\infty$ performance (10.32) is met if and only if there exist a scalar $\alpha$, a matrix $W_\infty$, and an $X_\infty > 0$ such that

$$\begin{bmatrix} \Psi(X_\infty, W_\infty) & B_2 & (C_\infty X_\infty + D_{\infty 1} W_\infty)^{\mathrm{T}} & (E_1 X_\infty + E_2 W_\infty)^{\mathrm{T}} \\ B_2^{\mathrm{T}} & -\gamma_\infty I & D_{\infty 2}^{\mathrm{T}} & 0 \\ C_\infty X_\infty + D_{\infty 1} W_\infty & D_{\infty 2} & -\gamma_\infty I & 0 \\ E_1 X_\infty + E_2 W_\infty & 0 & 0 & -\alpha I \end{bmatrix} < 0,$$

$$(10.34)$$

where

$$\Psi(X_\infty, W_\infty) = \langle A X_\infty + B_1 W_\infty \rangle_s + \alpha H H^{\mathrm{T}}.$$

When this condition is met, such a feedback gain can be taken as

$$K = K_\infty = W_\infty X_\infty^{-1}.$$

### 10.3.2.2 Condition for the Robust $H_2$ Performance

It follows from Theorem 8.6 that the $H_2$ performance (10.33) is met if and only if there exist a scalar $\beta$, a matrix $W_2$, two symmetric matrices $Z$ and $X_2$ such that

$$\begin{cases} \begin{bmatrix} \langle A X_2 + B_1 W_2 \rangle_s + B_2 B_2^{\mathrm{T}} + \beta H H^{\mathrm{T}} & (E_1 X_2 + E_2 W_2)^{\mathrm{T}} \\ E_1 X_2 + E_2 W_2 & -\beta I \end{bmatrix} < 0 \\ \begin{bmatrix} -Z & C_2 X_2 + D_{21} W_2 \\ (C_2 X_2 + D_{21} W_2)^{\mathrm{T}} & -X_2 \end{bmatrix} < 0 \\ \mathrm{trace}\,(Z) < \gamma_2^2. \end{cases}$$

$$(10.35)$$

When this condition is met, such a feedback gain can be taken as

$$K = K_2 = W_2 X_2^{-1}.$$

### 10.3.2.3 Condition for the Robust Closed-Loop Pole Location

It follows from Theorem 7.5 that the pole placement constraint is satisfied if and only if there exist a matrix $W_D$ and an $X_D > 0$ such that

$$L \otimes X_D + M \otimes (A X_D + B_1 W_D) + M^{\mathrm{T}} \otimes (A X_D + B_1 W_D)^{\mathrm{T}} < 0. \quad (10.36)$$

When this condition is met, such a feedback gain matrix can be taken as

$$K = K_D = W_D X_D^{-1}.$$

### 10.3.2.4 Solution

Following the same lines as in Section 10.2, let us define three sets based on the three conditions (10.34) through (10.36):

$$\Pi_\infty = \left\{ (X_\infty, W_\infty) \mid X_\infty > 0, \ W_\infty \in \mathbb{R}^{m \times n}, \exists \alpha > 0, \text{s.t. (10.34) holds} \right\},$$

$$\Pi_2 = \left\{ (X_2, W_2) \mid X_2 > 0, \ W_2 \in \mathbb{R}^{r \times n}, \exists \beta > 0, Z \in \mathbb{S}^n \text{ s.t. (10.35) holds} \right\},$$

and

$$\Pi_D = \left\{ (X_D, W_D) \mid X_D > 0, \ W_D \in \mathbb{R}^{r \times n}, \text{s.t. (10.36) holds} \right\}.$$

Also, we denote the intersection of the three sets by $\Pi_0$, that is,

$$\Pi_0 = \Pi_\infty \cap \Pi_2 \cap \Pi_D.$$

Since our problem is to seek a single feedback gain $K$ such that all the three requirements (10.34) through (10.36) are simultaneously met, in view of the three expressions for the gain matrix $K$, it is clear that the three requirements (10.34) through (10.36) are linked by

$$W_\infty X_\infty^{-1} = W_2 X_2^{-1} = W_D X_D^{-1}. \tag{10.37}$$

Based on the aforementioned reasoning, the following result is clearly true.

**Proposition 10.2** Problem 10.4 has a solution if and only if the following parameter set

$$\mathbb{F} = \left\{ (X_\infty, W_\infty, X_2, W_2, X_D, W_D) \ \middle| \ \begin{array}{l} \left\{ \begin{array}{l} (X_\infty, W_\infty) \in \Pi_\infty \\ (X_2, W_2) \in \Pi_2 \\ (X_D, W_D) \in \Pi_D \end{array} \right. \\[6pt] W_\infty X_\infty^{-1} = W_2 X_2^{-1} = W_D X_D^{-1} \end{array} \right\}$$

is not null, and in this case, a feedback gain is given by

$$K = W_\infty X_\infty^{-1} = W_2 X_2^{-1} = W_D X_D^{-1}.$$

Obviously, since (10.37) is involved in the definition of the parameter set $\mathbb{F}$, finding a $(X_\infty, W_\infty, X_2, W_2, X_D, W_D) \in \mathbb{F}$ is not an LMI problem, and hence can be very difficult. For convenience, let us set

$$X_D = X_\infty = X_2 \triangleq X,$$

and

$$W_D = W_\infty = W_2 \triangleq W,$$

then the parameter set $\mathbb{F}$ reduces to

$$\begin{aligned}
\mathbb{F}_0 &= \{(X, W, X, W, X, W) \mid (X, W) \in \Pi_\infty, (X, W) \in \Pi_2, (X, W) \in \Pi_D\} \\
&= \{(X, W, X, W, X, W) \mid (X, W) \in \Pi_\infty \cap \Pi_2 \cap \Pi_D\} \\
&= \{(X, W, X, W, X, W) \mid (X, W) \in \Pi_0\}.
\end{aligned}$$

Therefore, to avoid the difficulty, we now may find a $(X, W) \in \Pi_0$ instead of a $(X_\infty, W_\infty, X_2, W_2, X_D, W_D) \in \mathbb{F}$, and compute the gain matrix by

$$K = WX^{-1}. \tag{10.38}$$

This leads to the following result for solution to Problem 10.4.

**Theorem 10.4**    Problem 10.4 has a solution if there exist scalars $\alpha$, $\beta$, two symmetric matrices $X$, $Z$ and a matrix $W$, satisfying

$$\begin{bmatrix}
\Psi(X, W) & B_2 & (C_\infty X + D_{\infty 1} W)^T & (E_1 X + E_2 W)^T \\
B_2^T & -\gamma_\infty I & D_{\infty 2}^T & 0 \\
C_\infty X + D_{\infty 1} W & D_{\infty 2} & -\gamma_\infty I & 0 \\
E_1 X + E_2 W & 0 & 0 & -\alpha I
\end{bmatrix} < 0, \tag{10.39}$$

$$\begin{cases}
\begin{bmatrix}
\langle AX + B_1 W \rangle_s + B_2 B_2^T + \beta HH^T & (E_1 X + E_2 W)^T \\
E_1 X + E_2 W & -\beta I
\end{bmatrix} < 0 \\[2ex]
\begin{bmatrix}
-Z & C_2 X + D_{21} W \\
(C_2 X + D_{21} W)^T & -X
\end{bmatrix} < 0 \\[2ex]
\text{trace}\,(Z) < \gamma_2^2,
\end{cases} \tag{10.40}$$

and

$$L \otimes X + M \otimes (AX + B_1 W) + M^T \otimes (AX + B_1 W)^T < 0, \tag{10.41}$$

where

$$\Psi(X, W) = \langle AX + B_1 W \rangle_s + \alpha HH^T.$$

In this case, a feedback gain can be taken as in (10.38).

Based on the aforementioned theorem, the following optimization problem can be easily formulated, which minimizes a weighted $H_\infty/H_2$ index:

$$\begin{cases} \min & c_2 \gamma_2^2 + c_\infty \gamma_\infty \\ \text{s.t.} & (10.39)–(10.41), \end{cases} \tag{10.42}$$

where $c_2$ and $c_\infty$ are the weighting factors. The nonlinearity in the aforementioned problem can be avoided simply by introducing the simple variable substitution $\rho = \gamma_2^2$.

### Example 10.4

Consider a linear system in the form of (10.28) with the following parameters

$$A = \begin{bmatrix} 2 & 1 & -2 \\ 1 & -1 & -3 \\ 4 & 0 & -1 \end{bmatrix}, \quad B_1 = \begin{bmatrix} 1 & 0 \\ 0 & 3 \\ 3 & 1 \end{bmatrix}, \quad B_2 = \begin{bmatrix} -1 & 0 \\ 2 & 1 \\ 1 & -5 \end{bmatrix},$$

$$C_\infty = \begin{bmatrix} 2 & 1 & -0.5 \end{bmatrix}, \quad C_2 = \begin{bmatrix} 1 & 0 & 1 \end{bmatrix},$$

$$D_{\infty1} = \begin{bmatrix} 0 & 1 \end{bmatrix}, \quad D_{\infty2} = \begin{bmatrix} 0.01 & -0.02 \end{bmatrix}, \quad D_{21} = \begin{bmatrix} 1 & 0 \end{bmatrix},$$

$$H = \begin{bmatrix} 0 & 0 & -0.2 \\ 0 & -0.1 & 0.1 \\ 0.2 & 0 & 0.1 \end{bmatrix},$$

$$E_1 = \begin{bmatrix} 0.1 & 0.2 & 0.1 \\ 0.1 & 0.2 & 0.3 \\ 0.2 & 0.1 & 0.1 \end{bmatrix}, \quad E_2 = \begin{bmatrix} -0.1 & 0.5 \\ 0.2 & 0.1 \\ -0.2 & 0 \end{bmatrix}.$$

Design a state feedback control law such that

1. The closed-loop system is asymptotically stable, and the following $H_\infty$ and $H_2$ performances

$$\| G_{z_\infty w}(s) \|_\infty < \gamma_\infty,$$

$$\| G_{z_2 w}(s) \|_2 < \gamma_2,$$

are satisfied for possibly small positive scalars $\gamma_\infty$ and $\gamma_2$.
2. The closed-loop poles locate in the LMI region

$$\mathbb{H}_{0.5,10} = \left\{ x + yj \, | x, y \in R, \, -10 < x < -0.5 \right\}.$$

Choosing

$$c_2 = 1, \quad c_\infty = 5,$$

and using the function `mincx` in the MATLAB LMI toolbox, we obtain for the optimization problem (10.42) the following set of solutions:

$$X = 10^8 \times \begin{bmatrix} 2.3553 & -2.6201 & 0.4250 \\ -2.6201 & 6.6853 & 1.0505 \\ 0.4250 & 1.0505 & 1.0143 \end{bmatrix} > 0,$$

$$W = 10^8 \times \begin{bmatrix} -2.7803 & 1.5696 & -1.4393 \\ -1.8779 & -0.9198 & -1.3935 \end{bmatrix},$$

$$Z = 1.1777 \times 10^{-4},$$

and

$$\gamma_\infty = 0.0316, \quad \gamma_2^2 = 0.0002,$$

$$\alpha = 2.6639 \times 10^8, \quad \beta = 2.6618 \times 10^8,$$

and the state feedback gain is then given by

$$K = WX^{-1} = \begin{bmatrix} -1.0000 & 0.0000 & -1.0000 \\ -2.0000 & -1.0000 & 0.5000 \end{bmatrix}.$$

Thus, the set of the closed-loop poles can be obtained as

$$\lambda(A + B_1 K) = \{-0.5342 \pm 0.9546i, \; -5.4316\}.$$

To verify the effect of disturbance attenuation, we have plotted the output $z_\infty$ and $z_2$ with $w(t) = \begin{bmatrix} 0 & 0 \end{bmatrix}^T$ and $w(t) = \begin{bmatrix} \sin t & 2\sin 2t \end{bmatrix}^T$ in Figure 10.4, for the cases

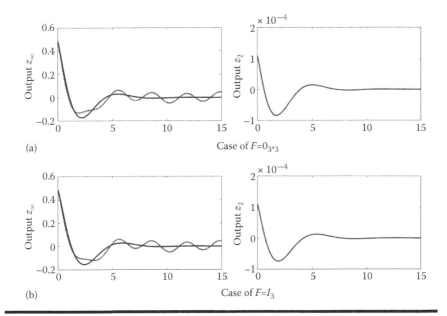

**Figure 10.4** The output $z_\infty$ and $z_2$, Example 10.4.

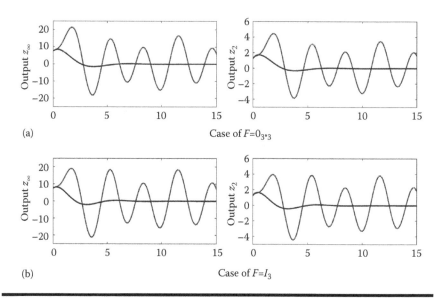

**Figure 10.5   System output corresponding to gain $K_1$, Example 10.4.**

of $F = 0_{3 \times 3}$ and $F = I_3$. It is seen from this figure that the uncertainty indeed has very small affection on the system output.

For comparison, using the function `place` in MATLAB control system toolbox to the same system, we obtain a feedback gain

$$K_1 = \begin{bmatrix} -0.7525 & 0.0200 & -0.8291 \\ -0.6031 & -1.3010 & 0.6426 \end{bmatrix},$$

which assigns the same set of closed-loop poles, that is, $\lambda(A + B_1 K)$, as the gain $K$. The closed-loop system output corresponding to the aforementioned feedback gain $K_1$ is also shown, in Figure 10.5. It is obvious to see from these two figures that the designed robust $H_\infty/H_2$ controller is much better because it gives responses with smaller affection from both the disturbance and the system parameter uncertainty.

# 10.4  Notes and References

## 10.4.1  Summary of Main Results

This chapter studies multiple objective design in control linear systems by LMI techniques. In Section 10.1, we introduce the problems of insensitive strip region and the disk region designs with minimum gain. The LMI conditions for these problems are given in Table 10.1. We remark that the requirements in these problems are

**Table 10.1  Criteria for Insensitive Region Designs**

| Regions | Criteria | Controllers |
|---------|----------|-------------|
| Strip | $\begin{bmatrix} -\gamma I & K \\ K^T & -\gamma I \end{bmatrix} < 0$ <br><br> $2\gamma_1 I < (A + BKC)^T + (A + BKC) < 2\gamma_2 I$ | $K$ |
| Disk | $\begin{bmatrix} -\gamma I & K \\ K^T & -\gamma I \end{bmatrix} < 0$ <br><br> $\begin{bmatrix} -\gamma_0 I & (A + BKC + qI) \\ (A + BKC + qI)^T & -\gamma_0 I \end{bmatrix} < 0$ | $K$ |

strong. Nevertheless, once we find a solution to a specific problem, we get a system with a very nice robustness property.

Section 10.2 deals with the problem of mixed $H_\infty/H_2$ design with a desired LMI pole region, in which the following requirements

$$||G_{z_\infty w}(s)||_\infty < \gamma_\infty,$$

$$||G_{z_2 w}(s)||_2 < \gamma_2,$$

and

$$\lambda (A + B_1 K) \subset \mathbb{D}$$

are satisfied simultaneously. The LMI conditions for the problem of mixed $H_\infty/H_2$ design with a desired LMI pole region are obtained as

$$
\begin{cases}
\text{trace} (Z) < \gamma_2^2 \\
\langle AX + B_1 W \rangle_s + B_2 B_2^T < 0 \\
\begin{bmatrix} -Z & * \\ (C_2 X + D_{21} W)^T & -X \end{bmatrix} < 0 \\
\begin{bmatrix} (AX + B_1 W)^T + AX + B_1 W & * & * \\ B_2^T & -\gamma_\infty I & * \\ C_\infty X + D_{\infty 1} W & D_{\infty 2} & -\gamma_\infty I \end{bmatrix} < 0 \\
L \otimes X + \langle M \otimes (AX + B_2 W) \rangle_s < 0.
\end{cases}
$$

In this case, a feedback gain can be taken as

$$K = WX^{-1}.$$

Section 10.3 deals with the problem of mixed robust $H_\infty/H_2$ design with a desired LMI pole region, in which the $H_\infty$, $H_2$, and the closed-loop eigenvalue performance requirements are satisfied simultaneously. The LMI conditions for the problem of mixed robust $H_\infty/H_2$ design with a desired LMI pole region are obtained as

$$\begin{cases}
\text{trace}\,(Z) < \gamma_2^2 \\[4pt]
\begin{bmatrix} -Z & C_2X + D_{21}W \\ (C_2X + D_{21}W)^{\mathrm{T}} & -X \end{bmatrix} < 0 \\[10pt]
\begin{bmatrix} \langle AX + B_1W \rangle_s + B_2B_2^{\mathrm{T}} + \beta HH^{\mathrm{T}} & * \\ E_1X + E_2W & -\beta I \end{bmatrix} < 0 \\[10pt]
L \otimes X + \langle M \otimes (AX + B_2W) \rangle_s < 0 \\[4pt]
\begin{bmatrix} \Psi(X, W) & * & * & * \\ B_2^{\mathrm{T}} & -\gamma_\infty I & * & * \\ C_\infty X + D_{\infty 1}W & D_{\infty 2} & -\gamma_\infty I & * \\ E_1X + E_2W & 0 & 0 & -\alpha I \end{bmatrix} < 0,
\end{cases}$$

where

$$\Psi(X, W) = \langle AX + B_1W \rangle_s + \alpha HH^{\mathrm{T}}.$$

In this case, a feedback gain can be taken as

$$K = WX^{-1}.$$

## 10.4.2 Further Remarks

Most control system design techniques have only paid attention to the optimal solution for one specific performance index, for example, $H_\infty$ norm of a closed-loop system transfer function (Scherer et al. 1997), the quadratic functional (Scherer et al. 1997), or the eigenvalue sensitivity indices (Duan 1992, 1993, Duan and Patton 1999, Duan and Wu 2005, Duan and Yu 2008). However, many practical control systems are required to have the ability to fit simultaneously different and often conflicting performance objectives as best as possible. Therefore, multiple objective designs are very important in practical control applications.

One type of approaches for multiple objective control is based on eigenstructure assignment, which provides complete parametrization of certain type of controllers (see Duan 1994, 1995, 1998, 1999, 2003, 2004b, Duan and Liu 2002). The degrees of freedom beyond eigenstructure assignment can be used to meet other system design requirements (see Duan 1992, 1993, Duan and Patton 1999, Duan et al. 2000, He et al. 2004a, Duan and Wu 2005, Bachelier et al. 2006, Duan and Yu 2008, Duan et al. 2009b). The advantage of this type of approaches lies in its flexibility in treating the various system design objectives, while the drawback is that often the formulated optimization problem is nonconvex and hence the optimal solution cannot be obtained usually.

In this chapter, we have presented the LMI approaches for multiple objective designs through a few typical problems. The advantage of this type of approaches is that a globally optimal and numerically reliable solution can always be obtained as long as the finally formulated LMI problem has a solution. Along this direction, there have been many reported results, for example, multiobjective control synthesis in uncertain fuzzy systems (He and Duan 2005), robust reliable $H_\infty$ control for uncertain systems with pole constraints (Gao et al. 2012b), substructural multiobjective $H_\infty$ controller design for large flexible structures (Toker and Sunar 2005), and $H_\infty$ and $H_2$ guaranteed cost gain-scheduling quadratic stabilization of linear time-varying polytopic systems (Montagner et al. 2007). Particularly, Scherer et al. (1997) investigated multiple objective output feedback control via LMI optimization, in which altogether 9 different design specifications have been considered, including $H_\infty$ and $H_2$ performances and the desired LMI pole region requirement.

# Exercises

**10.1** Consider a linear system in the form of (10.3) with the following parameters (Duan (2003)):

$$A = \begin{bmatrix} 0.5 & 0 & 0 \\ 0 & -2 & 10 \\ 0 & 1 & -2 \end{bmatrix}, \quad B = \begin{bmatrix} 1 & 0 \\ -2 & 2 \\ 0 & 1 \end{bmatrix}, \quad C = \begin{bmatrix} 1 & 0 & 0 \\ 0 & 0 & 1 \end{bmatrix}.$$

Solve the problem of insensitive strip region design with minimum gain for the case of $\gamma_1 = -4$ and $\gamma_2 = -1$.

**10.2** Consider a discrete-time linear system in the form of (10.7) with the following parameters:

$$A = \begin{bmatrix} 0.9696 & 0.0202 \\ 0.0404 & 0.9898 \end{bmatrix}, \quad B = \begin{bmatrix} 50500 \\ 50500 \end{bmatrix}, \quad C = \begin{bmatrix} 1 & 1 \end{bmatrix}.$$

This is a model describing the population distribution in a certain country (Zheng (2002), page 25–27). The system variables are

- $x_1(k)$, city population in the $k$th year;
- $x_2(k)$, rural population in the $k$th year;

■ $y(k)$, total population in the $k$th year; and
■ $u(k)$, control policy taken in the $k$th year.

Note that

$$\lambda(A) = \{0.9494, 1.0100\},$$

the system is not stable, and the purpose of control is to make the system robustly stable with minimal control effort. In doing so, we suffice to solve the problem of insensitive disk region design with minimum gain for the case of $q = 0$ and $\gamma_0 = 1$.

**10.3** Consider the linear system in the form of (10.11) with

$$A = \begin{bmatrix} 2 & 1 & -2 \\ 1 & -1 & -3 \\ 4 & 0 & -1 \end{bmatrix}, \quad B_1 = \begin{bmatrix} 1 & 0 \\ 0 & 3 \\ 3 & 1 \end{bmatrix}, \quad B_2 = \begin{bmatrix} 1 \\ 0.2 \\ -0.5 \end{bmatrix},$$

$$C_\infty = \begin{bmatrix} 2 & 1 & -0.5 \end{bmatrix}, \quad D_{\infty 1} = \begin{bmatrix} 0.2 & 0.5 \end{bmatrix}, \quad D_{\infty 2} = 0.05,$$

$$C_2 = \begin{bmatrix} 0.1 & 0 & -0.1 \\ 0 & 0.2 & 0.3 \end{bmatrix}, \quad D_{21} = \begin{bmatrix} 0.1 & 0.1 \\ 0 & 0.1 \end{bmatrix}.$$

Design a state feedback control law, such that

1. the closed-loop poles locate in the LMI region

$$\mathbb{D} = \left\{ x + yj \,\middle|\, x, y \in R,\ (x+2)^2 + \frac{1}{4}y^2 < 1 \right\};$$

2. the following objective is minimized:

$$J = \left\| G_{z_\infty w}(s) \right\|_\infty + \left\| G_{z_2 w}(s) \right\|_2^2,$$

where

$$G_{z_\infty w}(s) = (C_\infty + D_{\infty 1} K)(sI - (A + B_1 K))^{-1} B_2 + D_{\infty 2},$$

$$G_{z_2 w}(s) = (C_2 + D_{21} K)(sI - (A + B_1 K))^{-1} B_2.$$

**10.4** Consider a linear system in the form of (10.28) with the following parameters

$$A = \begin{bmatrix} 1 & 2 & -1 \\ 2 & 0 & -1 \\ 1 & 1 & 0 \end{bmatrix}, \quad B_1 = \begin{bmatrix} 0 & 1 \\ 1 & 2 \\ -1 & 1 \end{bmatrix}, \quad B_2 = \begin{bmatrix} 1 & -2 \\ 0 & -1 \\ -1 & 5 \end{bmatrix},$$

$$C_\infty = \begin{bmatrix} 0.2 & 1 & -1 \end{bmatrix}, \quad D_{\infty 1} = \begin{bmatrix} 0 & 0.1 \end{bmatrix}, \quad D_{\infty 2} = \begin{bmatrix} 0.01 & -0.02 \end{bmatrix},$$

$$C_2 = \begin{bmatrix} 0.1 & 0 & 0.2 \end{bmatrix}, \quad D_{21} = \begin{bmatrix} -0.1 & 0 \end{bmatrix},$$

and the following parameter uncertainties

$$
\begin{bmatrix} \Delta A & \Delta B_1 \end{bmatrix} = HF \begin{bmatrix} E_1 & E_2 \end{bmatrix}, \quad F^{\mathrm{T}} F \le I,
$$

$$
H = \begin{bmatrix} 0 & 0 & -0.2 \\ 0 & -0.1 & 0.1 \\ 0.2 & 0 & 0.1 \end{bmatrix},
$$

$$
E_1 = \begin{bmatrix} 0.1 & 0.2 & 0.1 \\ 0.1 & 0.2 & 0.3 \\ 0.2 & 0.1 & 0.1 \end{bmatrix}, \quad E_2 = \begin{bmatrix} -0.1 & 0.5 \\ 0.2 & 0.1 \\ -0.2 & 0 \end{bmatrix}.
$$

Design a state feedback control law, such that

1. the closed-loop poles locate in the LMI region

$$
\mathbb{D} = \left\{ x + yj \,\middle|\, x, y \in R, \ (x+3)^2 + 4y^2 < 4 \right\};
$$

2. the following objective is minimized:

$$
J = 5 \left\| \tilde{G}_{z_\infty w}(s) \right\|_\infty + \left\| \tilde{G}_{z_2 w}(s) \right\|_2^2,
$$

where

$$
\tilde{G}_{z_\infty w}(s) = (C_\infty + D_{\infty 1} K) (sI - A_c)^{-1} B_2 + D_{\infty 2},
$$
$$
\tilde{G}_{z_2 w}(s) = (C_2 + D_{21} K) (sI - A_c)^{-1} B_2,
$$

with

$$
A_c = (A + \Delta A) + (B_1 + \Delta B_1) K.
$$

# APPLICATIONS

# Chapter 11

# Missile Attitude Control

As one of the applications treated in the book, in this chapter we consider the attitude control of missiles. Two types of missiles are involved, one is the type of so-called nonrotating missiles, in which rotating is restricted during flight, the other is the type of BTT (bank-to-turn) missiles, which makes use of rotation in their turns.

In the first section, the simplified attitude system models of the two types of missiles are introduced, and then in the second section the design of the attitude control system of the type of nonrotating missiles is treated, where mixed $H_2/H_\infty$ performance is considered in order to attenuate the disturbance, and the closed-loop eigenvalues are restricted within a desired region in order to ensure the closed-loop dynamical properties. The third section investigates the attitude control system design in a BTT missile using techniques of simultaneous stabilization based on LMIs.

## 11.1 Dynamical Model

The dynamical model of a missile attitude system is generally very complicated. In practical applications, usually simplified models are used. The simplified dynamical model of the attitude systems of a type of missiles can be written as follows:

$$
\begin{cases}
\dot{\alpha}(t) = \omega_z(t) - \frac{1}{57.3}\omega_x(t)\beta(t) - a_4(t)\alpha(t) - a_5(t)\delta_z(t) \\[2ex]
\dot{\beta}(t) = \omega_y(t) + \frac{1}{57.3}\omega_x(t)\alpha(t) - b_4(t)\beta(t) - b_5(t)\delta_y(t) + a_6(t)\delta_z(t) \\[2ex]
\dot{\phi}(t) = \omega_x(t) \\[2ex]
\dot{\omega}_z(t) = -a_1(t)\omega_z(t) + a_1'(t)\dot{\alpha}(t) - a_2(t)\alpha(t) - a_3(t)\delta_z(t) + \frac{J_x - J_y}{57.3J_z}\omega_x(t)\omega_y(t) \\[2ex]
\dot{\omega}_y(t) = -b_1(t)\omega_y(t) - b_1'(t)\dot{\beta}(t) - b_2(t)\beta(t) - b_3(t)\delta_y(t) + \frac{J_z - J_x}{57.3J_y}\omega_x(t)\omega_z(t) \\[2ex]
\dot{\omega}_x(t) = -c_1(t)\omega_x(t) + c_2(t)\beta(t) - c_3(t)\delta_x(t) + c_4(t)\delta_y(t) + \frac{J_y - J_z}{57.3J_x}\omega_y(t)\omega_z(t) \\[2ex]
n_y(t) = \frac{V(t)}{57.3g}(a_4(t)\alpha(t) + a_5(t)\delta_z(t) + b_7(t)\beta(t)) \\[2ex]
n_z(t) = -\frac{V(t)}{57.3g}(b_4(t)\beta(t) + b_5(t)\delta_y(t) + b_6(t)\delta_z(t)),
\end{cases}
$$

$$(11.1)$$

where

- $\alpha$, $\beta$ and $\phi$ are the attack angle, the sideslip angle and the roll angle, respectively
- $\omega_x$, $\omega_y$ and $\omega_z$ are the roll angular velocity, the yaw angular velocity and the pitch angular velocity, respectively
- $n_y$ and $n_z$ are respectively the overloads on the side direction and the normal direction
- $J_x$, $J_y$ and $J_z$ are the rotary inertia of the missile corresponding to the body coordinates
- $g = 9.81 \mathrm{m/s}^2$ is the acceleration of gravity
- $V$ is the speed of the missile
- $a_1(t) \sim a_6(t)$, $b_1(t) \sim b_7(t)$, $a_1'(t)$, $b_1'(t)$ and $c_1(t) \sim c_4(t)$ are the system parameters
- $\delta_x$, $\delta_y$ and $\delta_z$ are the aileron deflection, the rudder deflection and the elevator deflection, respectively

The actual actuator deflections are governed by the dynamic behaviors of the actuators described as follows:

$$
\begin{cases}
\dot{\delta}_x(t) = \dfrac{1}{\tau_x}(\delta_{xc}(t) - \delta_x(t)) \\[2ex]
\dot{\delta}_y(t) = \dfrac{1}{\tau_y}(\delta_{yc}(t) - \delta_y(t)) \\[2ex]
\dot{\delta}_z(t) = \dfrac{1}{\tau_z}(\delta_{zc}(t) - \delta_z(t)),
\end{cases}
$$

$$(11.2)$$

where

- $\delta_{xc}$, $\delta_{yc}$ and $\delta_{zc}$ stand for the input actuator deflections
- $\tau_x$, $\tau_y$ and $\tau_z$ are the actuator time constants

The model (11.1)–(11.2) is a slightly generalized version of that in Hu (1997).

In the following, we further classify two different types of models for two different types of missiles.

## 11.1.1 Models for Nonrotating Missiles

Since, for the nonrotating missile considered in this section, the coupling between the yaw and the roll channels is relatively larger, for simplicity in design, in the following models for the pitch channel and the yaw/roll channel are established separately.

### 11.1.1.1 Pitch Channel Model

Denoting the state, control, output, and disturbance vectors, respectively, as

$$x = \begin{bmatrix} \alpha \\ \omega_z \\ \delta_z \end{bmatrix}, \quad u = \delta_{zc}, \quad y = \begin{bmatrix} \alpha \\ n_y \end{bmatrix}, \quad d = \begin{bmatrix} \beta \\ \omega_y \end{bmatrix}, \qquad (11.3)$$

we can obtain the state space model for the pitch channel from (11.1) and (11.2) as follows:

$$\begin{cases} \dot{x}(t) = A(t)x(t) + B_1 u(t) + B_2(t)d(t) \\ y(t) = C(t)x(t) + D_1 u(t) + D_2(t)d(t), \end{cases} \qquad (11.4)$$

where

$$A(t) = \begin{bmatrix} -a_4(t) & 1 & -a_5(t) \\ -d_1'(t)a_4(t) - a_2(t) & d_1'(t) - a_1(t) & -d_1'(t)a_5(t) - a_3(t) \\ 0 & 0 & -\frac{1}{\tau_z} \end{bmatrix}, \quad (11.5)$$

$$B_1 = \begin{bmatrix} 0 \\ 0 \\ \frac{1}{\tau_z} \end{bmatrix}, \quad B_2 = \frac{\omega_x(t)}{57.3} \begin{bmatrix} -1 & 0 \\ -d_1'(t) & \frac{J_x - J_y}{J_z} \\ 0 & 0 \end{bmatrix}, \qquad (11.6)$$

$$C(t) = \frac{1}{57.3g} \begin{bmatrix} 57.3g & 0 & 0 \\ V(t)a_4(t) & 0 & V(t)a_5(t) \end{bmatrix}, \qquad (11.7)$$

$$D_1(t) = 0, \quad D_2(t) = \frac{1}{57.3g} \begin{bmatrix} 0 & 0 \\ V(t)b_7(t) & 0 \end{bmatrix}. \tag{11.8}$$

### 11.1.1.2 Yaw/Roll Channel Model

Note that for the type of nonrotating missiles, we aim to achieve $\omega_x = 0$, yet are not interested in the value of the roll angle $\phi$, hence the equation $\dot{\phi}(t) = \omega_x(t)$ in this case becomes an identity and can be thus neglected.

Choosing the state vector, the input vector, and the output vector, respectively, as

$$x = \begin{bmatrix} \beta & \omega_y & \omega_x & \delta_x & \delta_y \end{bmatrix}^{\mathrm{T}},$$

$$u = \begin{bmatrix} \delta_{xc} & \delta_{yc} \end{bmatrix}^{\mathrm{T}}, \quad y = \begin{bmatrix} n_z & \omega_x \end{bmatrix}^{\mathrm{T}},$$

and the disturbance vector as

$$d = \delta_z,$$

we have from (11.1) and (11.2) the following state space representation of the yaw/roll attitude system as

$$\begin{cases} \dot{x}(t) = A(t)x(t) + B_1(t)u(t) + B_2(t)d(t) \\ y(t) = C(t)x(t) + D_1(t)u(t) + D_2(t)d(t), \end{cases} \tag{11.9}$$

where

$$A(t) = \begin{bmatrix} A_{11}(t) & A_{12}(t) \\ 0 & A_{22}(t) \end{bmatrix},$$

with

$$A_{11}(t) = \begin{bmatrix} -b_4(t) & 1 & \frac{\alpha(t)}{57.3} \\ b_1'(t)b_4(t) - b_2(t) & -b_1(t) - b_1'(t) & \frac{J_y - J_z}{57.3J_x}\omega_z(t) - \frac{b_1'(t)\alpha(t)}{57.3} \\ c_2(t) & \frac{J_y - J_z}{57.3J_x}\omega_z(t) & -c_1(t) \end{bmatrix},$$

$$A_{12}(t) = \begin{bmatrix} 0 & -b_5(t) \\ 0 & b_1'(t)b_5(t) - b_3(t) \\ -c_3(t) & c_4(t) \end{bmatrix}, \quad A_{22}(t) = -\frac{1}{\tau_x \tau_y}\begin{bmatrix} \tau_y & 0 \\ 0 & \tau_x \end{bmatrix},$$

and

$$
B_1(t) = \begin{bmatrix} 0 & 0 \\ 0 & 0 \\ 0 & 0 \\ \frac{1}{\tau_x} & 0 \\ 0 & \frac{1}{\tau_y} \end{bmatrix}, \quad B_2(t) = \begin{bmatrix} a_6(t) \\ -b_1'(t)a_6(t) \\ 0 \\ 0 \\ 0 \end{bmatrix},
$$

$$
C(t) = -\frac{1}{57.3g} \begin{bmatrix} V(t)b_4(t) & 0 & 0 & 0 & V(t)b_5(t) \\ 0 & 0 & -57.3g & 0 & 0 \end{bmatrix},
$$

$$
D_1(t) = 0, \quad D_2(t) = -\frac{V(t)}{57.3g} \begin{bmatrix} b_6(t) \\ 0 \end{bmatrix}.
$$

## 11.1.2 Models for BTT Missiles

The dynamical model of the attitude system of the type of considered missiles is again given by (11.1). However, assuming that this type of considered missiles are axis symmetrically designed, then we have

$$
J_y = J_z,
$$

and in this case,

$$
a_6 = b_6 = b_7 = c_2 = c_4 = 0. \tag{11.10}
$$

Further, we assume that the actuator time constants $\tau_x$, $\tau_y$ and $\tau_z$ are all small enough, and hence the actuator dynamics (11.2) can be neglected.

### 11.1.2.1 Rolling Channel Model

With the aforementioned preparation, it can be observed from (11.1) that the rolling channel model is independent and appears as

$$
\begin{cases} \dot{\omega}_x(t) = -c_1(t)\omega_x(t) - c_3(t)\delta_x(t) \\ \dot{\phi}(t) = \omega_x(t), \end{cases}
$$

which can be written in the following state space form:

$$
\begin{bmatrix} \dot{\omega}_x(t) \\ \dot{\phi}(t) \end{bmatrix} = \begin{bmatrix} -c_1(t) & 0 \\ 1 & 0 \end{bmatrix} \begin{bmatrix} \omega_x(t) \\ \phi(t) \end{bmatrix} + \begin{bmatrix} -c_3(t) \\ 0 \end{bmatrix} \delta_x(t). \tag{11.11}
$$

### 11.1.2.2 Pitch/Yaw Channel Model

It can be also observed from (11.1) that the model for the pitch/yaw channel is as follows:

$$
\begin{cases}
\dot{\alpha}(t) = \omega_z(t) - \omega_x(t)\beta(t)/57.3 - a_4(t)\alpha(t) - a_5(t)\delta_z(t) \\[2mm]
\dot{\beta}(t) = \omega_y(t) + \omega_x(t)\alpha(t)/57.3 - b_4(t)\beta(t) - b_5(t)\delta_y(t) \\[2mm]
\dot{\omega}_z(t) = -a_1(t)\omega_z(t) + a_1'(t)\dot{\alpha}(t) - a_2(t)\alpha(t) - a_3(t)\delta_z(t) + \frac{J_x - J_y}{57.3 J_z}\omega_x(t)\omega_y(t) \\[2mm]
\dot{\omega}_y(t) = -b_1(t)\omega_y(t) - b_1'(t)\dot{\beta}(t) - b_2(t)\beta(t) - b_3(t)\delta_y(t) + \frac{J_z - J_x}{57.3 J_y}\omega_x(t)\omega_z(t) \\[2mm]
n_y(t) = \frac{V(t)}{57.3g}(a_4(t)\alpha(t) + a_5(t)\delta_z(t)) \\[2mm]
n_z(t) = -\frac{V(t)}{57.3g}(b_4(t)\beta(t) + b_5(t)\delta_y(t)).
\end{cases}
$$

$$(11.12)$$

For the aforementioned pitch/yaw channel model, taking the state vector as

$$
x = \begin{bmatrix} \omega_z & \alpha & \omega_y & \beta \end{bmatrix}^{\mathrm{T}},
$$

and the input and output vectors, respectively, as

$$
u = \begin{bmatrix} \delta_z \\ \delta_y \end{bmatrix}, \quad y = \begin{bmatrix} n_z \\ n_y \end{bmatrix},
$$

then we get the state-space model for the yaw/pitch channel as

$$
\begin{cases}
\dot{x}(t) = A(t, \omega_x)x(t) + B(t)u(t) \\
y(t) = C(t)x + D(t)u(t),
\end{cases}
$$

$$(11.13)$$

where

$$
A(t, \omega_x) = \begin{bmatrix} A_{11}(t) & A_{12}(t, \omega_x) \\ A_{21}(t, \omega_x) & A_{22}(t) \end{bmatrix},
$$

with

$$
A_{11}(t) = \begin{bmatrix} a_1'(t) - a_1(t) & -a_1'(t)a_4(t) - a_2(t) \\ 1 & -a_4(t) \end{bmatrix},
$$

$$
A_{22}(t) = \begin{bmatrix} -b_1(t) - b_1'(t) & b_1'(t)b_4(t) - b_2(t) \\ 1 & -b_4(t) \end{bmatrix},
$$

$$A_{12}(t,\omega_x) = \frac{\omega_x(t)}{57.3} \begin{bmatrix} \frac{J_x-J_y}{J_z} & -d_1'(t) \\ 0 & -1 \end{bmatrix},$$

$$A_{21}(t,\omega_x) = \frac{\omega_x(t)}{57.3} \begin{bmatrix} \frac{J_z-J_x}{J_y} & -b_1'(t) \\ 0 & 1 \end{bmatrix},$$

and

$$B(t) = \begin{bmatrix} -d_1'(t)a_5(t) & 0 \\ -a_5(t) & 0 \\ 0 & b_1'(t)b_5(t) - b_3(t) \\ 0 & -b_5(t) \end{bmatrix},$$

$$C(t) = \frac{V(t)}{57.3g} \begin{bmatrix} 0 & 0 & 0 & -b_4(t) \\ 0 & a_4(t) & 0 & 0 \end{bmatrix}, \quad D(t) = \frac{V(t)}{57.3g} \begin{bmatrix} 0 & -b_5(t) \\ a_5(t) & 0 \end{bmatrix}.$$

## 11.2 Attitude Control of Nonrotating Missiles

In this section, we consider the attitude control of a type of nonrotating missiles using LMI techniques.

### 11.2.1 Problem

It is clearly seen that the pitch channel system (11.4) and the yaw/roll channel system (11.9) do have a mutual affection on each other. This affection really comes from the coupling of the two systems. From the practical application point of view, such an affection needs to be attenuated at a minimum level, which actually gives rise to the following problem of disturbance attenuation via $H_\infty$ control.

**Problem 11.1** For the attitude system (11.4) or (11.9), find a state feedback control law $u = Kx$ such that

1. The closed-loop system

$$\begin{cases} \dot{x} = (A + B_1K)x + B_2d \\ z = (C + D_1K)x + D_2d \end{cases}$$

    is stable.
2. The transfer function matrix

$$G_{zd}(s) = (C + D_1K)(sI - (A + B_1K))^{-1}B_2 + D_2$$

satisfies

$$\|G_{zd}(s)\|_\infty < \gamma$$

for a minimum positive scalar $\gamma$.

## 11.2.2 Solution

It directly follows from Theorem 8.1 that the proposed problem can be solved via the following optimization

$$
\begin{cases}
\min & \gamma \\
\text{s.t.} & X > 0 \\
& \begin{bmatrix} (AX + B_1 W)^{\mathrm{T}} + AX + B_1 W & B_2 & (CX + D_1 W)^{\mathrm{T}} \\ B_2^{\mathrm{T}} & -\gamma I & D_2^{\mathrm{T}} \\ CX + D_1 W & D_2 & -\gamma I \end{bmatrix} < 0,
\end{cases}
$$

$$(11.14)$$

or via the following feasibility problem

$$
\begin{cases}
X > 0 \\
\begin{bmatrix} (AX + B_1 W)^{\mathrm{T}} + AX + B_1 W & B_2 & (CX + D_1 W)^{\mathrm{T}} \\ B_2^{\mathrm{T}} & -\gamma I & D_2^{\mathrm{T}} \\ CX + D_1 W & D_2 & -\gamma I \end{bmatrix} < 0,
\end{cases} \quad (11.15)
$$

with a satisfactory disturbance attenuation level $\gamma$. When a pair of parameter matrices $W$ and $X$ to the aforementioned optimization (11.14) or the feasibility problem (11.15) are found, a solution to the problem can be given as

$$K = WX^{-1}.$$

Note that the attitude control system design problems for the pitch channel and the yaw/roll channel can be solved exactly in the same way; here, we only consider the case of the yaw/roll channel.

The design of the yaw/roll channel system at a certain single operating point is considered, at which $\alpha = 0$ and $\omega_z = 0$. The system parameter matrices at this operating point are given as follows (Jiang 1987, Duan and Wang 1992):

$$A = \begin{bmatrix} -0.5 & 1 & 0 & 0 & 0 \\ -62 & -0.16 & 0 & 0 & 30 \\ 10 & 0 & -50 & 40 & 5 \\ 0 & 0 & 0 & -40 & 0 \\ 0 & 0 & 0 & 0 & -20 \end{bmatrix}, \tag{11.16}$$

$$B_1 = \begin{bmatrix} 0 & 0 \\ 0 & 0 \\ 0 & 0 \\ 40 & 0 \\ 0 & 20 \end{bmatrix}, \quad B_2 = \begin{bmatrix} 0.001 \\ -0.05 \\ 0 \\ 0 \\ 0 \end{bmatrix},$$

$$C = \begin{bmatrix} 1.5 & 0 & 0 & 0 & 0.8 \\ 0 & 0 & 1 & 0 & 0 \end{bmatrix},$$

$$D_1 = 0, \quad D_2 = \begin{bmatrix} 0.01 \\ 0 \end{bmatrix}.$$

## 11.2.2.1 Optimal Solution

By using the function `mincx` in the MATLAB® LMI toolbox, the optimization problem (11.14) is solved and the optimal parameters are obtained as

$\gamma = 0.010,$

$$X = \begin{bmatrix} 5660061.4412 & 273814.4899 & -0.3022 & -88416.7582 & -10612613.1920 \\ 273814.4899 & 667358290.1678 & -0.0699 & -4274.4210 & -513400.8165 \\ -0.3022 & -0.0699 & 23.0243 & -1936.0966 & 0.5645 \\ -88416.7582 & -4274.4210 & -1936.0966 & 206851137.4282 & 165782.9759 \\ -10612613.1920 & -513400.8165 & 0.5645 & 165782.9759 & 19898751.5367 \end{bmatrix} > 0,$$

and

$$W = \begin{bmatrix} -92491.3889 & -266484.2871 & -206743444.9669 & 1.95793341.1380 & 109040.3609 \\ -11167641.7233 & -63320089.9897 & 377.4004 & 54520.1804 & -1451947.9947 \end{bmatrix}.$$

Thus, the state feedback gain is given by

$$K = WX^{-1}$$

$$= \begin{bmatrix} -1495.1248 & -0.0011 & -8986351.4887 & -83.1652 & -796.4424 \\ -397683.9112 & -0.0937 & -3.8796 & 0.0010 & -212097.0763 \end{bmatrix}.$$

Note that the attenuation level $\gamma$ is very small, the disturbance $d$ is effectively attenuated. However, with this optimal solution, the magnitude of the feedback gain matrix $K$ is obviously large, which is not desirable in applications. Therefore, in the following, we will seek a feasible solution instead.

## 11.2.2.2 Feasible Solution

Limiting the attenuation level by $\gamma < 0.0160$, and applying the function feasp in the MATLAB LMI toolbox to (11.15), we obtain the following feasible solutions

$$X = \begin{bmatrix} 4.5994 & 0.0380 & -0.0046 & -0.0672 & -8.6780 \\ 0.0380 & 545.3889 & -0.0078 & -0.0501 & 0.1566 \\ -0.0046 & -0.0078 & 0.9584 & -0.6916 & 0.0111 \\ -0.0672 & -0.0501 & -0.6916 & 192.4742 & 0.1568 \\ -8.6780 & 0.1566 & 0.0111 & 0.1568 & 17.7772 \end{bmatrix} > 0,$$

$$W = \begin{bmatrix} -0.0766 & -0.2812 & -192.8017 & 189.6008 & 0.9696 \\ -8.9753 & -52.4897 & -0.3344 & -1.4936 & 9.3408 \end{bmatrix},$$

with $\gamma = 0.0135$. In this case, the state feedback gain turns out to be

$$K = WX^{-1}$$

$$= \begin{bmatrix} 1.5717 & -0.0038 & -200.9755 & 0.2627 & 0.9452 \\ -12.1268 & -0.0939 & -0.3511 & -0.0089 & -5.3931 \end{bmatrix}.$$

For this yaw/roll channel attitude system, Duan (Duan and Wang 1992) has also given a stabilization gain as follows:

$$K_p = \begin{bmatrix} -3.6844 & -1.5266 & -1.2500 & -10 & -5.9703 \\ 1.9463 & -0.5378 & 0 & 0 & -1.0671 \end{bmatrix}.$$

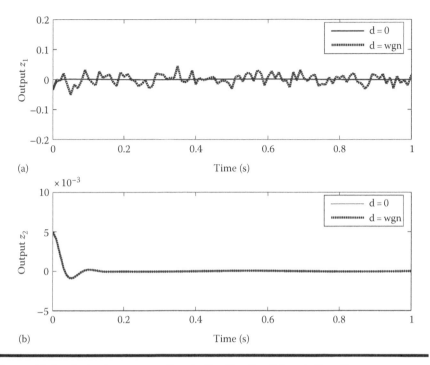

**Figure 11.1** Output corresponding to the H$_\infty$ feasible solution *K*.

We have plotted the output of the closed-loop systems in Figures 11.1 and 11.2, which respectively correspond to the H$_\infty$ feasible solution *K* and the stabilization solution $K_p$. To show the effect of disturbance attenuation, we have given simulation results for the zero disturbance case, represented by the solid lines, and the case of white Gaussian noise disturbance with power of 5 dBW, represented by the dash lines. It is seen from the two figures that the disturbance indeed has very small affection to the system output in both the two cases. But it is obvious that the settling time of the output of the H$_\infty$ control system is less than that of the stabilization system.

## 11.3 Attitude Control of BTT Missiles

In this section, we consider the attitude control of a type of BTT (bank-to-turn) missiles using LMI techniques. Theory of quadratic stabilization is applied.

### 11.3.1 Problem

Obviously, the rolling channel system (11.11) can be designed independently. Since the model of the rolling channel is very simple and easy to handle, here in this section only the design problem of the pitch/yaw channel is considered.

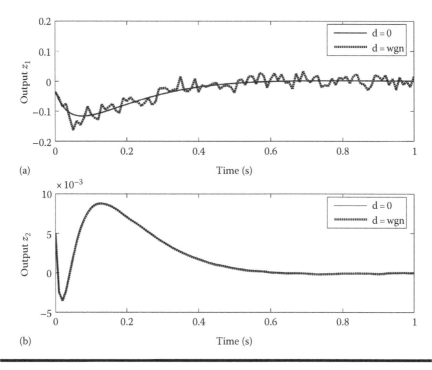

(a)

(b)

**Figure 11.2 Output corresponding to the stabilization solution $K_p$.**

### 11.3.1.1 Trajectory and Operating Points

In the flying process, the system parameters $a_1(t)$–$a_5(t)$, $b_1(t)$–$b_5(t)$, $a'_1(t)$, $b'_1(t)$, $c_1(t)$, $c_2(t)$ all change with the variation of the flying height and flying speed, and their varying rules are very complicated and hard to express analytically. These parameters are generally assumed to be continuous and uniformly bounded.

To deal with this complicated time-varying system, as is done in practice, we choose seven operating points on the whole trajectory, as shown in Figure 11.3, which correspond to several important flight moments, as shown in Table 11.1. The values of the parameters $a_1(t)$–$a_5(t)$, $a'_1(t)$, $b_1(t)$–$b_5(t)$, $b'_1(t)$, $c_1(t)$, $c_2(t)$, at the seven operating points are known and are given in Table 11.2 (Hu 1997).

### 11.3.1.2 Rolling Rate $\omega_x$

Although the rolling channel system is independent, it does affect the yaw/pitch channel through the rolling rate $\omega_x$, which directly appears in the system matrix $A$ of the yaw/pitch channel model, and this makes the yaw/pitch channel attitude system an uncertain one.

The rolling rate $\omega_x$ is assumed to be differentiable and uniformly bounded. In this design, it is assumed to alter within the interval $[-400°/s, 400°/s]$. Therefore,

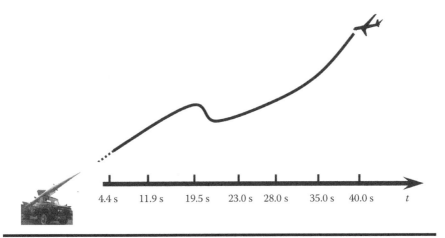

**Figure 11.3   Operating points along trajectory.**

**Table 11.1   Operating Points**

| $t_1$ | $t_2$ | $t_3$ | $t_4$ | $t_5$ | $t_6$ | $t_7$ |
|-------|-------|-------|-------|-------|-------|-------|
| 4.4s | 11.9s | 19.5s | 23.0s | 28.0s | 35.0s | 40.0s |

**Table 11.2   Parameter Values at the Seven Operating Points**

| Time (s) | $t_1$ | $t_2$ | $t_3$ | $t_4$ | $t_5$ | $t_6$ | $t_7$ |
|----------|-------|-------|-------|-------|-------|-------|-------|
| $a_1(t)$ | 1.593 | 1.485 | 1.269 | 1.130 | 0.896 | 0.559 | 0.398 |
| $a_1'(t)$ | 0.285 | 0.192 | 0.147 | 0.118 | 0.069 | 0.055 | 0.043 |
| $a_2(t)$ | 260.559 | 266.415 | 196.737 | 137.385 | 129.201 | 66.338 | 51.003 |
| $a_3(t)$ | 185.488 | 182.532 | 176.932 | 160.894 | 138.591 | 78.404 | 53.840 |
| $a_4(t)$ | 1.506 | 1.295 | 1.169 | 1.130 | 1.061 | 0.599 | 0.421 |
| $a_5(t)$ | 0.298 | 0.243 | 0.217 | 0.191 | 0.165 | 0.105 | 0.078 |
| $b_1(t)$ | 1.655 | 1.502 | 1.269 | 1.130 | 0.896 | 0.559 | 0.398 |
| $b_1'(t)$ | 0.295 | 0.195 | 0.147 | 0.118 | 0.069 | 0.055 | 0.043 |
| $b_2(t)$ | 39.988 | −24.627 | −31.452 | −41.425 | −68.165 | −21.448 | −9.635 |
| $b_3(t)$ | 159.974 | 170.532 | 182.030 | 184.093 | 154.608 | 89.853 | 59.587 |
| $b_4(t)$ | 0.771 | 0.652 | 0.680 | 0.691 | 0.709 | 0.360 | 0.243 |
| $b_5(t)$ | 0.254 | 0.191 | 0.188 | 0.182 | 0.162 | 0.102 | 0.072 |

at each operating point shown in Table 11.1, the pitch/yaw system (11.13) is an interval system with a varying parameter $\omega_x \in [-400, 400]$.

### 11.3.1.3 Problem

With the aforementioned description, we can now state our problem of attitude control system design as follows.

**Problem 11.2**  For the attitude system (11.13), find a constant state feedback

$$u = Kx + v(t), \tag{11.17}$$

with $v$ being an external input such that the closed-loop system

$$\dot{x}(t) = A_c(t, \omega_x)x(t) + B(t)v(t),$$

where

$$A_c(t, \omega_x) = A(t, \omega_x) + B(t)K$$

is uniformly asymptotically stable.

## 11.3.2 Quadratic Stabilization

### 11.3.2.1 Family of Systems

It follows from the aforementioned description that, at each operating point $t_k$, the system is a constant one but with an uncertain parameter $\omega_x$, which is denoted, for simplicity, by

$$\Pi_k(\omega_x) = (A(t_k, \omega_x), B(t_k)). \tag{11.18}$$

Note that the matrix $A(t, \omega_x)$ is linear in $\omega_x$, thus, in order to stabilize the system $(A(t, \omega_x), B(t))$ at each operating point $t_k$ for arbitrary parameters $\omega_x \in [-400, 400]$, it suffices to stabilize $\Pi_k(\omega_x) = (A(t_k, \omega_x), B(t_k))$ at the two endpoints of the parameter $\omega_x$, that is, $\omega_{x1} = 400°/s$ and $\omega_{x2} = -400°/s$. Therefore, at each operating point $t_k$, we need to consider stabilization of the following two systems

$$\Pi_{Lk} = (A(t_k, \omega_{x1}), B(t_k)),$$

and

$$\Pi_{Rk} = (A(t_k, \omega_{x2}), B(t_k)).$$

This tells us that in order to realize stabilization of the attitude system at all the operating points, we need to realize simultaneous stabilization of 14 systems, namely, $\Pi_{Lk}$ and $\Pi_{Rk}$, $k = 1, 2, \ldots, 7$.

For simplicity, the 14 systems can be further uniformly defined as

$$\Pi_i = (A_i, B_i), \tag{11.19}$$

where

$$(A_i, B_i) = \begin{cases} (A(t_k, \omega_{x1}), B(t_k)), & i = 2k - 1 \\ (A(t_k, \omega_{x2}), B(t_k)), & i = 2k, \end{cases} \tag{11.20}$$

$$k = 1, 2, \ldots, 7.$$

Further, with these 14 systems as vertexes, we can form a polytopic system (see Figure 11.4)

$$\Pi = (A(\delta(t)), B(\delta(t))),$$

that is,

$$\dot{x} = A(\delta(t))x + B(\delta(t))u,$$

where

$$\begin{cases} A(\delta(t)) = \delta_1(t)A_1 + \delta_2(t)A_2 + \cdots + \delta_{14}(t)A_{14} \\ B(\delta(t)) = \delta_1(t)B_1 + \delta_2(t)B_2 + \cdots + \delta_{14}(t)B_{14}, \end{cases} \tag{11.21}$$

and

$$\delta = \begin{bmatrix} \delta_1(t) & \delta_2(t) & \cdots & \delta_{14}(t) \end{bmatrix} \in \Delta_P,$$

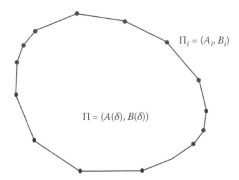

**Figure 11.4  The polytopic system.**

with

$$\Delta_P = \left\{ \delta(t) = \begin{bmatrix} \delta_1(t) & \cdots & \delta_{14}(t) \end{bmatrix} \middle| \sum_{i=1}^{14} \delta_i(t) = 1, \ \delta_i(t) \geq 0, \quad i = 1, 2, \ldots, 14 \right\}.$$

(11.22)

The set of extremes of $\Delta_P$ is

$$\Delta_E = \left\{ \begin{pmatrix} 1 & 0 & \cdots & 0 \end{pmatrix}, \begin{pmatrix} 0 & 1 & \cdots & 0 \end{pmatrix}, \ldots, \begin{pmatrix} 0 & 0 & \cdots & 1 \end{pmatrix} \right\}.$$

### 11.3.2.2 Quadratic Stabilization

Let

$$\Pi = \{(A(\delta(t)), B(\delta(t))) \mid \delta(t) \in \Delta_P\},$$

(11.23)

then $\Pi$ is a polytopic system, with vertices being $(A_i, B_i)$, $i = 1, 2, \ldots, 14$. When the 14 operating points are properly chosen, we may assume

$$(A(t, \omega_x), B(t)) \in \Pi, \ \forall t > 0, \ \omega_x \in [-400, 400].$$

Further note that quadratic stability implies uniform asymptotical stability. To solve Problem 11.2, we suffice to solve the following one.

**Problem 11.3** Let $A_i$, $B_i$, $i = 1, 2, \ldots, 14$, be defined by (11.20), and $(A(\delta(t)), B(\delta(t)))$ defined by (11.21). Find a constant state feedback

$$u = Kx,$$

(11.24)

which quadratically stabilizes the set of system $\Pi$ defined by (11.18).

According to the theory introduced in Section 7.8, to solve Problem 11.3, it suffices only to find a state feedback in the form of (11.24) which quadratically stabilizes the extreme points of $\Pi$, that is,

$$\Pi_E = \{(A(\delta(t)), B(\delta(t))) \mid \delta(t) \in \Delta_E\}$$
$$= \{(A_i, B_i), \quad i = 1, 2, \ldots, 14\}.$$

According to Theorem 7.8, this is equivalent to solving a pair of matrices $W$ and $P > 0$ satisfying the following set of LMIs:

$$A_i P + B_i W + P A_i^T + W^T B_i^T < 0, \quad i = 1, 2, \ldots, 14.$$

(11.25)

Once a pair of solutions $W$ and $P > 0$ to the aforementioned set of LMIs are obtained, the gain matrix is given by

$$K = WP^{-1}.$$

## 11.3.3 Numerical Results and Simulation

By using the MATLAB function `feasp` in the MATLAB LMI toolbox, the corresponding LMI problem (11.25) is solved and the two parameter matrices are obtained as

$$W = \begin{bmatrix} 0.4539 & -0.0285 & 0 & 0 \\ 0 & 0 & 0.0627 & 0.0234 \end{bmatrix},$$

and

$$P = \begin{bmatrix} 2.9241 & -0.3080 & 0 & 0 \\ -0.3080 & 0.0393 & 0 & 0 \\ 0 & 0 & 3.0039 & -0.2937 \\ 0 & 0 & -0.2937 & 0.0373 \end{bmatrix} > 0.$$

Thus, the feedback control gain matrix can be obtained as

$$K = WP^{-1} = \begin{bmatrix} 0.4510 & 2.8077 & 0 & 0 \\ 0 & 0 & 0.3571 & 3.4389 \end{bmatrix}.$$

The corresponding closed-loop poles at each operating point with different values of $\omega_x$ are listed in Table 11.3.

The designed stabilizing controller is applied to the linear time-varying attitude system (11.13) to validate its effect.

For system (11.13), the parameters $a_1(t) \sim a_5(t)$, $b_1(t) \sim b_5(t)$, $a_1'(t)$, $b_1'(t)$ are obtained by linear interpolation between the adjacent operating points, that is,

$$a_k(t) = \frac{t_{i+1} - t}{t_{i+1} - t_i} a_k(t_i) + \frac{t - t_i}{t_{i+1} - t_i} a_k(t_{i+1}),$$

$$b_k(t) = \frac{t_{i+1} - t}{t_{i+1} - t_i} b_k(t_i) + \frac{t - t_i}{t_{i+1} - t_i} b_k(t_{i+1}),$$

$$a_1'(t) = \frac{t_{i+1} - t}{t_{i+1} - t_i} a_1'(t_i) + \frac{t - t_i}{t_{i+1} - t_i} a_1'(t_{i+1}),$$

$$b_1'(t) = \frac{t_{i+1} - t}{t_{i+1} - t_i} b_1'(t_i) + \frac{t - t_i}{t_{i+1} - t_i} b_1'(t_{i+1}),$$

$$t_i \le t < t_{i+1}, \quad i = 1, 2, \ldots, 6, \quad k = 1, 2, \ldots, 5.$$

**Table 11.3    Closed-Loop Poles**

| $\omega_x$ | $t_k$ | $s_1$ | $s_2$ | $s_3$ | $s_4$ |
|---|---|---|---|---|---|
| | $t_1$ | −76.4467 | −47.3087 | −11.4674 | −13.3857 |
| | $t_2$ | −74.4519 | −52.3237 | −11.5453 | −11.5658 |
| | $t_3$ | −72.3643 | −56.3288 | −10.6407 | −11.4067 |
| 0 | $t_4$ | −65.3712 | −57.0581 | −11.2387 | −10.1164 |
| | $t_5$ | −54.4027 | −46.5242 | −10.9134 | −10.5963 |
| | $t_6$ | −24.3356 | −12.5350 | −16.7046 ± 4.6079$i$ | |
| | $t_7$ | −12.6822 ± 7.0736$i$ | | −11.1045 ± 8.8081$i$ | |
| | $t_1$ | −75.5712 | −48.2827 | −12.3764 ± 7.3558$i$ | |
| | $t_2$ | −73.2926 | −53.5192 | −11.5366 ± 7.3109$i$ | |
| | $t_3$ | −70.6459 | −58.0807 | −11.0061 ± 7.2847$i$ | |
| 400 | $t_4$ | −61.2303 ± 2.7003$i$ | | −10.6611 ± 7.2751$i$ | |
| | $t_5$ | −50.4879 ± 2.8356$i$ | | −10.7297 ± 7.3870$i$ | |
| | $t_6$ | −22.8872 | −20.4674 | −13.4623 ± 9.8123$i$ | |
| | $t_7$ | −10.5999 ± 14.2593$i$ | | −13.1865 ± 1.5925$i$ | |
| | $t_1$ | −75.5712 | −48.2827 | −12.3764 ± 7.3558$i$ | |
| | $t_2$ | −73.2926 | −53.5192 | −11.5366 ± 7.3109$i$ | |
| | $t_3$ | −70.6459 | −58.0807 | −11.0061 ± 7.2847$i$ | |
| −400 | $t_4$ | −61.2303 ± 2.7003$i$ | | −10.6611 ± 7.2751$i$ | |
| | $t_5$ | −50.4879 ± 2.8356$i$ | | −10.7297 ± 7.3870$i$ | |
| | $t_6$ | −22.8872 | −20.4674 | −13.4623 ± 9.8123$i$ | |
| | $t_7$ | −10.5999 ± 14.2593$i$ | | −13.1865 ± 1.5925$i$ | |

The rolling angular velocity $\omega_x$ is taken as

$$\omega_x(t) = 400 \sin(0.2\pi t).$$

In order to realize tracking of given overload signals, here we employ the following feedforward tracking scheme based on the model reference theory (Duan et al. 1994, 2001):

$$u(t) = K(t)x + G(t)y_r(t), \tag{11.26}$$

with $y_r$ being the overload signal to be followed, and

$$G(t) = H(t) - K(t)Z(t), \tag{11.27}$$

where the coefficient matrices $H(t)$ and $Z(t)$ are derived by

$$\begin{bmatrix} Z(t) \\ H(t) \end{bmatrix} = \begin{bmatrix} A(t) & B(t) \\ C(t) & D(t) \end{bmatrix}^{-1} \begin{bmatrix} 0 \\ I \end{bmatrix}. \tag{11.28}$$

When $y_r(t)$ is a step function, and $K(t)$ stabilizes system (11.13), the output of the closed-loop system satisfies

$$\lim_{t \to \infty} (y(t) - y_r) = 0. \tag{11.29}$$

Using the aforementioned scheme we have considered tracking of the following sine wave

$$y_{r1}(t) = y_{r2}(t) = 0.5 \sin (0.5t), \quad t \geq 4.4,$$

and also the following trapezoidal wave

$$y_{r1}(t) = y_{r2}(t) = \begin{cases} -0.5, & 4.4 \leq t < 4.875 \\ 4t - 20, & t_0(k) \leq t < t_1(k) \\ 0.5, & t_1(k) \leq t < t_2(k) \\ -4t + 60, & t_2(k) \leq t < t_3(k) \\ -0.5, & t_3(k) \leq t < t_4(k), \end{cases}$$

where

$$\begin{cases} t_0(k) = 40k + 4.875 \\ t_1(k) = 40k + 5.125 \\ t_2(k) = 40k + 14.875 \\ t_3(k) = 40k + 15.125 \\ t_4(k) = 40k + 24.875, \quad k = 0, 1, \ldots . \end{cases}$$

The results are shown in Figures 11.5 and 11.6, which demonstrate that the designed time varying attitude control system works perfectly and offers good dynamical performance, bearing in mind that we have used only one single-state feedback gain to cope with all the variations in the 12 system parameters shown in Table 11.2.

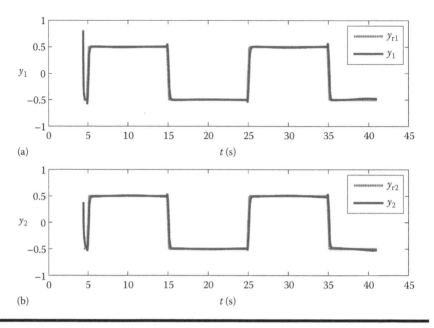

(a)

(b)

**Figure 11.5    Tracking a square wave.**

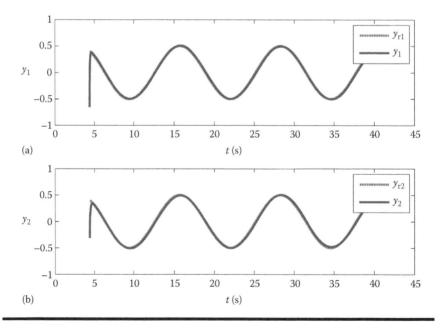

(a)

(b)

**Figure 11.6    Tracking a sine wave.**

## 11.4 Notes and References

In this chapter, the attitude control of two types of missiles is considered. Specifically, one is the design of the attitude control systems of a type of nonrotating missiles, where mixed $H_2/H_\infty$ performance is considered in order to attenuate the disturbance, and the closed-loop eigenvalues are restricted within desired region in order to ensure the closed-loop dynamical properties; the other is the attitude control system design of a BTT missile using techniques of simultaneous stabilization based on LMIs.

The data for the model of the nonrotating missile are taken from Duan et al. (1994), while those for the BTT model of missiles are taken from Duan et al. (2009b) and Duan (2012). The authors are grateful to the Chinese Ministry of Aerospace for providing the data through project funding.

Probably the most difficult theoretical problem existing in missile attitude control systems design is to guarantee the closed-loop stability. This is because the plant is highly nonlinear and time varying! To follow the desired trajectory, normally several controllers are designed and they are switched as the plant model changes. With such a scheme, closed-loop stability is difficult to be guaranteed, at least in theory. But in Section 11.3, the problem is solved via quadratic stabilization. It is shown that, instead of designing a local controller at each characteristic point and operating the switching, a single controller for the whole trajectory is sufficient. What is more, closed-loop stability is well guaranteed by the quadratic stabilization theory, and the time-varying property as well as the system uncertainty are all well coped with.

For missile control systems design using advanced control theory and techniques, there are numerous reported results in the literature. The author's work along this direction also includes Cai et al. (2011) and Cai et al. (2010), both investigating the control of airbreathing hypersonic vehicles, and Tan and Duan (2008), which is about designing globally stabilizing controllers for BTT missiles as well, while Hou and Duan (2008) studied integrated guidance and control of homing missiles against ground fixed targets.

## Exercises

**11.1** Find out the open-loop stability of the non-rotating missile attitude system (11.9) with matrix $A$ given in (11.16).

**11.2** For the following linear system

$$\begin{cases} \dot{x} = Ax + Bu \\ y = Cx + Du, \end{cases} \tag{11.30}$$

a controller is designed in the form of (11.26), with $K$ enabling the asymptotical stability of the system

$$\dot{x} = (A + BK)x$$

and $G$ given by (11.27) and (11.28). Show that the asymptotical tracking relation (11.29) holds.

**11.3** Consider the pitch channel attitude model (11.3)-(11.8) of the type of non-rotating missiles with

$$A = \begin{bmatrix} -1.2950 & 1 & -0.2430 \\ 0.1920 & -1.6770 & -0.0467 \\ 0 & 0 & 20 \end{bmatrix},$$

$$B_1 = \begin{bmatrix} 0 \\ 0 \\ 1 \end{bmatrix}, \quad B_2 = \begin{bmatrix} -0.1745 & 0 \\ 0.0335 & -0.1309 \\ 0 & 0 \end{bmatrix},$$

$$C = \begin{bmatrix} 1 & 0 & 0 \\ -3.2253 & 0 & -0.6052 \end{bmatrix},$$

$$D_1 = 0, \quad D_2 = \begin{bmatrix} 0 & 0 \\ -0.0025 & 0 \end{bmatrix}.$$

Find a state feedback control law $u = Kx$ such that the closed-loop system is stable, and the transfer function matrix $G_{yd}(s)$ satisfies

$$\|G_{yd}(s)\|_\infty < \gamma$$

for a minimum positive scalar $\gamma$.

**11.4** Let $A, A_0 \in \mathbb{R}^{n \times n}$. Show that

$$A(\omega) = A + A_0\omega, \quad \omega \in [\alpha, \beta],$$

is quadratically stable if $\{A + A_0\alpha, \ A + A_0\beta\}$ is quadratically stable. (Note: This is related to Subsection 11.3.2, in which the quadratic stabilization of system (11.18) with $\omega_x \in [-400, 400]$ is treated.)

**11.5** The closed-loop poles at each operating point with different values of $\omega_x$ are listed in Table 11.3, and they can be observed to be all stable. Is this fact sufficient for the stability of the designed closed-loop missile attitude system? Why?

**11.6** Consider the rolling channel model (11.11) of the type of BTT missiles. The seven operating points along the whole trajectory are the same as in Table 11.1, while the parameters of the system at these operating points are given in Table 11.4. Following the lines in Section 11.3, complete the attitude control system design for this rolling channel using the quadratic stabilization approach.

**Table 11.4  Parameters at Characteristic Points**

| Times (s) | $t_1$ | $t_2$ | $t_3$ | $t_4$ | $t_5$ | $t_6$ | $t_7$ |
|---|---|---|---|---|---|---|---|
| $c_1(t)$ | 1.264 | 1.600 | 1.636 | 1.635 | 1.607 | 0.936 | 0.644 |
| $c_3(t)$ | 1787.048 | 1832.067 | 2128.877 | 2231.985 | 3045.292 | 1329.481 | 818.706 |

# Chapter 12

# Satellite Control

The motion of satellites in orbit performs an important role in military, civil, and scientific activities. To meet the various objectives, attitude control of a satellite must be required, which includes attitude maneuvering and attitude stabilization.

A satellite attitude system is a complex one subject to various uncertainties and disturbances. External disturbances may include the aerodynamic moments, sunlight pressure torques, gravity gradient torques, and magnetic moments, while the internal disturbances may include parameter uncertainties, the influence resulted from fuel and some carried facilities, such as the two-dimensional scanners. In practical engineering, the attitude fast maneuvering and accurate pointing are required in the presence of all kinds of disturbances and parameter uncertainties. Therefore, robust attitude controller design is of great importance and challenge.

In this chapter, we treat the problem of satellite attitude control. In Section 12.1, we present the model of the attitude system, while in the second section we treat the problem of $H_\infty$ and $H_2$ control. In the third section, attitude stabilization is considered where disturbance attenuation and desired closed-loop eigenvalue location are achieved based on LMI techniques.

## 12.1 System Modeling

Based on the well-known moment of momentum theorem, the attitude dynamics of a satellite in the inertial coordinate system can be generally expressed as

$$\dot{H} = T_c + T_g + T_d, \tag{12.1}$$

where

- $T_c$, $T_g$ and $T_d$ are the flywheel torque, the gravitational torque and the disturbance torque, respectively
- $H$ is the total momentum acting on the satellite

The total momentum $H$ is given by (Sidi (1997), Zhang (1998))

$$H = I_b\omega,$$

where

- $I_b$ is the inertia matrix
- $\omega$ is the angular velocity

While the derivative of the total momentum $H$ can be given as follows (Liu and Zhao (2003))

$$\dot{H} = I_b\dot{\omega} + \omega \times (I_b\omega), \tag{12.2}$$

where $\times$ represents the cross product of vectors. Combining (12.2) and (12.1), yields the following dynamical equation:

$$I_b\dot{\omega} + \omega \times (I_b\omega) = T_c + T_g + T_d. \tag{12.3}$$

Choose

$$I_b = \text{diag}(I_x, I_y, I_z),$$

and let

$$T_c = \begin{bmatrix} T_{cx} \\ T_{cy} \\ T_{cz} \end{bmatrix}, \quad T_g = \begin{bmatrix} T_{gx} \\ T_{gy} \\ T_{gz} \end{bmatrix}, \quad T_d = \begin{bmatrix} T_{dx} \\ T_{dy} \\ T_{dz} \end{bmatrix},$$

then we can expand Equation (12.3) into

$$\begin{cases} I_x\dot{\omega}_x + (I_z - I_y)\omega_y\omega_z = T_{cx} + T_{gx} + T_{dx} \\ I_y\dot{\omega}_y + (I_x - I_z)\omega_z\omega_x = T_{cy} + T_{gy} + T_{dy} \\ I_z\dot{\omega}_z + (I_y - I_x)\omega_x\omega_y = T_{cz} + T_{gz} + T_{dz}. \end{cases} \tag{12.4}$$

Under small angle approximation, the angular velocity of the satellite in the inertial coordinate system represented in the body coordinate system is given by (Zhang 1998)

$$\omega = \begin{bmatrix} \omega_x \\ \omega_y \\ \omega_z \end{bmatrix} = \begin{bmatrix} \dot{\varphi} - \omega_0 \psi \\ \dot{\theta} - \omega_0 \\ \dot{\psi} + \omega_0 \varphi \end{bmatrix}, \tag{12.5}$$

where
$\omega_0 = 7.292115 \times 10^{-5} \, \text{rad/s}$, is the rotational-angular velocity of the earth
$\varphi, \theta$, and $\psi$ are the three Euler angles

By substituting the kinematic equations (12.5) into (12.4), the attitude dynamics of the satellite can be obtained as

$$\begin{cases} I_x \ddot{\varphi} + (I_y - I_z)\omega_0^2 \varphi + (I_y - I_z - I_x)\omega_0 \dot{\psi} = T_{cx} + T_{gx} + T_{dx} \\ I_y \ddot{\theta} = T_{cy} + T_{gy} + T_{dy} \\ I_z \ddot{\psi} + (I_x + I_z - I_y)\omega_0 \dot{\varphi} + (I_y - I_x)\omega_0^2 \psi = T_{cz} + T_{gz} + T_{dz}, \end{cases} \tag{12.6}$$

where the gravitational torques are easily shown to be given by

$$\begin{cases} T_{gx} = -3\omega_0^2 (I_y - I_z)\varphi \\ T_{gy} = -3\omega_0^2 (I_x - I_z)\theta \\ T_{gz} = 0. \end{cases} \tag{12.7}$$

Substituting the aforementioned relation into (12.6) gives finally the dynamical model of the satellite as follows:

$$\begin{cases} I_x \ddot{\varphi} + 4(I_y - I_z)\omega_0^2 \varphi + (I_y - I_z - I_x)\omega_0 \dot{\psi} = T_{cx} + T_{dx} \\ I_y \ddot{\theta} + 3\omega_0^2 (I_x - I_z)\theta = T_{cy} + T_{dy} \\ I_z \ddot{\psi} + (I_x + I_z - I_y)\omega_0 \dot{\varphi} + (I_y - I_x)\omega_0^2 \psi = T_{cz} + T_{dz}. \end{cases} \tag{12.8}$$

## 12.1.1 Second-Order System Form

Define vectors

$$q = \begin{bmatrix} \varphi \\ \theta \\ \psi \end{bmatrix}, \quad u = \begin{bmatrix} T_{cx} \\ T_{cy} \\ T_{cz} \end{bmatrix}, \quad d = \begin{bmatrix} T_{dx} \\ T_{dy} \\ T_{dz} \end{bmatrix},$$

then the equations in (12.8) can be written compactly in the following second-order matrix form:

$$M\ddot{q} + H\dot{q} + Gq = L_1 u + L_2 d, \tag{12.9}$$

where

$$M = \text{diag}\left(I_x,\ I_y,\ I_z\right), \qquad L_1 = L_2 = I_{3\times3},$$

$$H = \omega_0(I_y - I_x - I_z)\begin{bmatrix} 0 & 0 & 1 \\ 0 & 0 & 0 \\ -1 & 0 & 0 \end{bmatrix},$$

$$G = \text{diag}\left(4\omega_0^2(I_y - I_z),\ 3\omega_0^2(I_x - I_z),\ \omega_0^2(I_y - I_x)\right).$$

## 12.1.2 State-Space Form

Define

$$x = \begin{bmatrix} q & \dot{q} \end{bmatrix}^{\mathrm{T}}, \qquad z_\infty = 10^{-3}M\ddot{q}, \qquad z_2 = q,$$

and introduce the notations

$$I_{\alpha\beta} = I_\alpha - I_\beta, \qquad I_{\alpha\beta\gamma} = I_\alpha - I_\beta - I_\gamma,$$

where $\alpha$, $\beta$, and $\gamma$ stand for any element in $\{x, y, z\}$. Then, we can rewrite the system (12.9) in the following state-space form:

$$\begin{cases} \dot{x} = Ax + B_1 u + B_2 d \\ z_\infty = C_1 x + D_1 u + D_2 d \\ z_2 = C_2 x, \end{cases} \tag{12.10}$$

where

$$A = \begin{bmatrix} 0 & 0 & 0 & 1 & 0 & 0 \\ 0 & 0 & 0 & 0 & 1 & 0 \\ 0 & 0 & 0 & 0 & 0 & 1 \\ -\dfrac{4\omega_0^2 I_{yz}}{I_x} & 0 & 0 & 0 & 0 & -\omega_0\dfrac{I_{yzx}}{I_x} \\ 0 & -\dfrac{3\omega_0^2 I_{xz}}{I_y} & 0 & 0 & 0 & 0 \\ 0 & 0 & -\dfrac{\omega_0^2 I_{yx}}{I_z} & \omega_0\dfrac{I_{yxz}}{I_z} & 0 & 0 \end{bmatrix}, \tag{12.11}$$

$$B_1 = B_2 = \begin{bmatrix} 0 & 0 & 0 \\ 0 & 0 & 0 \\ 0 & 0 & 0 \\ \dfrac{1}{I_x} & 0 & 0 \\ 0 & \dfrac{1}{I_y} & 0 \\ 0 & 0 & \dfrac{1}{I_z} \end{bmatrix}, \tag{12.12}$$

$$C_1 = 10^{-3} \times \begin{bmatrix} -4\omega_0^2 I_{yz} & 0 & 0 & 0 & 0 & -\omega_0 I_{yxz} \\ 0 & -3\omega_0^2 I_{xz} & 0 & 0 & 0 & 0 \\ 0 & 0 & -\omega_0^2 I_{yx} & \omega_0 I_{yxz} & 0 & 0 \end{bmatrix},$$

(12.13)

$$C_2 = \begin{bmatrix} I_{3\times3} & 0_{3\times3} \end{bmatrix},$$

(12.14)

$$D_1 = 10^{-3} \times L_1, \quad D_2 = 10^{-3} \times L_2.$$

(12.15)

## 12.2 H₂ and H∞ Feedback Control

First, let us investigate the H∞ feedback control of the satellite system.

### 12.2.1 H∞ Control

#### 12.2.1.1 Problem and Solution

For the aforementioned linear system in the state-space form (12.10), design a state feedback control law

$$u = Kx,$$

such that the closed-loop system is stable and the transfer function matrix

$$G_{z_\infty d}(s) = (C_1 + D_1 K)\,(sI - (A + B_1 K))^{-1}\, B_2 + D_2$$

satisfies

$$\|G_{z_\infty d}(s)\|_\infty < \gamma_\infty,$$

for a minimal positive scalar $\gamma_\infty$.

According to Theorem 8.1, the aforementioned problem with a minimal attenuation level $\gamma_\infty$ can be sought via the following optimization problem:

$$\begin{cases} \min & \gamma_\infty \\ \text{s.t.} & X > 0 \\ & \begin{bmatrix} (AX + B_1 W)^{\mathrm{T}} + AX + B_1 W & B_2 & (C_1 X + D_1 W)^{\mathrm{T}} \\ B_2^{\mathrm{T}} & -\gamma_\infty I & D_2^{\mathrm{T}} \\ C_1 X + D_1 W & D_2 & -\gamma_\infty I \end{bmatrix} < 0. \end{cases}$$

(12.16)

## 12.2.1.2 Numerical Results

Taking

$$I_x = 1030.17\text{kg} \cdot \text{m}^2, \quad I_y = 3015.65\text{kg} \cdot \text{m}^2, \quad I_z = 3030.43\text{kg} \cdot \text{m}^2, \quad (12.17)$$

and using the MATLAB® function `mincx` in MATLAB LMI toolbox to the aforementioned problem (12.16), we obtain $\gamma_\infty = 0.0010$ and

$$X = \begin{bmatrix} 3358849.2116 & 0 & 0.0001 \\ 0 & 3412059.8084 & 0 \\ 0.0001 & 0 & 3417761.6655 \\ -1386.3746 & 0 & 0 \\ 0 & -684.7144 & 0 \\ 0 & 0 & -683.2073 \end{bmatrix}$$

$$\begin{matrix} -1386.3746 & 0 & 0 \\ 0 & -684.7144 & 0 \\ 0 & 0 & -683.2073 \\ 1.1419 & 0 & 0 \\ 0 & 0.2742 & 0 \\ 0 & 0 & 0.2725 \end{matrix} \Bigg] > 0,$$

$$W = \begin{bmatrix} -1.0559 & 0 & 52.0597 & -0.9703 & 0 & -0.0208 \\ 0 & -108.8758 & 0 & 0 & -0.3098 & 0 \\ -105.6403 & 0 & 36.0840 & 0.0870 & 0 & -0.3372 \end{bmatrix},$$

and the corresponding state feedback gain is given by

$$K_0 = WX^{-1}$$

$$= \begin{bmatrix} -0.0007 & 0 & 0.0000 & -1.7040 & 0 & -0.0762 \\ 0 & -0.0005 & 0 & 0 & -2.4249 & 0 \\ 0.0000 & 0 & -0.0005 & 0.0762 & 0 & -2.4273 \end{bmatrix}.$$

Note that the attenuation level $\gamma_\infty = 0.0010$ is very small, eventually the effect of the disturbance $d$ must be effectively attenuated. However, with this solution, the closed-loop system poles are

$$-0.000863, \; -0.000791, \; -0.000417, \; -0.000384,$$

$$-0.000419, \; -0.000385,$$

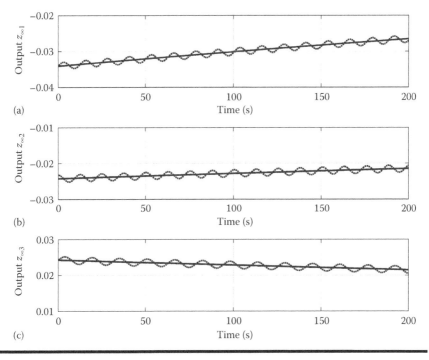

**Figure 12.1  Closed-loop output $z_\infty$ corresponding to $K_0$.**

which are all stable, but are all very close to the imaginary axis. Therefore, the system response must be too slow and must have a long settling time. This has been verified by plotting the output $z_\infty$ of the system in Figure 12.1, where the solid lines represent the response of the system for the case of $d = 0$ and the dash lines for the case of

$$d(t) = \left[\sin \tfrac{t}{200} \quad \cos \tfrac{t}{200} \quad \sin \tfrac{t}{250}\right]^{\mathrm{T}}.$$

It is seen in this figure that within 200 seconds the system responses are still far from settling down.

To prevent such a phenomenon, instead of using the second LMI constraint in (12.16), we use

$$-I < \begin{bmatrix} (AX + B_1 W)^{\mathrm{T}} + AX + B_1 W & B_2 & (C_1 X + D_1 W)^{\mathrm{T}} \\ B_2^{\mathrm{T}} & -\gamma_\infty I & D_2^{\mathrm{T}} \\ C_1 X + D_1 W & D_2 & -\gamma_\infty I \end{bmatrix} < 0,$$

$$(12.18)$$

which is obtained by adding at the left-hand side a simple constraint. With this altered condition, using again the `feasp` function in MATLAB LMI toolbox, we obtain $\gamma_\infty = 0.0010$ and

$$W = \begin{bmatrix} -0.1295 & 0 & 0.0229 & -0.9720 & 0 & -0.0004 \\ 0 & -13.1714 & 0 & 0 & -0.3320 & 0 \\ -0.0229 & 0 & 4.3581 & 0.0010 & 0 & -0.3304 \end{bmatrix}.$$

The corresponding state feedback gain is given by

$$K = WX^{-1}$$

$$= \begin{bmatrix} -0.0001 & 0 & 0 & -70.7320 & 0 & -0.0762 \\ 0 & -0.0001 & 0 & 0 & -70.5375 & 0 \\ 0 & 0 & 0 & 0.0762 & 0 & -70.5371 \end{bmatrix}.$$

With this feedback gain $K$, we also plotted the closed-loop system output for the same disturbance choices, which is shown in Figure 12.2. Comparing these two

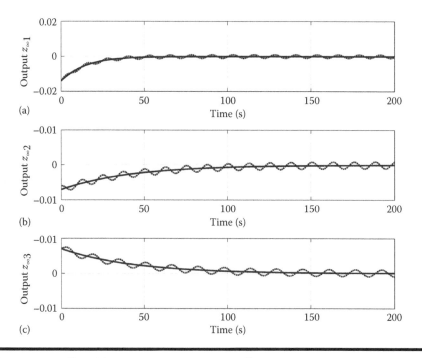

**Figure 12.2   Closed-loop output $z_\infty$ corresponding to $K$.**

figures, we can find that this new solution $K$ is much better in the rapidness and yet does not lose any disturbance attenuation level.

## 12.2.2 $H_2$ Control

### 12.2.2.1 Problem and Solution

For the satellite attitude system (12.10), design a state feedback control law

$$u = Kx,$$

such that the closed-loop system is stable and the transfer function matrix

$$G_{z_2 w}(s) = C_2 (sI - (A + B_1 K))^{-1} B_2$$

satisfies

$$||G_{z_2 w}(s)||_2 < \gamma_2,$$

for a minimal positive scalar $\gamma_2$.

According to Theorem 8.3, the aforementioned problem with a minimal attenuation level $\gamma_2$ can be sought via the following optimization problem:

$$
\begin{cases}
\min \quad \rho \\
\text{s.t.} \quad AX + B_1 W + (AX + B_1 W)^{\mathrm{T}} + B_2 B_2^{\mathrm{T}} < 0 \\
\qquad \begin{bmatrix} -Z & C_2 X \\ (C_2 X)^{\mathrm{T}} & -X \end{bmatrix} < 0 \\
\qquad \operatorname{trace}(Z) < \rho,
\end{cases}
\tag{12.19}
$$

where $\rho = \gamma_2^2$.

### 12.2.2.2 Numerical Results

Using the MATLAB function `mincx` in the MATLAB LMI toolbox to the aforementioned problem (12.19), we obtain $\rho = 3.3706 \times 10^{-11}$, and

$$X = \begin{bmatrix} 0.0000 & 0 & -0.0000 \\ 0 & 0.0000 & 0 \\ -0.0000 & 0 & 0.0000 \\ -0.6678 & 0 & -0.0000 \\ 0 & -0.4631 & 0 \\ 0.0000 & 0 & -0.4615 \end{bmatrix}$$

$$\begin{bmatrix} -0.6678 & 0 & 0.0000 \\ 0 & -0.4631 & 0 \\ -0.0000 & 0 & -0.4615 \\ 456742.4312 & 0 & -0.0011 \\ 0 & 146337.9739 & 0 \\ -0.0011 & 0 & 145578.4508 \end{bmatrix} > 0,$$

$$Z = 10^{-10} \times \begin{bmatrix} 0.0690 & 0 & -0.0000 \\ 0 & 0.1072 & 0 \\ -0.0000 & 0 & 0.1070 \end{bmatrix},$$

$$W = \begin{bmatrix} -470520069.0898 & 0 & 1.1871 \\ 0 & -441299747.4459 & 0 \\ 3.3326 & 0 & -441160935.7371 \end{bmatrix}$$

$$\begin{bmatrix} -280354806.5468 & 0 & 351.6082 \\ 0 & -281251955.9531 & 0 \\ 119.5264 & 0 & -281256221.9998 \end{bmatrix}.$$

The corresponding state feedback gain is given by

$$K_0 = WX^{-1}$$

$$= \begin{bmatrix} -875925392039229 & 0 & -536523549 \\ 0 & -113922040007657 & 0 \\ -497126691 & 0 & -114527952325858 \end{bmatrix}$$

$$\begin{bmatrix} -1280625924 & 0 & 6282 \\ 0 & -360501424 & 0 \\ -885 & 0 & -363086358 \end{bmatrix}.$$

Note that the attenuation level $\rho$ is very small, the disturbance $d$ is indeed effectively attenuated. However, with this optimal solution, the feedback gain matrix $K_0$ is very large, which is not desirable in applications. The reason for this is that in the optimal solution the matrix $X$ is nearly singular. To make a modification, we add another LMI condition

$$X > 1000I$$

in problem (5.4) and using the MATLAB function `mincx` in the MATLAB LMI toolbox to the new problem, we obtain $\rho = 0.0030$, and

$$X = \begin{bmatrix} 1000.7433 & 0 & -0.0000 \\ 0 & 1002.9883 & 0 \\ -0.0000 & 0 & 1002.9901 \\ -515.5386 & 0 & 0.0005 \\ 0 & -403.5612 & 0 \\ 0.0004 & 0 & -402.6945 \end{bmatrix}$$

$$\begin{bmatrix} -515.5386 & 0 & 0.0004 \\ 0 & -403.5612 & 0 \\ 0.0005 & 0 & -402.6945 \\ 452378.3015 & 0 & 0.0270 \\ 0 & 148013.1738 & 0 \\ 0.0270 & 0 & 147288.82317 \end{bmatrix} > 0,$$

$$Z = \begin{bmatrix} 0.0010 & 0 & -0.0000 \\ 0 & 0.0010 & 0 \\ -0.0000 & 0 & 0.0010 \end{bmatrix},$$

$$W = \begin{bmatrix} -464248215.3113 & 0 & 3.0511 \\ 0 & -442473079.0442 & 0 \\ -121.9329 & 0 & -442455256.4189 \end{bmatrix}$$

$$\begin{bmatrix} -280426475.1312 & 0 & 311.6877 \\ 0 & -281293587.9586 & 0 \\ 105.9557 & 0 & -281297771.4290 \end{bmatrix},$$

and the corresponding state feedback gain is given by

$$K = WX^{-1}$$

$$= \begin{bmatrix} -464495.4185 & 0 & 0.0017 \\ 0 & -442404.7666 & 0 \\ -0.1235 & 0 & -442388.6057 \end{bmatrix}$$

$$\begin{bmatrix} -1149.2412 & 0 & 0.0035 \\ 0 & -3106.6896 & 0 \\ 0.0008 & 0 & -3119.3489 \end{bmatrix}.$$

Corresponding to this feedback gain, we have plotted the output $z_2$ in Figure 12.3, where the solid lines represent the case of $d = 0$ and the dash lines the case of

$$d(t) = 10 \times \begin{bmatrix} \sin(t) & \cos(t) & \sin(2t) \end{bmatrix}^{\mathrm{T}}.$$

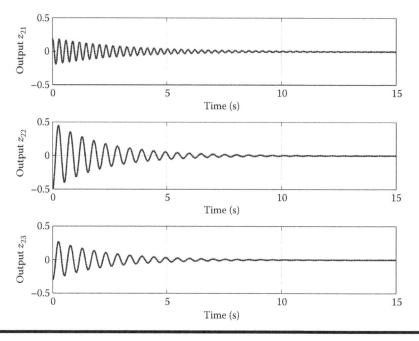

**Figure 12.3    The output $z_2$ of the closed-loop system.**

It can be seen from this figure that the effect of the disturbance to the system output has been greatly attenuated. This is consistent with the theory since the restriction level $\gamma = \sqrt{\rho} = 0.0548$ is very small.

## 12.3 Mixed H₂/H∞ Feedback Control

This section discusses controller design for the satellite attitude system (12.10) through (12.15) with external disturbances and parameter uncertainties. The mixed $H_2/H_\infty$ control design method introduced in Chapter 10 is applied.

### 12.3.1 Problem and Solution

For the design of the satellite attitude control system, we stress on two types of requirements.

#### 12.3.1.1 Pole Region Assignment

In order that the designed satellite attitude system has desired dynamical properties, the closed-loop poles are restricted to be within certain desired LMI region

$$\mathbb{D} = \{s|\ s \in \mathbb{C},\ L + sM + \bar{s}M^{\mathrm{T}} < 0\}, \tag{12.20}$$

where
  $L$ and $M$ are matrices of appropriate dimensions
  $L$ is symmetric

## 12.3.1.2 Disturbance Attenuation

Clearly, the disturbance $d$ has an affection on both output vectors $z_\infty$ and $z_2$. Such an affection needs to be minimized.

With the aforementioned analysis, a multiobjective design problem for the satellite attitude system (12.10) can now be stated as follows.

**Problem 12.1**  For the linear system (12.10), design a state feedback control law

$$u = Kx$$

such that

1. The closed-loop eigenvalues are all located in $\mathbb{D}$, that is

$$\lambda(A + BK) \subset \mathbb{D}, \tag{12.21}$$

  with $\mathbb{D}$ being given by (12.20).
2. The following $H_\infty$ and $H_2$ performance conditions

$$\left\| G_{z_\infty d} \right\|_\infty = \left\| (C_1 + N_2K)(sI - (A + B_1K))^{-1}B_2 + N_1 \right\|_\infty \le \gamma_\infty \tag{12.22}$$

  and

$$\left\| G_{z_2 d} \right\|_2 = \left\| C_2(sI - (A + B_1K))^{-1}B_2 \right\|_\infty \le \gamma_2 \tag{12.23}$$

  are satisfied with some possibly small $\gamma_\infty$ and $\gamma_2$.

Based on Theorem 10.3 and the optimization problem (10.27), we can formulate the aforementioned problem into the following optimization problem with LMI constraints:

$$
\left\{
\begin{aligned}
&\text{min} \quad c_\infty \gamma_\infty + c_2 \rho \\[4pt]
&\text{s.t.} \quad
\begin{bmatrix} -Z & C_2 X \\ X C_2^{\mathrm{T}} & -X \end{bmatrix} < 0 \\[6pt]
&\quad\quad \text{trace}\,(Z) < \rho \\[4pt]
&\quad\quad AX + B_1 W + (AX + B_1 W)^{\mathrm{T}} + BB^{\mathrm{T}} < 0 \\[4pt]
&\quad\quad L \otimes X + M \otimes (AX + B_1 W) + M^{\mathrm{T}} \otimes (AX + B_1 W)^{\mathrm{T}} < 0 \\[6pt]
&\quad\quad
\begin{bmatrix}
(AX + BW)^{\mathrm{T}} + AX + B_1 W & B_1 & (C_1 X + D_2 W)^{\mathrm{T}} \\
B & -\gamma_\infty I & D_1^{\mathrm{T}} \\
(C_1 X + D_2 W) & D_1 & -\gamma_\infty I
\end{bmatrix} < 0,
\end{aligned}
\right.
$$

$$(12.24)$$

and once a set of solutions $\gamma_\infty$, $\gamma_2^2$, $W$, $Z$ and $X > 0$ are obtained, the state feedback gain matrix can be constructed as $K = WX^{-1}$, and the corresponding $\gamma_2$ is given by $\sqrt{\rho}$.

## 12.3.2 Numerical and Simulation Results

### 12.3.2.1 Controller

For the specific satellite, we have $\omega_0 = 7.292115 \times 10^{-5}$ rad/s, and

$$
I_x = 1030.17 \text{kg·m}^2, \quad I_y = 3015.65 \text{kg·m}^2, \quad I_z = 3030.43 \text{kg·m}^2,
$$

the same as in (12.17). Further, the LMI region in the design is taken to be the strip region $\mathbb{H}_{1,3}$, which has the following characteristic function

$$
F_{\mathbb{H}_{1,3}}(s) = L + sM + \bar{s}M^{\mathrm{T}},
$$

with

$$
L = \begin{bmatrix} 2 & 0 \\ 0 & -6 \end{bmatrix}, \quad M = \begin{bmatrix} 1 & 0 \\ 0 & -1 \end{bmatrix}.
$$

To solve the optimization problem (12.24), we choose the weighting factors as $\alpha_1 = \alpha_2 = 1$. With the function `mincx` in the MATLAB LMI toolbox, we obtain the following set of solutions:

$\gamma_\infty = 0.001, \quad \gamma_2 = 0.0048,$

$X = 10^{-3} \times$

$$\begin{bmatrix} 0.0164 & 0.0000 & 0.0000 & -0.0481 & 0.0000 & 0.0000 \\ 0.0000 & 0.0027 & 0.0000 & 0.0000 & -0.0069 & 0.0000 \\ 0.0000 & 0.0000 & 0.0027 & 0.0000 & 0.0000 & -0.0069 \\ -0.0481 & 0.0000 & 0.0000 & 0.3006 & 0.0000 & -0.0000 \\ 0.0000 & -0.0069 & 0.0000 & 0.0000 & 0.0386 & 0.0000 \\ 0.0000 & 0.0000 & -0.0069 & -0.0000 & 0.0000 & 0.0383 \end{bmatrix} > 0,$$

and

$$W = \begin{bmatrix} -0.0155 & 0.0000 & 0.0000 & -0.9230 & 0.0000 & 0.0000 \\ 0.0000 & -0.0014 & 0.0000 & 0.0000 & -0.3281 & 0.0000 \\ 0.0000 & 0.0000 & -0.0014 & -0.0000 & 0.0000 & -0.3265 \end{bmatrix}.$$

The state feedback matrix is then computed to be

$K = WX^{-1} = 10^4 \times$

$$\begin{bmatrix} -1.8722 & 0.0000 & 0.0000 & -0.6068 & 0.0000 & 0.0000 \\ 0.0000 & -3.9822 & 0.0000 & 0.0000 & -1.5611 & 0.0000 \\ 0.0000 & 0.0000 & -3.9883 & -0.0000 & 0.0000 & -1.5665 \end{bmatrix}.$$

With this gain matrix, the closed-loop poles are

$$\begin{cases} s_{1,2} = -2.9453 \pm j3.0820 \\ s_{3,4} = -2.5846 \pm j2.5457 \\ s_{5,6} = -2.5883 \pm j2.5507. \end{cases} \tag{12.25}$$

Using the function norm in the MATLAB $H_\infty$ toolbox, we have for the closed-loop system corresponding to the feedback matrix $K$ the following:

$$\|G_{z_\infty d}\|_\infty = 0.0010, \quad \|G_{z_2 d}\|_2 = 7.75 \times 10^{-5}. \tag{12.26}$$

In addition, for comparison, by some pole assignment technique we obtained the following state feedback gain matrix

$K_p = 10^6$

$$\times \begin{bmatrix} 0.3241 & -0.3230 & -0.0932 & 0.0815 & -0.0562 & -0.0583 \\ -0.0776 & 0.0024 & 0.0623 & -0.0034 & -0.0228 & 0.0144 \\ 1.4482 & -1.3762 & 0.4510 & 0.3677 & -0.2356 & -0.2660 \end{bmatrix}.$$

Corresponding to this gain matrix, using the function `norm` in the MATLAB $H_\infty$ toolbox again, we have, correspondingly,

$$\left\| G_{z_\infty d} \right\|_\infty = 0.1114, \quad \left\| G_{z_2 d} \right\|_2 = 0.0054,$$

which are clearly larger than those given in (12.26).

### 12.3.2.2 Numerical Simulation

In this section, the numerical simulation is developed to show the validity of the mixed $H_2/H_\infty$ feedback controller. The initial state is taken as

$$x(0) = \begin{bmatrix} 0.05 & 0.05 & 0.05 & 0.01 & 0.01 & 0.01 \end{bmatrix}^T,$$

where the unit of angle is rad and the unit of angular velocity is rad/s. Meanwhile, the disturbance torques are chosen as

$$d = \begin{bmatrix} 10^{-3} \sin(\omega_0 t) \text{ Nm} \\ 10^{-3} \sin(\omega_0 t) \text{ Nm} \\ 10^{-3} \sin(\omega_0 t) \text{ Nm} \end{bmatrix}.$$

For comparison, using feedback matrices $K$ and $K_p$, respectively, we obtain the simulation results as shown in Figures 12.4 and 12.5. Figure 12.4 shows $z_\infty$ signals, and Figure 12.5 shows $z_2$ signals. In these figures, the solid lines correspond to the controller $K$, while the dotted lines correspond to the controller $K_p$.

From the simulation results, we can see that, with the mixed $H_2/H_\infty$ feedback controller, the closed-loop system is not only stable but also maintains better disturbance attenuation ability.

## 12.4 Notes and References

In this chapter, attitude stabilization in a type of satellites is considered. Disturbance attenuation and desired closed-loop eigenvalue location are achieved based on LMI techniques.

### 12.4.1 Optimality versus Feasibility

In Section 12.2, the problems of $H_\infty$ and $H_2$ feedback control of the satellite attitude system are carried out. It is interesting to note that, in both cases, the optimal solutions directly derived from the theories are not satisfactory. In the case of $H_\infty$ feedback control, the obtained optimal gain matrix gives very slow response transfer property, while in the case of $H_2$ feedback control the optimal gain matrix

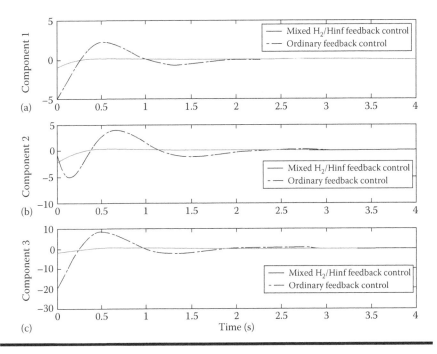

**Figure 12.4  System output $z_\infty$.**

has a very large magnitude. Through some proper adjustments for both cases, we have obtained satisfactory solutions. Such a phenomena indicates the following points:

▪ In the original designs, only the $H_\infty$ and $H_2$ requirements are considered, while the response transfer property and the control energy problem are neglected. The adjusted conditions have taken these into consideration.

▪ From a practical application point of view, the originally formulated $H_\infty$ and $H_2$ feedback control problems are not appropriate, they should have taken the response transfer and the control energy into account.

▪ With many practical applications, often, many trials and tests need to be carried out. This is the case with research, do not just give up easily!

### 12.4.2 Problem of Space Rendezvous

Besides the problems addressed in this chapter, there are also many others related to satellite control. Particularly, the authors and their collaborators have also tackled the problems of attitude maneuvering (Zhang and Duan 2012b, Zhang et al. 2012), orbital transfer and rendezvous (Duan et al. 2010, Zhou et al. 2011a, 2012a, Gao et al. 2011a, 2012a), and soft lunar landing control (Liu et al. 2008, Zhang

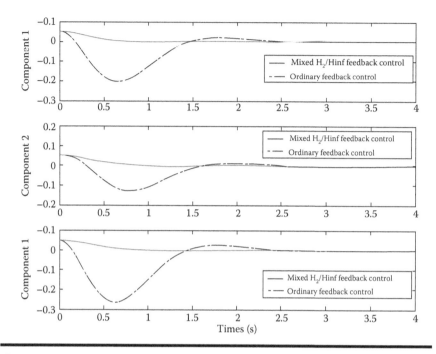

**Figure 12.5   System output $z_2$.**

and Duan 2012a). We remark that most of these problems can also be solved via LMI techniques. For a demonstration, here we give the formulation of the space rendezvous problem.

It is seen from Example 7.14 that, under certain conditions, the relative orbital dynamic model of spacecraft rendezvous can be described by (7.70). By denoting

$$x = \begin{bmatrix} r_x & r_y & r_z & \dot{r}_x & \dot{r}_y & \dot{r}_z \end{bmatrix}^{\mathrm{T}}, \quad y = \begin{bmatrix} r_x & r_y & r_z \end{bmatrix}^{\mathrm{T}},$$

$$u = \begin{bmatrix} T_x & T_y & T_z \end{bmatrix}^{\mathrm{T}}, \quad d = \begin{bmatrix} d_x & d_y & d_z \end{bmatrix}^{\mathrm{T}},$$

the second-order linear system (7.70) can be converted into the following state-space form:

$$\begin{cases} \dot{x} = Ax + B_1 u + B_2 d \\ y = Cx, \end{cases}$$

with

$$A = \begin{bmatrix} 0 & 0 & 0 & 1 & 0 & 0 \\ 0 & 0 & 0 & 0 & 1 & 0 \\ 0 & 0 & 0 & 0 & 0 & 1 \\ 3\omega_0^2 & 0 & 0 & 0 & 2\omega_0 & 0 \\ 0 & 0 & 0 & -2\omega_0 & 0 & 0 \\ 0 & 0 & -\omega_0^2 & 0 & 0 & 0 \end{bmatrix}, \tag{12.27}$$

$$B_1 = B_2 = \begin{bmatrix} 0_{3\times3} \\ I_3 \end{bmatrix}, \tag{12.28}$$

and the observation distribution matrix is taken, assuming that all the three Euler angles are measurable, as

$$C = \begin{bmatrix} I_3 & 0_{3\times3} \end{bmatrix},$$

The transfer function of the system is obviously

$$G_{yd}(s) = C(sI - A - B_1K)^{-1}B_2.$$

The problem of spacecraft rendezvous is to find a controller for the system in the form of

$$u = Kx,$$

such that $\|G_{yd}(s)\|_\infty$ or $\|G_{yd}(s)\|_2$ is minimized.
Of course, a consequence of this design is the achievement of closed-loop stability, which produces

$$\lim_{t\to\infty} x(t) = 0.$$

This ensures the rendezvous requirement since the relative distance between the object and the chaser tends to zero.
Such a problem obviously can be solved via LMI technique in a similar way.

## Exercises

**12.1** For the satellite system (12.6), show that the gravity torques are given by the formulas in (12.7).

**12.2** With the parameters given in (12.17),

1. check the stability of the open-loop system (12.10)–(12.15);
2. find out the value of $\left\| G_{z_\infty d}^o \right\|_\infty$ and $\left\| G_{z_2 d}^o \right\|_2$, with

$$G_{z_\infty d}^o = C_1 \left( sI - A \right)^{-1} B_2 + D_2,$$

and

$$G_{z_2 d}^o = C_2 \left( sI - A \right)^{-1} B_2,$$

and compare the obtained values with those corresponding to the closed-loop system designed in Section 12.2.

**12.3** It is seen from Subsection 12.2.1 that the obtained optimal solution does not give satisfactory response performance, while a feasible solution to the altered LMI in (12.18) does. Find another solution for the problem by reasonably altering the original LMI condition.

**12.4** Let $\omega_0 = \pi/12 (\text{rad/h})$. Solve the $H_\infty$ spacecraft rendezvous control problem described in Section 12.4.

**12.5** Let $\omega_0 = \pi/12 (\text{rad/h})$. Solve the $H_2$ spacecraft rendezvous problem described in Section 12.4.

# APPENDICES

# Appendix A: Proofs of Theorems

## A.1 Proof of Theorem 4.1

This proof follows the same lines in Chilali and Gahinet (1996) and Yu (2002) but with all the details fully explained.

### A.1.1 Preliminaries

In order to prove the theorem, we first give two preliminary lemmas.

**Lemma A.1** Let $A \in \mathbb{R}^{n \times n}$, and $\lambda(A) = \{s_1, s_2, \ldots, s_n\}$. Then, there exist a sequence of nonsingular matrices $T_k, k = 1, 2, \ldots$, such that

$$\lim_{k \to \infty} T_k^{-1} A T_k = \Lambda \triangleq \text{diag}\,(s_1, s_2, \ldots, s_n). \tag{A.1}$$

*Proof* When the Jordan matrix of $A$ is diagonal, it follows from Jordan decomposition that there exists a nonsingular matrix $T$ satisfying

$$T^{-1} A T = \Lambda.$$

In this case, we may simply take $T_k = T, k = 1, 2, \ldots$.

Now consider the case that the Jordan matrix $J$ of $A$ is not diagonal. Without loss of generality, we may only demonstrate with the case that $J$ is composed of only one Jordan block, that is,

$$J = \begin{bmatrix} s & 1 & & \\ & s & \ddots & \\ & & \ddots & 1 \\ & & & s \end{bmatrix}.$$

Again, by Jordan decomposition, there exists a nonlinear matrix $V_0$ such that

$$V_0^{-1} A V_0 = J. \tag{A.2}$$

Choose

$$V_k = \text{diag}\left(k^{n-1}, k^{n-2}, \ldots, k, 1\right),$$

then we can easily verify that

$$V_k^{-1} J V_k = \Theta(s, k),$$

with

$$
\Theta(s, k) =
\begin{bmatrix}
s & \frac{1}{k} & & & \\
& s & \frac{1}{k} & & \\
& & \ddots & \ddots & \\
& & & s & \frac{1}{k} \\
& & & & s
\end{bmatrix}.
$$

Premultiplying by $V_k^{-1}$ and postmultiplying by $V_k$ both sides of (A.2) produces

$$V_k^{-1} V_0^{-1} A V_0 V_k = V_k^{-1} J V_k = \Theta(s, k).$$

Since

$$\lim_{k \to \infty} \Theta(s, k) = \text{diag}(s, s, \ldots, s),$$

it is obvious that

$$T_k = V_0 V_k, \quad k = 1, 2, \ldots$$

are a sequence of matrices satisfying (A.1).  ■

**Lemma A.2**  Let $P = X + jY$, with $X, Y \in \mathbb{R}^{n \times n}$, be a Hermite matrix. Then, $P$ is positive definite, that is,

$$z^H P z > 0, \quad \forall z \in \mathbb{C}^n,$$

if and only if

$$Q = \begin{bmatrix} X & -Y \\ Y & X \end{bmatrix} > 0.$$

*Proof*  Since $P$ is Hermite, we have

$$X^T = X, \quad Y^T = -Y. \tag{A.3}$$

Therefore,

$$u^T Y u = 0, \quad \forall u \in \mathbb{R}^n. \tag{A.4}$$

Choose

$$z = x + jy \in \mathbb{C}^n,$$

where $x, y \in \mathbb{R}^n$. Using (A.3) and (A.4), we have

$$
\begin{aligned}
z^H P z &= (x + jy)^H (X + jY)(x + jy) \\
&= (x^T - jy^T)(X + jY)(x + jy) \\
&= (x^T - jy^T)[Xx - Yy + j(Xy + Yx)] \\
&= x^T Xx - x^T Yy + y^T Xy + y^T Yx \\
&\quad + j(-y^T Xx + x^T Xy + x^T Yx + y^T Yy) \\
&= x^T (Xx - Yy) + y^T (Yx + Xy) \\
&= [x^T \quad y^T]\begin{bmatrix} X & -Y \\ Y & X \end{bmatrix}\begin{bmatrix} x \\ y \end{bmatrix}.
\end{aligned}
$$

Therefore, the conclusion follows in view of the arbitrariness of vectors $x$, $y$, and $z$.  ∎

It clearly follows from Lemma A.2 that the real part of a Hermite positive definite matrix is symmetric positive definite. Such a fact will be used in the following proof.

## A.1.2  Sufficiency

Suppose $R_\mathbb{D}(A, P) < 0$ holds for some symmetric matrix $P$, and let $s$ be an arbitrary eigenvalue of $A$, and $v \in \mathbb{C}^n$ a nonzero vector satisfying $v^H A = s v^H$. Then, it is easy to derive

$$
\begin{aligned}
(I \otimes v^H)&R_\mathbb{D}(A, P)(I \otimes v) \\
&= (I \otimes v^H)\left(L \otimes P + M \otimes (AP) + M^T \otimes (AP)^T\right)(I \otimes v) \\
&= L \otimes (v^H P v) + M \otimes (v^H A P v) + M^T \otimes \left(v^H (AP)^T v\right) \\
&= v^H P v(L + sM + \bar{s}M^T) \\
&= v^H P v F_\mathbb{D}.
\end{aligned}
$$

Since $R_{\mathbb{D}}(A, P) < 0$, we have from the aforementioned relation that

$$v^{\mathrm{H}} P v F_{\mathbb{D}} < 0,$$

which, together with $P > 0$, gives $F_{\mathbb{D}} < 0$. Thus, $A$ is $\mathbb{D}$-stable.

## A.1.3 Necessity

Suppose $A$ is $\mathbb{D}$-stable. We need to show that there exists a symmetric positive definite matrix $P$ such that $R_{\mathbb{D}}(A, P) < 0$. The proof is divided into three steps.

*Step 1. The diagonal case*
Let us consider the case that $A$ is diagonal, that is,

$$A = \mathrm{diag}\,(s_1, \ldots, s_n) \triangleq \Delta,$$

where $s_i \in \mathbb{D}$, $i = 1, 2, \ldots, n$. Then, we have

$$R_{\mathbb{D}}(\Delta, I) = L \otimes I + M \otimes \Delta + M^{\mathrm{T}} \otimes \Delta^{\mathrm{H}}$$
$$= U^{\mathrm{T}} \mathrm{diag}\,(F_{\mathbb{D}}(s_1), \ldots, F_{\mathbb{D}}(s_n))\,U,$$

where $U$ is the product of some elementary matrices. It follows from $s_i \in \mathbb{D}$ that $F_{\mathbb{D}} < 0$ holds. Therefore, the aforementioned relation gives $R_{\mathbb{D}}(\Delta, I) < 0$. In this case, we have $P = I$ satisfying $R_{\mathbb{D}}(\Delta, P) < 0$.

*Step 2. The general case*
Without loss of generality, we assume that the diagonal elements of $\Delta$ are composed of the eigenvalues of $A$. Similarly, we can get $R_{\mathbb{D}}(\Delta, I) < 0$. Following from Lemma A.1, we can construct a matrix sequence $\{T_k\}$ such that

$$\lim_{k \to \infty} T_k^{-1} A T_k = \Delta.$$

Since $R_{\mathbb{D}}(Y, I)$ is continuous with respect to $Y$, using the aforementioned relation we get

$$\lim_{k \to \infty} R_{\mathbb{D}}(T_k^{-1} A T_k, I) = R_{\mathbb{D}}(\Delta, I) < 0.$$

Therefore, there exists a sufficient large number $q > 0$ such that $R_{\mathbb{D}}(T_q^{-1} A T_q, I) < 0$. Taking $T = T_q$, gives

$$R_{\mathbb{D}}(T^{-1} A T, I) < 0, \tag{A.5}$$

then

$$(I \otimes T)R_{\mathbb{D}}(T_q^{-1}AT_q, I)(I \otimes T^{\mathrm{H}})$$
$$= (I \otimes T)[L \otimes I + M \otimes (T^{-1}AT) + M^{\mathrm{T}} \otimes (T^{-1}AT)^{\mathrm{H}}](I \otimes T^{\mathrm{H}})$$
$$= L \otimes (TT^{\mathrm{H}}) + M \otimes (ATT^{\mathrm{H}}) + M^{\mathrm{T}} \otimes (ATT^{\mathrm{H}})^{\mathrm{H}}$$
$$= R_{\mathbb{D}}(A, TT^{\mathrm{H}}). \tag{A.6}$$

Using this relation and (A.5), we obtain

$$R_{\mathbb{D}}(A, TT^{\mathrm{H}}) < 0.$$

Letting $P = TT^{\mathrm{H}}$, we have $R_{\mathbb{D}}(A, P) < 0$.

*Step 3. Finding a real P*
If the matrix $P = TT^{\mathrm{H}}$ is real, then $P = TT^{\mathrm{H}} > 0$ holds. Therefore, the proof is done. Now suppose that $P = TT^{\mathrm{H}}$ is not real, note that

$$R_{\mathbb{D}}(A, P)$$
$$= L \otimes [\mathrm{Re}(P) + j\,\mathrm{Im}(P)] + M \otimes [A(\mathrm{Re}(P) + j\,\mathrm{Im}(P))]$$
$$+ M^{\mathrm{T}} \otimes [A(\mathrm{Re}(P) + j\,\mathrm{Im}(P))]^{\mathrm{H}}$$
$$= R_{\mathbb{D}}(A, \mathrm{Re}(P))$$
$$+ j[L \otimes \mathrm{Im}(P) + M \otimes (A\,\mathrm{Im}(P)) + M^{\mathrm{T}} \otimes (A\,\mathrm{Im}(P))^{\mathrm{T}}],$$

we have

$$\mathrm{Re}[R_{\mathbb{D}}(A, P)] = R_{\mathbb{D}}(A, \mathrm{Re}(P)).$$

Further, it follows from (A.6) that

$$R_{\mathbb{D}}(A, TT^{\mathrm{H}}) = L \otimes (TT^{\mathrm{H}}) + M \otimes (ATT^{\mathrm{H}}) + M^{\mathrm{T}} \otimes (ATT^{\mathrm{H}})^{\mathrm{H}}, \tag{A.7}$$

we can easily show that $R_{\mathbb{D}}(A, TT^{\mathrm{H}})$ is Hermite. Therefore, it follows from Lemma A.2 that

$$R_{\mathbb{D}}(A, P) < 0 \implies \mathrm{Re}(R_{\mathbb{D}}(A, P)) < 0,$$

we obtain

$$R_{\mathbb{D}}(A, \mathrm{Re}(P)) < 0.$$

Finally, using again Lemma A.2 to $P = TT^{\mathrm{H}}$, we know $\mathrm{Re}(P) > 0$. Therefore, $\mathrm{Re}(P)$ is the desired real symmetric positive definite matrix. The proof is then completed.

## A.2 Proof of Theorem 5.1

The proof is divided into two subsections.

### A.2.1 Inequality

In this step, we show that

$$\frac{\|G(s)u(s)\|_2}{\|u(s)\|_2} \le \|G(s)\|_\infty. \tag{A.8}$$

Notice that

$$\frac{\|G(s)u(s)\|_2}{\|u(s)\|_2}\Bigg|_{\|u(s)\|_2 \ne 0,\, u(s) \in \mathcal{H}_2}$$

$$= \frac{\frac{1}{\|u(s)\|_2}\|G(s)u(s)\|_2}{\frac{1}{\|u(s)\|_2}\|u(s)\|}\Bigg|_{\|u(s)\|_2 \ne 0,\, u(s) \in \mathcal{H}_2}$$

$$= \frac{\left\|G(s)\frac{u(s)}{\|u(s)\|_2}\right\|_2}{\left\|\frac{u(s)}{\|u(s)\|_2}\right\|}\Bigg|_{\|u(s)\|_2 \ne 0,\, u(s) \in \mathcal{H}_2}$$

$$= \|G(s)v(s)\|_2 \big|_{\|v(s)\|_2 = 1,\, v(s) \in \mathcal{H}_2},$$

we then have

$$\sup_{\|u(s)\|_2 \ne 0,\, u(s) \in \mathcal{H}_2}\left\{\frac{\|G(s)u(s)\|_2}{\|u(s)\|_2}\right\} = \sup_{\|v(s)\|_2 = 1,\, v(s) \in \mathcal{H}_2}\{\|G(s)v(s)\|_2\}.$$

Hence, we need only to prove the first relation in the theorem.
Furthermore, note that

$$G^H(j\omega)G(j\omega) \le \lambda_{\max}\{G^H(j\omega)G(j\omega)\}I,$$

by definition, we have

$$\|G(s)u(s)\|_2^2 = \frac{1}{2\pi}\int_{-\infty}^{\infty} u^H(j\omega)G^H(j\omega)G(j\omega)u(j\omega)\,d\omega$$

$$\le \frac{1}{2\pi}\int_{-\infty}^{\infty} u^H(j\omega)\lambda_{\max}\{G^H(j\omega)G(j\omega)\}u(j\omega)\,d\omega$$

$$\le \frac{1}{2\pi}\int_{-\infty}^{\infty} u^H(j\omega)\sup_{\omega \in \mathbf{R}}(\lambda_{\max}\{G^H(j\omega)G(j\omega)\})u(j\omega)\,d\omega$$

$$= \sup_{\omega \in \mathbf{R}} \left( \sigma_{max}^2 \left\{ G\left(j\omega\right) \right\} \right) \frac{1}{2\pi} \int_{-\infty}^{\infty} u^H \left(j\omega\right) u \left(j\omega\right) d\omega$$

$$= \|G(s)\|_{\infty}^2 \|u(s)\|_2^2 ,$$

that is, (A.8) holds.

## A.2.2  Tightness

In this step, we need only to show that $\|G(s)\|_\infty$ is the least upper bound of

$$\frac{\|G(s)u(s)\|_2}{\|u(s)\|_2}.$$

To do this, we need only to prove, for any given $\epsilon > 0$, there exists a signal $u_\epsilon(s) \in \mathcal{H}_2$ with $\|u_\epsilon(s)\|_2 = 1$ such that

$$\|G(s)u_\epsilon(s)\|_2 \geq \|G(s)\|_\infty - \epsilon. \tag{A.9}$$

*Step 1. Constructing $u_\epsilon(s)$*
Let $\omega_0$ be a frequency such that $\sigma_{max}\left\{ G\left(j\omega\right) \right\} = \sigma_{max}$. Then, by singular value decomposition, there exist vectors $u_i$, $v_i$ with unit length and

$$u_i u_j = \delta\left(i - j\right), \quad v_i v_j = \delta\left(i - j\right),$$

such that

$$G\left(j\omega_0\right) = \sigma_{max} u_1\left(j\omega_0\right) v_1^H\left(j\omega_0\right) + \sum_{i=2}^{r} \sigma_i u_i\left(j\omega_0\right) v_i^H\left(j\omega_0\right) \tag{A.10}$$

with $r = \mathrm{rank}\left(G\left(j\omega_0\right)\right)$ and $\sigma_i \leq \sigma_{max}, i \geq 2$.
    If $0 \leq \omega_0 < \infty$, we can write $v_1\left(j\omega_0\right)$ as

$$v_1\left(j\omega_0\right) = \begin{bmatrix} \alpha_1 e^{j\theta_1} \\ \alpha_2 e^{j\theta_2} \\ \vdots \\ \alpha_q e^{j\theta_q} \end{bmatrix},$$

where $\alpha_i \in \mathbb{R}$ is such that $\theta_i \in [-\pi, 0)$. Clearly, we have

$$\sum_{i=1}^{q} \alpha_i^2 = 1.$$

Let $\beta_i \geq 0$ be such that

$$\theta_i = \arg \left( \frac{\beta_i - j\omega_0}{\beta_i + j\omega_0} \right) \in [-\pi, 0), \tag{A.11}$$

and let $u_s(s)$ be given by

$$u_s(s) = \begin{bmatrix} \alpha_1 \frac{\beta_1 - s}{\beta_1 + s} \\ \alpha_2 \frac{\beta_2 - s}{\beta_2 + s} \\ \vdots \\ \alpha_q \frac{\beta_q - s}{\beta_q + s} \end{bmatrix} u_\varepsilon(s) \triangleq \phi(s) u_\varepsilon(s), \tag{A.12}$$

where the scalar function $u_\varepsilon(s)$ is taken as

$$u_\varepsilon(j\omega) = \begin{cases} \sqrt{\frac{\pi}{2\varepsilon}}, & \omega \in \{\omega : |\omega - \omega_0| < \varepsilon\} \bigcup \{\omega : |\omega + \omega_0| < \varepsilon\}, \\ 0, & \text{otherwise}, \end{cases}$$

with $\varepsilon$ being a sufficiently small scalar. Clearly, by construction, we have $\phi(j\omega_0) = v_1(j\omega_0)$. Then, we can derive

$$\left\| u_\varepsilon(j\omega) \right\|_2^2 = \frac{1}{2\pi} \int_{-\infty}^{\infty} u_\varepsilon^{\mathrm{H}}(j\omega) u_\varepsilon(j\omega) \, d\omega = \frac{1}{2\pi} \frac{\pi}{2\varepsilon} 4\varepsilon = 1,$$

from which it follows that

$$\left\| u_s(j\omega) \right\|_2^2 = \frac{1}{2\pi} \int_{-\infty}^{\infty} u_s^{\mathrm{H}}(j\omega) u_s(j\omega) \, d\omega$$

$$= \frac{1}{2\pi} \int_{-\infty}^{\infty} \sum_{i=1}^{q} \left( \alpha_i^2 \left| \frac{\beta_i - j\omega}{\beta_i + j\omega} \right|^2 \right) u_\varepsilon^2(j\omega) \, d\omega$$

$$= \frac{1}{2\pi} \int_{-\infty}^{\infty} \sum_{i=1}^{q} \alpha_i^2 u_\varepsilon^2(j\omega) \, d\omega$$

$$= \frac{1}{2\pi} \left( \sum_{i=1}^{q} \alpha_i^2 \right) \int_{-\infty}^{\infty} u_\varepsilon^2(j\omega) \, d\omega$$

$$= \frac{1}{2\pi} \int_{-\infty}^{\infty} u_\varepsilon^2(j\omega) \, d\omega$$

$$= \left\| u_\varepsilon \left( j\omega \right) \right\|_2^2$$

$$= 1.$$

Hence, $u_s(s) \in \mathcal{H}_2$ and has unit $H_2$ norm.

*Step 2. Showing (A.9)*
With the $u_s(s)$ obtained in Step 1, we have

$$\| G(s) u_s(s) \|_2^2$$

$$= \frac{1}{2\pi} \int\limits_{-\infty}^{\infty} u_s^H(j\omega) G^H(j\omega) G(j\omega) u_s(j\omega) d\omega$$

$$= \frac{1}{2\pi} \int\limits_{\omega_0-\varepsilon}^{\omega_0+\varepsilon} u_\varepsilon^H(j\omega) \phi^H(j\omega) G^H(j\omega) G(j\omega) \phi(j\omega) u_\varepsilon(j\omega) d\omega$$

$$+ \frac{1}{2\pi} \int\limits_{-\omega_0-\varepsilon}^{-\omega_0+\varepsilon} u_\varepsilon^H(j\omega) \phi^H(j\omega) G^H(j\omega) G(j\omega) \phi(j\omega) u_\varepsilon(j\omega) d\omega. \qquad \text{(A.13)}$$

Clearly, there exists a scalar $\gamma > 0$ such that for any $\omega \in \{\omega : |\omega - \omega_0| < \varepsilon\}$,

$$\phi^H(j\omega_0) G^H(j\omega_0) G(j\omega_0) \phi(j\omega_0)$$

$$\leq \phi^H(j\omega) G^H(j\omega) G(j\omega) \phi(j\omega) + \gamma |\omega - \omega_0|$$

$$\leq \phi^H(j\omega) G^H(j\omega) G(j\omega) \phi(j\omega) + \gamma \varepsilon, \qquad \text{(A.14)}$$

and for any $\omega \in \{\omega : |\omega + \omega_0| < \varepsilon\}$,

$$\phi^H(j\omega_0) G^H(j\omega_0) G(j\omega_0) \phi(j\omega_0)$$

$$\leq \phi^H(-j\omega) G^H(-j\omega) G(-j\omega) \phi(-j\omega) + \gamma |\omega + \omega_0|$$

$$\leq \phi^H(-j\omega) G^H(-j\omega) G(-j\omega) \phi(-j\omega) + \gamma \varepsilon. \qquad \text{(A.15)}$$

Hence, using (A.13) through (A.15), and the notation

$$H(s) = G^H(s) G(s),$$

we get

$$\| G(s) u_s(s) \|_2^2$$

$$\geq \frac{1}{2\pi} \int\limits_{\omega_0-\varepsilon}^{\omega_0+\varepsilon} u_\varepsilon^H(j\omega) \phi^H(j\omega_0) G^H(j\omega_0) G(j\omega_0) \phi(j\omega_0) u_\varepsilon(j\omega)\, d\omega$$

$$+ \frac{1}{2\pi} \int\limits_{-\omega_0-\varepsilon}^{-\omega_0+\varepsilon} u_\varepsilon^H(j\omega) \phi^H(-j\omega_0) H(-j\omega_0) \phi(-j\omega_0) u_\varepsilon(j\omega)\, d\omega$$

$$- \gamma\varepsilon \frac{1}{2\pi} \left( \int\limits_{\omega_0-\varepsilon}^{\omega_0+\varepsilon} u_\varepsilon^H(j\omega) u_\varepsilon(j\omega)\, d\omega + \int\limits_{-\omega_0-\varepsilon}^{-\omega_0+\varepsilon} u_\varepsilon^H(j\omega) u_\varepsilon(j\omega)\, d\omega \right)$$

$$= \frac{1}{2\pi} \int\limits_{\omega_0-\varepsilon}^{\omega_0+\varepsilon} u_\varepsilon^H(j\omega) v_1^H(j\omega_0) G^H(j\omega_0) G(j\omega_0) v_1(j\omega_0) u_\varepsilon(j\omega)\, d\omega$$

$$+ \frac{1}{2\pi} \int\limits_{-\omega_0-\varepsilon}^{-\omega_0+\varepsilon} u_\varepsilon^H(j\omega) v_1^H(-j\omega_0) H(-j\omega_0) v_1(-j\omega_0) u_\varepsilon(j\omega)\, d\omega$$

$$- \gamma\varepsilon \| u_\varepsilon(j\omega) \|_2^2$$

$$= \frac{1}{2\pi} \int\limits_{\omega_0-\varepsilon}^{\omega_0+\varepsilon} \sigma_{\max}^2 u_\varepsilon^H(j\omega) u_\varepsilon(j\omega)\, d\omega$$

$$+ \frac{1}{2\pi} \int\limits_{-\omega_0-\varepsilon}^{-\omega_0+\varepsilon} \sigma_{\max}^2 u_\varepsilon^H(j\omega) u_\varepsilon(j\omega)\, d\omega - \gamma\varepsilon \| u_\varepsilon(j\omega) \|_2^2$$

$$= \sigma_{\max}^2 \| u_\varepsilon(j\omega) \|_2^2 - \gamma\varepsilon \| u_\varepsilon(j\omega) \|_2^2$$

$$= \sigma_{\max}^2 - \gamma\varepsilon$$

$$= \| G(s) \|_\infty^2 - \gamma\varepsilon,$$

which clearly implies (A.9).

If $\omega_0 = \infty$, we can simply choose $u_s(s) = v_1 u_\varepsilon(s)$. The proof is finished.

## A.3   Proof of Theorem 5.2

This proof is also composed of subsections, and in a way it is similar to the one given in Section A.2.

## A.3.1 Inequality

In this step, we show that

$$\| G(s)u(s) \|_P^2 \leq \| G(s) \|_\infty^2 \| u(s) \|_P^2. \tag{A.16}$$

Define the autocorrelation matrix of $u(t)$ as

$$R_{uu}(\tau) = \lim_{T \to \infty} \frac{1}{2T} \int_{-T}^{T} u(t + \tau)u^H(t)dt,$$

if the limit exists and is finite for all $\tau$. The Fourier transform is called the spectral density of $u$, denoted by $S_{uu}(j\omega)$

$$S_{uu}(j\omega) = \int_{-\infty}^{\infty} R_{uu}(\tau)e^{-j\omega\tau} d\tau.$$

Then, the power of the signal $u(t)$ is

$$\| u(t) \|_P = (\text{trace}\, (R_{uu}(0)))^{1/2}.$$

It is also easy to show that

$$\| u(t) \|_P^2 = \frac{1}{2\pi} \int_{-\infty}^{\infty} \text{trace}\, (S_{uu}(j\omega))\, d\omega. \tag{A.17}$$

Moreover, if $G(s)$ is a stable $p \times q$ transfer function with (5.10), then

$$S_{yy}(j\omega) = G(j\omega)S_{uu}(j\omega)G^H(j\omega). \tag{A.18}$$

Similar to the proof of Theorem 5.1, we need only to prove the first relation. By (A.17) and (A.18), we have

$$\| G(s)u(s) \|_P^2 = \frac{1}{2\pi} \int_{-\infty}^{\infty} \text{trace}\, (G(j\omega)S_{uu}(j\omega)G^H(j\omega))\, d\omega$$

$$= \frac{1}{2\pi} \int_{-\infty}^{\infty} \text{trace}\, (S_{uu}(j\omega)G^H(j\omega)G(j\omega))\, d\omega$$

$$\leq \sup_{\omega \in \mathbf{R}} (\lambda_{\max}(G^H(j\omega)G(j\omega))) \frac{1}{2\pi} \int_{-\infty}^{\infty} \text{trace}\, (S_{uu}(j\omega))\, d\omega$$

$$= \| G(s) \|_\infty^2 \| u(s) \|_P^2.$$

That is, the relation in (A.16) holds.

## A.3.2 Tightness

In this step, we only need to prove that there exists a $u(s) \in \mathcal{P}$ with $\|u(t)\|_P = 1$ such that the equality relation in the inequality (A.16) holds.

*Step 1. Constructing u(s)*
Similar to the proof of Theorem 5.1, we let (A.10) and (A.11) hold. If $0 \leq \omega_0 < \infty$, then we choose $u_s(s)$ as (A.12) where

$$u_\varepsilon(s) = u_{\omega_0}(s) = \sqrt{2}\sin\omega_0 s.$$

Then,

$$R_{u_{\omega_0} u_{\omega_0}}(\tau) = \cos(\omega_0\tau)$$

and

$$\left\| u_{\omega_0}(s) \right\|_P = R_{u_{\omega_0} u_{\omega_0}}(0) = 1.$$

Also, we have

$$S_{u_{\omega_0} u_{\omega_0}}(j\omega) = \pi\left(\delta(\omega - \omega_0) + \delta(\omega + \omega_0)\right).$$

Then, by (A.18), we have

$$S_{u_s u_s}(j\omega) = \phi(j\omega)S_{u_{\omega_0} u_{\omega_0}}(j\omega)\phi^H(j\omega),$$

from which it follows that

$$\begin{aligned}
\|u_s\|_P^2 &= \frac{1}{2\pi}\int_{-\infty}^{\infty} \text{trace}\left(S_{u_s u_s}(j\omega)\right)d\omega \\
&= \frac{1}{2\pi}\int_{-\infty}^{\infty} \text{trace}\left(\phi(j\omega)S_{u_{\omega_0} u_{\omega_0}}(j\omega)\phi^H(j\omega)\right)d\omega \\
&= \frac{1}{2\pi}\int_{-\infty}^{\infty} S_{u_{\omega_0} u_{\omega_0}}(j\omega)\phi^H(j\omega)\phi(j\omega)d\omega \\
&= \frac{\pi}{2\pi}\left(\phi^H(j\omega_0)\phi(j\omega_0) + \phi^H(-j\omega_0)\phi(-j\omega_0)\right) \\
&= 1.
\end{aligned}$$

*Step 2. Showing the equality relation in (A.16)*

With $u(s) = u_s(s)$ and by using (A.10), we have

$$\|G(s)u(s)\|_P^2 = \frac{1}{2\pi} \int\limits_{-\infty}^{\infty} \text{trace}\left(G(j\omega)S_{u_s u_s}(j\omega)G^H(j\omega)\right) d\omega$$

$$= \frac{1}{2\pi} \int\limits_{-\infty}^{\infty} \text{trace}\left(G(j\omega)\phi(j\omega)S_{u_{\omega_0} u_{\omega_0}}(j\omega)\phi^H(j\omega)G^H(j\omega)\right) d\omega$$

$$= \frac{1}{2\pi} \int\limits_{-\infty}^{\infty} S_{u_{\omega_0} u_{\omega_0}}(j\omega)\phi^H(j\omega)\, G^H(j\omega)\, (G(j\omega)\phi(j\omega))\, d\omega$$

$$= \frac{\pi}{2\pi}\left[\phi^H(j\omega_0)G^H(j\omega_0)\, (G(j\omega_0)\phi(j\omega_0))\right.$$
$$\left. + \phi^H(-j\omega_0)\, G^H(-j\omega_0)\, (G(-j\omega_0)\,\phi(-j\omega_0))\right]$$

$$= \frac{1}{2}\left[2\phi^H(j\omega_0)G^H(j\omega_0)\, (G(j\omega_0)\,\phi(j\omega_0))\right]$$

$$= \sigma_{\max}^2$$

$$= \|G(s)\|_\infty^2 .$$

The case $\omega_0 = \infty$ can be treated similarly. The proof is finished.

## A.4  Proof of Theorem 6.11

### A.4.1  *Necessity*

Let $\zeta \in \mathbb{R}$ satisfy $\det(i\zeta I - A) \neq 0$ and consider the input

$$u(t) = \exp(i\zeta t)u_0,$$

with $u_0 \in \mathbb{R}^r$. Then, it can be shown that the corresponding state response is given by

$$x(t) = \exp(i\zeta t)(i\zeta I - A)^{-1}Bu_0.$$

In fact, with $x(t)$ given earlier, we have

$$Ax + Bu = \left[A\exp(i\zeta t)(i\zeta I - A)^{-1} + \exp(i\zeta t)\right]Bu_0$$
$$= \left[A(i\zeta I - A)^{-1} + I\right]\exp(i\zeta t)Bu_0$$
$$= i\zeta(i\zeta I - A)^{-1}\exp(i\zeta t)Bu_0,$$

and

$$\dot{x}(t) = \left(\exp(i\zeta t)(i\zeta I - A)^{-1} B u_0\right)'$$

$$= (\exp(i\zeta t))'(i\zeta I - A)^{-1} B u_0$$

$$= i\zeta \exp(i\zeta t)(i\zeta I - A)^{-1} B u_0$$

$$= Ax + Bu,$$

thus this $x(t)$ is a state response of the system. Further, the corresponding output is

$$y(t) = Cx(t) + Du(t)$$

$$= C \exp(i\zeta t)(i\zeta I - A)^{-1} B u_0 + D \exp(i\zeta t) u_0$$

$$= \exp(i\zeta t) G(i\zeta) u_0.$$

With this pair of system input and output, we have

$$s(u, y) = \begin{bmatrix} \exp(i\zeta t) G(i\zeta) u_0 \\ \exp(i\zeta t) u_0 \end{bmatrix}^{\mathrm{H}} Q \begin{bmatrix} \exp(i\zeta t) G(i\zeta) u_0 \\ \exp(i\zeta t) u_0 \end{bmatrix}$$

$$= \exp(-i\zeta t) \exp(i\zeta t) u_0^{\mathrm{H}} \begin{bmatrix} G(i\zeta) \\ I \end{bmatrix}^{\mathrm{H}} Q \begin{bmatrix} G(i\zeta) \\ I \end{bmatrix} u_0$$

$$= u_0^{\mathrm{H}} \begin{bmatrix} G(i\zeta) \\ I \end{bmatrix}^{\mathrm{H}} Q \begin{bmatrix} G(i\zeta) \\ I \end{bmatrix} u_0, \tag{A.19}$$

which happens to be a constant for all time $t \in \mathbb{R}$.

For $T > 0$, we further have, using (A.19),

$$\int_0^T s(u(t), y(t)) dt = \int_0^T u_0^{\mathrm{H}} \begin{bmatrix} G(i\zeta) \\ I \end{bmatrix}^{\mathrm{H}} Q \begin{bmatrix} G(i\zeta) \\ I \end{bmatrix} u_0 d\zeta$$

$$= T u_0^{\mathrm{H}} \begin{bmatrix} G(i\zeta) \\ I \end{bmatrix}^{\mathrm{H}} Q \begin{bmatrix} G(i\zeta) \\ I \end{bmatrix} u_0.$$

Since $u_0$ and $T > 0$ are arbitrary, the relation (6.29) clearly implies

$$\begin{bmatrix} G(i\zeta) \\ I \end{bmatrix}^{\mathrm{H}} Q \begin{bmatrix} G(i\zeta) \\ I \end{bmatrix} \geq 0, \quad \forall \zeta \in \mathbb{R}. \tag{A.20}$$

A.20 can be easily shown to be equivalent to (6.33). In fact, using the maximum modulus principle (Lemma 2.14), we have

$$\begin{bmatrix} G(i\zeta) \\ I \end{bmatrix}^{\mathrm{H}} Q \begin{bmatrix} G(i\zeta) \\ I \end{bmatrix} \geq 0, \quad \forall \zeta \in \mathbb{R}$$

$$\Longleftrightarrow \sigma_{\min} \left( \begin{bmatrix} G(i\zeta) \\ I \end{bmatrix}^{\mathrm{H}} Q \begin{bmatrix} G(i\zeta) \\ I \end{bmatrix} \right) \geq 0, \quad \forall \zeta \in \mathbb{R}$$

$$\Longleftrightarrow \sigma_{\min} \left( \begin{bmatrix} G(s) \\ I \end{bmatrix}^{\mathrm{H}} Q \begin{bmatrix} G(s) \\ I \end{bmatrix} \right) \geq 0, \quad \forall s \in \mathbb{C}^+$$

$$\Longleftrightarrow \begin{bmatrix} G(s) \\ I \end{bmatrix}^{\mathrm{H}} Q \begin{bmatrix} G(s) \\ I \end{bmatrix} \geq 0, \; \forall s \in \mathbb{C}^+.$$

## A.4.2  Sufficiency

Let the relation (6.33) hold, hence (A.20) holds. Since the system is controllable, for an arbitrary positive number $T$, there exists a control $u$ which transfers $x(0) = 0$ to $x(T) = 0$ within the time interval $[0, T]$. Without loss of generality, we may assume that

$$u(t) = 0, \quad x(t) = 0, \quad \text{when } t < 0 \text{ and } t > T. \tag{A.21}$$

Hence,

$$y(t) = 0, \quad \text{when } t > T. \tag{A.22}$$

On the other hand, we have

$$Y(i\zeta) = C(Ii\zeta - A)^{-1} BU(i\zeta) = G(i\zeta)U(i\zeta). \tag{A.23}$$

Premultiplying by $U^{\mathrm{H}}(i\zeta)$ and postmultiplying by $U(i\zeta)$ both sides of (A.20), and using the relation (A.23), gives

$$\begin{bmatrix} Y(i\zeta) \\ U(i\zeta) \end{bmatrix}^{\mathrm{H}} Q \begin{bmatrix} Y(i\zeta) \\ U(i\zeta) \end{bmatrix} \geq 0, \quad \forall \zeta \in \mathbb{R}. \tag{A.24}$$

Let

$$V(i\zeta) = \begin{bmatrix} Y(i\zeta) \\ U(i\zeta) \end{bmatrix},$$

then (A.24) is equivalent to

$$V(i\zeta)^{\mathrm{H}} Q V(i\zeta) \geq 0, \quad \forall \zeta \in \mathbb{R}.$$

Taking integration yields

$$\langle V(i\zeta)\rangle_Q = \frac{1}{2\pi} \int_{-\infty}^{+\infty} V(i\zeta)^H Q V(i\zeta) d\zeta \geq 0.$$

Define the inverse Laplace transform of $V(s)$ as follows:

$$v(t) = \mathcal{L}^{-1}[V(s)] = \begin{bmatrix} y(t) \\ u(t) \end{bmatrix}, \tag{A.25}$$

and using the generalized Parseval lemma (Lemma 2.16), we have

$$\lfloor v(t) \rfloor_Q = \frac{1}{2\pi} \int_0^{+\infty} v(t)^H Q v(t) dt = \lfloor V(i\zeta) \rfloor_Q \geq 0. \tag{A.26}$$

Further, noticing (A.25), (A.21), and (A.22), we have

$$\int_0^{+\infty} v(t)^T Q v(t) \, dt = \int_0^{+\infty} \begin{bmatrix} y(t) \\ u(t) \end{bmatrix}^T Q \begin{bmatrix} y(t) \\ u(t) \end{bmatrix} dt$$

$$= \int_0^T \begin{bmatrix} y(t) \\ u(t) \end{bmatrix}^T Q \begin{bmatrix} y(t) \\ u(t) \end{bmatrix} dt,$$

which, together with (A.26), clearly implies (6.29).

## A.5   Direct Proof of Theorem 6.15

The bounded-real lemma was proved in Section 6.5 based on a result for dissipativity. Here, we give a direct proof, which makes use of Parseval lemma and Theorem 5.1.

Using the well-known Parseval lemma, we have

$$\|y(s)\|_2 = \|y(t)\|_2 = \left( \int_0^{+\infty} y^H(t) y(t) dt \right)^{\frac{1}{2}}, \tag{A.27}$$

$$\|u(s)\|_2 = \|u(t)\|_2 = \left( \int_0^{+\infty} u^H(t) u(t) dt \right)^{\frac{1}{2}}. \tag{A.28}$$

On the other hand, since

$$y(s) = G(s) u(s),$$

we have, following Corollary 5.1,

$$\|y(s)\|_2 \leq \|G(s)\|_\infty \|u(s)\|_2. \tag{A.29}$$

## A.5.1 Necessity

Now suppose that the linear system (6.27) is nonexpansive, then by definition we have for arbitrary $T \geq 0$, that

$$\int_0^T y^H(t)y(t)\,dt \leq \int_0^T u^H(t)u(t)\,dt. \tag{A.30}$$

Specially taking $T \to +\infty$ in the aforementioned relation yields

$$\int_0^{+\infty} y^H(t)y(t)\,dt \leq \int_0^{+\infty} u^H(t)u(t)\,dt. \tag{A.31}$$

With the help of (A.27) and (A.28), we get from (A.31),

$$\|y(s)\|_2 \leq \|u(s)\|_2. \tag{A.32}$$

On the other hand, it follows from Theorem 5.1 (and also its proof in Section A.2) that $\|G(s)\|_\infty$ is the least upper bound of

$$\frac{\|G(s)u(s)\|_2}{\|u(s)\|_2}.$$

Therefore, for arbitrary $\epsilon > 0$, there exists a $u_\epsilon(s) \in \mathcal{H}_2$ such that

$$\frac{\|G(s)u_\epsilon(s)\|_2}{\|u_\epsilon(s)\|_2} = \|G(s)\|_\infty - \epsilon. \tag{A.33}$$

Denote the output corresponding to $u_\epsilon(s)$ as $y_\epsilon(s)$, then

$$y_\epsilon(s) = G(s)u_\epsilon(s),$$

and we can get from the relation (A.33) the following relation

$$\frac{\|y_\epsilon(s)\|_2}{\|u_\epsilon(s)\|_2} = \|G(s)\|_\infty - \epsilon. \tag{A.34}$$

Since (A.32) holds by assumption, we also have

$$\frac{\|y_\epsilon(s)\|_2}{\|u_\epsilon(s)\|_2} \leq 1. \tag{A.35}$$

Combining the aforementioned two relations gives

$$\| G(s) \|_\infty \leq 1 + \epsilon.$$

Note that $\epsilon$ can be chosen arbitrarily small, we thus have $\| G(s) \|_\infty \leq 1$, that is, the transfer function of system (6.27) is bounded-real.

## A.5.2  Sufficiency

Suppose that the transfer function of system (6.27) is bounded-real, then it follows from Proposition 6.3 that (6.48) holds. This gives, in view of (A.29), the relation in (A.31).

Based on the pair of system input $u(t)$ and output $y(t)$ satisfying (A.31), define, for arbitrary $T \geq 0$, the following new system input

$$\tilde{u}(t) = \begin{cases} u(t), & t \leq T \\ 0, & t > T, \end{cases}$$

and define the corresponding output $\tilde{y}(t)$ by

$$\tilde{y}(t) = \mathcal{L}^{-1}[\tilde{y}(s)], \quad \tilde{y}(s) = G(s)\tilde{u}(s),$$

then we have for this pair of system input and output the following relation:

$$\int_0^{+\infty} \tilde{y}^H(t)\tilde{y}(t)\, dt \leq \int_0^{+\infty} \tilde{u}^H(t)\tilde{u}(t)\, dt. \tag{A.36}$$

Note that

$$\int_0^T y^H(t)y(t)\, dt = \int_0^T \tilde{y}^H(t)\tilde{y}(t)\, dt \tag{A.37}$$

and

$$\int_0^{+\infty} \tilde{u}^H(t)\tilde{u}(t)\, dt = \int_0^T u^H(t)u(t)\, dt, \tag{A.38}$$

we finally have from (A.36) that the relation in (A.30) holds for arbitrary $T \geq 0$, that is, the linear system (6.27) is nonexpansive. The proof is then completed.

## A.6  Proofs of Lemmas 7.3 and 7.4

Letting

$$z = \begin{bmatrix} x \\ \dot{x} \end{bmatrix}, \quad y = \begin{bmatrix} y_p \\ y_d \end{bmatrix},$$

we can convert the second-order linear system (7.51) into the following extended first-order state-space model

$$\begin{cases} \dot{z} = A_e z + B_e u \\ y = C_e z, \end{cases} \tag{A.39}$$

where

$$A_e = \begin{bmatrix} 0 & I \\ -A_2^{-1}A_0 & -A_2^{-1}A_1 \end{bmatrix}, \quad B_e = \begin{bmatrix} 0 \\ A_2^{-1}B \end{bmatrix}, \quad C_e = \text{diag}\left(C_p, C_d\right). \tag{A.40}$$

### A.6.1  Proof of Lemma 7.3

To prove this lemma, we need the following two results.

**Lemma A.3**  (Duan et al. 1989) A matrix $A \in \mathbb{R}^{n \times n}$ is Hurwitz stable if $A + A^{\mathrm{T}} < 0$.

This result can be easily deduced from the main result in Section 7.5.2. Here, we present a different proof.

*Proof*  Let $\lambda$ be an eigenvalue of $A$, $x$ be a corresponding eigenvector, then we have

$$Ax = \lambda x.$$

Premultiplying both sides of the aforementioned equation by $x^{\mathrm{H}}$ yields

$$x^{\mathrm{H}} A x = \lambda x^{\mathrm{H}} x. \tag{A.41}$$

Taking complex conjugate of (A.41) and adding it to (A.41) gives

$$x^{\mathrm{H}}(A^{\mathrm{T}} + A)x = (\lambda + \bar{\lambda})x^{\mathrm{H}}x.$$

Thus, $A^T + A < 0$ implies

$$\text{Re}\,\lambda(A) = \frac{\lambda + \bar{\lambda}}{2} < 0.$$

This completes the proof. ■

**Lemma A.4** (Duan et al. 1989) Let $A \in \mathbb{R}^{n \times n}$, $E \in \mathbb{S}^n$ and $E > 0$. Then, the matrix $EA$ is Hurwitz stable if

$$A + A^T \le 0, \tag{A.42}$$

and the matrix pair $(EA, C)$ is observable, where

$$C^T C = -(A + A^T). \tag{A.43}$$

*Proof* It can be easily verified that the following Lyapunov equality

$$P(EA) + (EA)^T P = -C^T C \le 0,$$

has a solution $P = E^{-1} > 0$. Then, matrix $EA$ is Hurwitz stable in view of the Lyapunov stability theory. ■

With the help of the aforementioned lemmas, we can give a proof of Lemma 7.3 as follows.

### Proof of Lemma 7.3

The second-order linear system (7.51) is Hurwitz stable if and only if $A_e$ in (A.40) is Hurwitz stable. Thus, we only need to prove the Hurwitz stability of $A_e$.

Partition the matrix $A_e$ as

$$A_e = EA,$$

where

$$E = \begin{bmatrix} A_0^{-1} & 0 \\ 0 & A_2^{-1} \end{bmatrix}, \quad A = \begin{bmatrix} 0 & A_0 \\ -A_0 & -A_1 \end{bmatrix}.$$

Then, we can verify that $E > 0$, and

$$A + A^T = \begin{bmatrix} 0 & 0 \\ 0 & -A_1 - A_1^T \end{bmatrix} \le 0,$$

in view of (7.54).

On the other hand, there exists a nonsingular matrix $G$ satisfying $G^{\mathrm{T}}G = A_1 + A_1^{\mathrm{T}} > 0$. Letting

$$C = \begin{bmatrix} 0 & G \end{bmatrix},$$

then we have

$$A + A^{\mathrm{T}} = -C^{\mathrm{T}}C.$$

Furthermore, in view of $\operatorname{rank}(A_2^{-1}A_0) = n$, $\operatorname{rank}(G) = n$, there holds

$$\operatorname{rank}\begin{bmatrix} sI - A_e \\ C \end{bmatrix} = \operatorname{rank}\begin{bmatrix} sI & -I \\ A_2^{-1}A_0 & sI + A_2^{-1}A_1 \\ 0 & G \end{bmatrix} = 2n, \quad \forall s \in \mathbb{C}. \qquad \text{(A.44)}$$

Thus, $(A_e, C)$ is observable. Therefore, the conclusion follows from Lemma A.4.

## A.6.2 Proof of Lemma 7.4

To prove this lemma, we suffice to show the following three results.

**Lemma A.5** Given $P \in \mathbb{S}^n$, $P > 0$, the following conclusions are equivalent:

1. $P + \Delta P > 0$.
2. $\det(P + r\Delta P) \neq 0$, $\forall r \in (0, 1]$.
3. $\lambda_i(P^{-1}\Delta P) > -1$.

*Proof* First we show the equivalence between the first two conclusions.
Let $P + \Delta P > 0$, then

$$P + r\Delta P = r(P + \Delta P) + (1 - r)P > 0, \quad \forall r \in (0, 1],$$

which means $\det(P + r\Delta P) \neq 0, \forall r \in (0, 1]$.
Conversely, denote

$$f(r) = \det(P + r\Delta P), \quad g_i(r) = \lambda_i(P + r\Delta P), \quad i = 1, 2, \ldots, n,$$

then functions $f(r)$ and $g_i(r)$, $i = 1, 2, \ldots, n$, are both continuous in the interval $r \in (0, 1]$. If the second conclusion holds, then

$$f(r) \neq 0, \quad \forall r \in (0, 1],$$

which implies

$$g_i(r) \neq 0, \quad i = 1, 2, \dots, n, \quad \forall r \in (0, 1].$$

Furthermore, we know $g_i(0) = \lambda_i(P) > 0$, then due to the continuity of $g_i(r)$, we have

$$g_i(r) > 0, \quad i = 1, 2, \dots, n, \quad \forall r \in (0, 1].$$

Specially letting $r = 1$ yields the first conclusion.

Next, we prove the equivalence between the second and the third conclusions. For arbitrary $r \in (0, 1]$,

$$\det(P + r\Delta P) \neq 0 \iff \det(I + rP^{-1}\Delta P) \neq 0$$

$$\iff \lambda_i(P^{-1}\Delta P) \neq -\frac{1}{r}$$

$$\iff \lambda_i(P^{-1}\Delta P) > -1.$$

With this we complete the proof. ■

**Lemma A.6** Let $P > 0$, then $P + \Delta P > 0$ if $\lambda_{\min}(P) > \|\Delta P\|_2$.

*Proof* In view of Lemma A.5, we have

$$\lambda_{\min}(P) > \|\Delta P\|_2 \iff \sigma_{\min}(P) > \|\Delta P\|_2$$

$$\iff \frac{1}{\sigma_{\max}(P^{-1})} > \sigma_{\max}(\Delta P)$$

$$\implies \sigma_{\max}(P^{-1}\Delta P) < 1$$

$$\implies \rho(P^{-1}\Delta P) < 1$$

$$\implies \lambda_{\min}(P^{-1}\Delta P) > -1$$

$$\implies P + \Delta P > 0.$$

Then, we complete the proof. ■

**Lemma A.7** Given $P > 0$, $P + \Delta P > 0$ if

$$\varepsilon \sqrt{l} < \lambda_{\min}(P),$$

where $l$ is the number of the nonzero elements of matrix $\Delta P$, and

$$\varepsilon = \max\{|\Delta p_{ij}|\}, \quad \Delta P = [\Delta p_{ij}]_{n \times n}.$$

*Proof*　In view of Lemma A.6, we have

$$\varepsilon \sqrt{l} < \lambda_{\min}(P) \Longrightarrow \|\Delta P\|_F < \lambda_{\min}(P)$$
$$\Longrightarrow \|\Delta P\|_2 < \lambda_{\min}(P)$$
$$\Longrightarrow P + \Delta P > 0.$$

Thus, the conclusion holds. ■

With the help of Lemmas 7.3, A.6, and A.7, we can prove Lemma 7.4 easily.

# Appendix B: Using the MATLAB® LMI Toolbox

LMIs have been shown to provide a powerful control design tool. For solution of LMIs, there are basically two types of effective methods, namely, the ellipsoid methods and the interior point methods.

Based on the interior-point method, Gahinet, Nemirovskii, and some other researchers produced a software package called LMI Toolbox, which is for use with MATLAB®. This LMI toolbox is designed to assist control engineers and researches with a user-friendly interactive environment. Through this environment, one can specify and solve many often encountered engineering problems involving LMIs.

In this appendix, we have given a very brief introduction to the MATLAB LMI toolbox to help the readers to get a quick start. For a much more detailed description, one needs to refer to the User's Guide for LMI Control Toolbox by Gahinet et al. (1995). One can also check with the help document for LMI Lab from the MATLAB window following the procedure:

$$\text{Help} \rightarrow \text{Product Help} \rightarrow \text{Robust Control Toolbox} \rightarrow \text{LMI Lab.}$$

This is convenient when you are really working on some LMI problems with MATLAB since it provides you with a sort of online help environment.

As is all know, for usage of a particular MATLAB command in the LMI toolbox, one can simply check with the help message in the command file simply by typing in the MATLAB working window "help M-file name."

## B.1  How to Describe LMIs?

The LMI toolbox can handle any system of LMIs of the form

$$N^{\mathrm{T}} L(X_1, \ldots, X_K) N < M^{\mathrm{T}} R(X_1, \ldots, X_K) M,$$

where $X_1, \ldots, X_K$ are the decision variables, $L(X_1, \ldots, X_K)$ and $R(X_1, \ldots, X_K)$ are linear functions and $M$, $N$ are matrices of appropriate dimensions.

417

The specification of an LMI system involves two steps:

1. Declare the dimensions and structure of each matrix variable.
2. Describe the term content of each LMI.

There are two ways of generating the internal description of a given LMI system:

■ **Command-based description** by a sequence of `lmivar`/`lmiterm` commands that build it incrementally
■ **LMI editor description** via the `lmiedit` where LMIs can be specified directly as symbolic matrix expressions

## B.1.1 Command-Based Description

The description of LMI problems in the MATLAB LMI toolbox involves several MATLAB commands.

### setlmis

The description of an LMI system should begin with `setlmis` and end with `getlmis`. The function `setlmis` initializes the LMI system description. It can be used in the following two ways:

■ `setlmis([])`: Resets the internal variables used for creating LMIs so that one can create a new system of LMIs.
■ `setlmis(lmisys0)`: Uses an existing LMI system `lmisys0` as the starting point for creating a new system of LMIs. Subsequent commands will add to the LMI system `lmisys0` to construct a new set of LMIs.

### getlmis

When the LMI system is completely specified, type

$$\text{lmisys = getlmis}$$

This returns the internal representation `lmisys` of this LMI system. The command `getlmis` must be used only once and after declaring all the matrix variables and all the LMI terms.

### lmivar

After initializing the description with the command `setlmis`, the matrix variables are declared one at a time with `lmivar` and are characterized by their structure. The function `lmivar` can be used as follows:

- X = lmivar(type, struct) adds a new matrix variable X to the LMI system currently specified. You can use the identifier X for subsequent references to the variable X in calls to lmiterm.
- [X,ndec,xdec] = lmivar(type, struct) also returns the total number of decision variables associated with X, ndec, and the entry-wise dependence of X on these decision variables, xdec.

In command lmivar, the first input specifies the structure type and the second input contains additional information about the structure of the variable.

*Inputs:*

type: Structure of X, which takes values of 1, 2, or 3, as defined next:
When type = 1, X is symmetric block diagonal.
When type = 2, X is full rectangular.
When type = 3, X takes some other form.

struct: Additional data on the structure of X, which varies with the variable type as follows:

  i. When type = 1, this second input struct is a matrix with two columns and as many rows as diagonal blocks. The first column lists the sizes of the diagonal blocks and the second column specifies their nature with the following convention:
1 stands for full symmetric block.
0 stands for scalar block.
−1 stands for zero block.

  ii. When type = 2, struct = [m,n] if X is an m×n matrix.

  iii. When type = 3, struct is a matrix of the same dimension as X, where each entry is
0 if X(i,j) = 0.
+n if X(i,j) = $n$th decision variable.
−n if X(i,j) = (−1) × $n$th decision variable.

*Outputs:*

X: Identifier for the new matrix variable. Its value is $k$ if $k − 1$ matrix variables have already been declared. This identifier is not affected by subsequent modifications of the LMI system.

nedc: Total number of decision variables.

xdec: Entry-wise dependence of X on the decision variables.

## lmiterm

lmiterm(termid,A,B,flag) adds one term to some LMI in the LMI system currently specified.

*Inputs:*

termid: four-entry vector specifying the term location and nature.

    i. The first entry specifies which LMI.

        +n: Left-hand side of the *n*th LMI.

        −n: right-hand side of the *n*th LMI.

    ii. The next two entries state which block. For outer factors, set termid (2:3) = [0 0]; otherwise, set termid (2:3) = [i j] if the term belongs to the (i,j) block of the LMI.

    iii. The last one is what type of term.

        0: Constant term.

        X: Variable term AXB.

        −X: variable term $AX^T B$.

        In the earlier points, X is a matrix variable by lmivar.

A: Value of the outer factor, constant term, or left coefficient in variable terms AXB or $AX^T B$.

B: Right coefficient in variable terms AXB or $AX^T B$.

flag: Quick way of specifying the expression $AXB+(AXB)^T$ in a diagonal block. Set flag = 's' to specify it with only one lmiterm command.

Two examples are given as follows:

**Example B.1**

Given

$$A_1 = \begin{bmatrix} -1 & 2 \\ 1 & -3 \end{bmatrix}, \quad A_2 = \begin{bmatrix} -0.8 & 1.5 \\ 1.3 & -2.7 \end{bmatrix}, \quad A_3 = \begin{bmatrix} -1.4 & 0.9 \\ 0.7 & -2.0 \end{bmatrix},$$

describe the following LMIs in MATLAB:

$$\begin{cases} A_1^T P + PA_1 < 0 \\ A_2^T P + PA_2 < 0 \\ A_3^T P + PA_3 < 0 \\ I < P. \end{cases}$$

The MATLAB file for solving the aforementioned problem is given as follows:

```
% specify parameter matrices
A1 = [-1 2;1 -3]; A2 = [-0.8 1.5;1.3 -2.7];
A3 = [-1.4 0.9;0.7 -2.0];
% define the LMI system
```

```
setlmis ([]) % initialing
P = lmivar (1, [2 1]); % declare the LMI variable
lmiterm ( [1 1 1 P],1, A1, 's'); % #1 LMI, left-hand side
lmiterm ( [2 1 1 P],1, A2, 's'); % #2 LMI, left-hand side
lmiterm ( [3 1 1 P],1, A3, 's'); % #3 LMI, left-hand side
lmiterm ( [-4 1 1 P],1,1); % #4 LMI, right-hand side
lmiterm ( [4 1 1 0],1); % #4 LMI, left-hand side
lmis1 = getlmis; %finishing description
```

**Example B.2**

Given

$$A = \begin{bmatrix} -1 & -2 & 1 \\ 3 & 2 & 1 \\ 1 & -2 & -1 \end{bmatrix}, \quad B = \begin{bmatrix} 1 \\ 0 \\ 1 \end{bmatrix},$$

describe the following LMIs in MATLAB:

$$\begin{cases} \begin{bmatrix} A^\mathsf{T} X + XA + Q & XB \\ B^\mathsf{T} X & -I \end{bmatrix} < 0 \\ 0 < X \\ 0 < Q. \end{cases}$$

The MATLAB file for solving the aforementioned problem is given as follows:
```
% specify parameter matrices
A = [-1 -2 1;3 2 1;1 -2 -1]; B = [1; 0; 1];
% define the LMI system
setlmis ([]) %initialing
X=lmivar (1, [3 1]); % declare the LMI variable
Q=lmivar (1, [3 1]); % declare the LMI variable
lmiterm ( [1 1 1 X],1, A, 's'); % #1 LMI, the (1,1) block
lmiterm ( [1 1 1 Q],1,1); % #1 LMI, (1,1) block
lmiterm ( [1 2 2 0],-1); % #1 LMI, the (2,2) block
lmiterm ([1 2 1 X],B',1); % #1 LMI, the (2,1) block
lmiterm ([-2 1 1 X],1,1); % #2 LMI, right-hand side
lmiterm ([-3 1 1 Q],1,1); % #3 LMI, right-hand side
lmis2 = getlmis; % finishing description
```

## B.1.2 LMI Editor Description

The LMI editor `lmiedit` is a graphical user interface (GUI) to specify LMI systems in a straightforward symbolic manner. Typing

```
lmiedit
```

calls up a window with several editable text areas and various push-buttons. To specify your LMI system,

1. Declare each matrix variable (name and structure) in the upper half of the worksheet. The structure is characterized by its type and by an additional "structure" matrix. This matrix contains specific information about the structure and corresponds to the second argument of `lmivar`.
2. Specify the LMIs as MATLAB expressions in the lower half of the worksheet.

Though somewhat less flexible and powerful than the command-based description, the LMI editor is more straightforward to use, hence particularly well-suited for beginners. Thanks to its coding and decoding capabilities, it also constitutes a good tutorial introduction to `lmivar` and `lmiterm`.

# B.2 How to Modify LMIs?

Once specified, a system of LMIs can be modified in several ways with the functions `dellmi`, `delmvar`, and `setmvar`.

### dellmi

The first possibility is to remove an entire LMI from the system with `dellmi`. The format is as follows:

```
newsys = dellmi(lmisys,lmid)
```

It removes the LMI with identifier `lmid` from the system of LMIs described in `lmisys`. `lmid` is the ranking of the LMI in the LMI system initially created with `setlmis/getlmis`. To easily keep track of LMIs after deletions, set `lmid` to the identifier returned by `newsys` when this LMI is created.

### delmvar

Another way of modifying an LMI system is to delete a matrix variable, that is, to remove all variable terms involving this matrix variable. This operation is performed by `delmvar`. Note that `delmvar` automatically removes all LMIs that depended only on the deleted matrix variable. The format is as follows:

$$newsys = delmvar(lmisys,xid)$$

It deletes the matrix variable with identifier `xid` from the LMI system `lmisys`. The updated LMI system, `newsys`, is returned. For safety and to easily keep track of modifications, set `xid` to the value returned by `lmivar` when the matrix variable was created.

## setmvar

The function `setmvar` is used to set a matrix variable to some given values. As a result, this variable is removed from the problem and all terms involving it become constant terms. The format is as follows:

$$newsys = setmvar(lmisys,xid,xval)$$

It sets the matrix variable with identifier `xid` to the value `xval`. All LMI terms involving this matrix variable are evaluated and added to the constant terms. Since `setmvar` does not alter the variable identifiers, the remaining matrix variables can still be referred to by their original identifier.

*Inputs:*

    `lmisys`: Array describing the system of LMIs.

    `xid`: Identifier of the variable matrix to be set.

    `xval`: Matrix value assigned to the matrix variable `xid` (a scalar $t$ is interpreted as $t * I$).

*Output:*

    `newsys`: Updated description of the LMI system.

# B.3 How to Retrieve Information?

Recalling that the full description of an LMI system is stored as a single vector called the internal representation, the user should not attempt to read or retrieve information directly from this vector. Three functions called `lmiinfo`, `lminbr`, and `matnbr` are provided to extract and display all relevant information in a user-readable format.

## lmiinfo

`Lmiinfo` is an interactive facility to retrieve qualitative information about LMI systems. This includes the number of LMIs, the number of matrix variables and their structure, the term content of each LMI block, etc. To invoke `lmiinfo`, enter

$$\texttt{lmiinfo(lmisys)}$$

where `lmisys` is the internal representation of the LMI system produced by `getlmis`.

### `lminbr` and `matnbr`

These two functions return the number of LMIs and the number of matrix variables in the system. To get the number of LMIs in LMI system `lmisys`, for instance, enter

$$\texttt{nlmis = lminbr(lmisys)}$$

To get the number of matrix variables in LMI system `lmisys`, type

$$\texttt{[nmvars,varid] = matnbr(lmisys)}$$

or

$$\texttt{nmvars = matnbr(lmisys)}$$

where `nmvars` is the number of matrix variables, and `varid` is the vector of matrix variable identifiers in the LMI system `lmisys`.

## B.4 How to Convert between Decision and Matrix Variables?

While LMIs are specified in terms of their matrix variables $X_1, \ldots, X_k$, the LMI solvers optimize the vector $x$ of free scalar entries of these matrices, called the decision variable. The two functions `mat2dec` and `dec2mat` perform the conversion between these two descriptions of the problem variables.

### `mat2dec`

Command `mat2dec` constructs decision variables vector from values of the matrix variables in the LMI systems. The format of this function is

$$\texttt{xdec = mat2dec(lmisys,x1,x2,...,xn)}$$

*Inputs:*

lmisys: The LMI system.

x1,x2,...,xn: The particular values of the matrix variables $X_1, X_2, \ldots, X_3$ of system lmisys. mat2dec accepts up to 20 matrix variables. An error is issued if some matrix variable remains unassigned.

*Outputs:*

xdec: The decision variables from matrix variable values.

## dec2mat

Command dec2mat extracts matrix variable value from vector of decision variables. The format of this function is

$$X = \text{dec2mat}(\text{lmisys}, \text{xdec}, \text{xid})$$

*Inputs:*

lmisys: The LMI system.

xdec: The vector of decision variables.

xid: The identifier of matrix variable of LMI system lmisys.

*Outputs:*

X: The corresponding values of matrix variables with the identifier xid.

# B.5    How to Solve LMIs?

With the MATLAB LMI toolbox, one can solve three types of basic LMI problems with commands feasp, mincx, and gevp, respectively.

## B.5.1    *feasp* Command

The function feasp solves the feasibility problem, that is, Problem 3.7, which can be stated as follows.

*Find a solution $x \in R^n$ satisfying the following LMI problem*

$$A(x) < B(x).$$

The corresponding function in MATLAB is

$$[\text{tmin}, \text{xfeas}] = \text{feasp}(\text{lmisys}, \text{options}, \text{target})$$

which solves the feasibility problem defined by the system `lmisys` of LMI constraints. When the problem is feasible, the output `xfeas` is a feasible value of the vector of (scalar) decision variables. The format is as follows:

*Inputs:*

`lmisys`: Array describing the system of LMI constraints.

`options`: (optional) Five-entry vector of control parameters. Default values are selected by setting options (*i*)=0:

`options (1)`: Not used.

`options (2)`: Maximum number of iterations (default = 100).

`options (3)`: Feasibility radius $R$. $R > 0$ constrains $x$ to $x^T x < R^2$ (default = $10^9$). $R < 0$ means "no bound."

`options (4)`: When set to an integer value $L > 1$, forces termination when $t$ has not decreased by more than 1% over the last $L$ iterations (default = 10).

`options (5)`: When nonzero, the trace of execution is turned off.

`target`: (optional) target for `tmin`. The code terminates as soon as $t < $ `target` (default = 0).

*Outputs:*

`tmin`: Value of $t$ upon termination. The LMI system is feasible if and only if `tmin` $\leq 0$.

`xfeas`: corresponding minimizer. If `tmin` $\leq 0$, `xfeas` is a feasible vector for the set of LMI constraints.

Given a feasibility problem of the form $A(x) < B(x)$, `feasp` solves the auxiliary convex programming:

$$\begin{cases} \min & t \\ \text{s.t.} & A(x) < B(x) + tI. \end{cases}$$

The system of LMIs is feasible if and only if the global minimum `tmin` is negative. The current best value of $t$ is displayed by `feasp` at each iteration.

The following example illustrates the use of the `feasp` solver.

**Example B.3**

Find a positive definite symmetric matrix $P$ such that the LMIs in Example B.1 hold.

The MATLAB commands for the LMI description (step 1) are given in Example B.1. Here, we only write the commands for the feasibility problem (step 2).

```
% Step 2. Solve this LMI problem and use dec2mat to
% get the corresponding matrix P
% solve the feasibility problem
[tmin,xfeas] = feasp(lmis1);
% extract the matrix P from the decision variable xfeas
P = dec2mat(lmis1,xfeas,P);
```

## B.5.2  *mincx Command*

The function mincx solves the convex minimization problem, that is, Problem 3.8, which has the following form:

$$\begin{cases} \min\limits_{x} & c^T x \\ \text{s.t.} & A(x) < B(x), \end{cases}$$

where

$x \in \mathbb{R}^n$ is the vector of decision variables

$c \in \mathbb{R}^n$ is a given vector

The corresponding function in MATLAB is

```
[copt,xopt] = mincx( lmisys,c,options,xinit,target)
```

*Inputs:*

lmisys: Description of the system of LMI constraints.

c: Vector of the same size as $x$. Use defcx to specify the objective $c^T x$ directly in terms of matrix variables.

options: (optional) Is similar to that in feasp.

xinit: (optional) Initial guess for $x$ ([] if none, ignored when unfeasible).

target: (optional) Target for the objective value. The code terminates as soon as a feasible $x$ is found such that $c^T x <$ target. (default = $-10^{20}$).

*Outputs:*

copt: Global minimum of the objective $c^T x$.

xopt: Minimizing value of the vector $x$ of decision variables.

The following example illustrates the use of the mincx solver.

**Example B.4**

Given the following linear system

$$\begin{cases} \dot{x} = Ax + B_2 u + B_1 w \\ z = Cx + Du, \end{cases}$$

where

$$A = \begin{bmatrix} -3 & -2 & 1 \\ 1 & 2 & 1 \\ 1 & -1 & -1 \end{bmatrix}, \quad B_1 = \begin{bmatrix} 1 \\ 0 \\ 1 \end{bmatrix}, \quad B_2 = \begin{bmatrix} 2 & 0 \\ 0 & 2 \\ 0 & 1 \end{bmatrix},$$

$$C = \begin{bmatrix} 1 & 0 & 1 \\ 0 & 1 & 0 \end{bmatrix}, \quad D = I_2,$$

design a state feedback control law $u = Kx$ and find the minimal positive scalar $\gamma_{min}$ such that the closed-loop transfer function matrix satisfies $\|G_{zw}(s)\|_2 < \gamma_{min}$.

According to (8.16), the $H_2$ control problem with minimum disturbance attenuation level $\gamma_{min}$ can be turned into the following convex minimization problem:

$$\begin{cases} \min_{X, Z, W, \rho} \rho \\ \text{s.t.} \quad (AX + B_2 W)^T + AX + B_2 W + B_1 B_1^T < 0 \\ \qquad \begin{bmatrix} -Z & CX + DW \\ (CX + DW)^T & -X \end{bmatrix} < 0 \\ \qquad \text{trace}(Z) < \rho, \end{cases}$$

where $X$, $W$, and $Z$ are matrix variables to be solved, and the minimal value of $\sqrt{\rho}$ equals to $\gamma_{min}$.

The MATLAB file for solving the aforementioned problem is given as follows:

```
% Step 1. Describe this LMIs in MATLAB
% specify parameter matrices
A = [-3 -2 1;1 2 1;1 -1 -1]; B1 = [1; 0; 1];
B2 = [2 0;0 2;0 1]; C = [1 0 1;0 1 0]; D = eye(2);
% define the LMI system
setlmis ([]) % initialing
% declare the LMI variable
X = lmivar(1, [3 1]); W = lmivar(2,[2,3]);
Z = lmivar(1, [2 1]); rou = lmivar(1, [1 1]);
% #1 LMI, left-hand side
lmiterm ( [1 1 1 X],A, 1, 's');
lmiterm ( [1 1 1 W],B2, 1, 's');
```

```
lmiterm ( [1 1 1 0],B1*B1');
% #2 LMI, left-hand side
lmiterm ( [2 1 1 Z],1,-1); %  the (1,1) block
lmiterm ( [2 1 2 W],D,1); %  the (1,2) block
lmiterm ( [2 1 2 X],C,1); %  the (1,2) block
lmiterm ( [2 2 2 X],1,-1); %  the (2,2) block
% #3 LMI, left-hand side
lmiterm ( [3 1 1 Z],[1 0],[1 0]');
lmiterm ( [3 1 1 Z],[0 1],[0 1]');
% #3 LMI, right-hand side
lmiterm ( [3 1 1 rou],1,-1);
lmis3=getlmis; % finishing description
% Step 2. Solve this LMI problem and use dec2mat to get the
% corresponding matrix X
c = mat2dec(lmis3,0,0,ones(2)); % obtain vector c
[copt,xopt] = mincx(lmis3,c); % solve the
  optimization problem
X = dec2mat(lmis3,xopt,1); % extract matrix X
W = dec2mat(lmis3,xopt,2); % extract matrix W
Z = dec2mat(lmis3,xopt,3); % extract matrix Z
K = W*inv(X);  % get the feedback gain
```

By the LMI Toolbox in MATLAB, we obtain the controller gain matrix

$$K = \begin{bmatrix} -1.0000 & -0.0000 & -1.0000 \\ -0.0000 & -1.0000 & 0.0000 \end{bmatrix}.$$

## B.5.3   *gevp* Command

The function gevp solves the generalized eigenvalue problem, that is, Problem 3.9, which has the following form:

$$\begin{cases} \min_{x,\lambda} & \lambda \\ \text{s.t.} & A(x) < \lambda B(x) \\ & 0 < B(x) \\ & C(x) < D(x). \end{cases}$$

The corresponding format in MATLAB is

```
[tmin,xopt] = gevp(lmisys,nlfc,options,t0,x0,target)
```

*Inputs:*

lmisys: Description of the system of LMI constraints.

nlfc: Number of linear fractional constraints (LMIs involving $t$).

options: (optional) Is similar to that in feasp.

t0, x0: (optional) Initial guesses for $t$, $x$ (ignored when unfeasible).

target: (optional) Target for tmin. The code terminates as soon as $t$ falls below this value (default $= -10^5$).

*Outputs:*

tmin: Minimal value of $t$.

xopt: Minimizing value of the vector $x$ of decision variables.

**Example B.5**

Consider the following optimization problem

$$
\begin{cases}
\min_{x} & \alpha \\
\text{s.t.} & I < P \\
& A_1^T P + P A_1 < \alpha P \\
& A_2^T P + P A_2 < \alpha P \\
& A_3^T P + P A_3 < \alpha P,
\end{cases}
$$

where $A_1$, $A_2$, and $A_3$ are given in Example B.1.

The MATLAB file for solving the aforementioned problem is given as follows:

```
%Step 1. Describe this LMIs in MATLAB
% specify parameter matrices
A1 = [-1 2;1 -3]; A2 = [-0.8 1.5;1.3 -2.7];
A3 = [-1.4 0.9;0.7 -2.0];
% define the LMI system
setlmis ([]) % initialing
P = lmivar (1, [2 1]); % declare the LMI variable
lmiterm ([1 1 1 0],1); % #1 LMI, left-hand
lmiterm ([-1 1 1 P],1,1); % #1 LMI, right-hand
lmiterm ([2 1 1 P],1, A1, 's'); % #2 LMI, left-hand
lmiterm ([-2 1 1 P],1,1); % #2 LMI, right-hand
lmiterm ([3 1 1 P],1, A2, 's'); % #3 LMI, left-hand
lmiterm ([-3 1 1 P],1,1); % #3 LMI, right-hand
lmiterm ([4 1 1 P],1, A3, 's'); % #4 LMI, left-hand
lmiterm ([-4 1 1 P],1,1); % #4 LMI, right-hand
lmis4 = getlmis % finishing description
```

```
%Step 2. Solve this LMI problem
[aa,xopt] = gevp(lmis4,3);
```

# B.6   How to Validate Results?

The LMI toolbox offers two functions to analyze and validate the results of an LMI optimization.

## evallmi

The function `evallmi` evaluates all variable terms in an LMI system for a given value of the vector of decision variables, for instance, the feasible or optimal vector returned by the LMI solvers. The format of this function is

$$\text{evalsys} = \text{evallmi}(\text{lmisys},\text{xdec})$$

where
    `lmisys` is a created LMI system
    `xdec` is the vector of decision variables

Recall that decision variables fully determine the values of the matrix variables. The "evaluation" consists of replacing all terms involving the matrix variables by their matrix value. The output `evalsys` is an LMI system containing only constant terms. The matrix values of the left and right sides of each LMI are then returned by `showlmi`.

## showlmi

Once the function `evallmi` is performed, the left- and right-hand sides of a particular LMI are returned by `showlmi`.

$$[\text{lhs},\text{rhs}] = \text{showlmi}(\text{evalsys},\text{n})$$

*Inputs:*
    `evalsys`: Array describing the set of evaluated LMIs (output of `evallmi`).
    `n`: Label of the selected LMI as returned by `newlmi`.
*Outputs:*
    `lhs`: The left-hand-side value of the $n$th LMI in the LMI system `lmisys`.
    `rhs`: The right-hand-side value of the $n$th LMI in the LMI system `lmisys`.

# References

Akgül, M. (1984). *Topics in Relaxation and Ellipsoidal Methods* (*Research Notes in Mathematics*, Vol. 97). Pitman, London.

Alizadeh, F. (1991). *Combinatorial optimization with interior point methods and semidefinite matrices*. PhD thesis, University of Minnesota, Minneapolis, MN.

Alizadeh, F. (1992a). Combinatorial optimization with semidefinite matrices. In *Proceedings of the Second Annual Integer Programming and Combinatorial Optimization Conference*, Carnegie-Mellon University, Pittsburgh, PA.

Alizadeh, F. (1992b). Optimization over the positive-definite cone: Interior point method and combinatorial applications. In: P. Pardalos (ed.), *Advances in Optimization and Parallel Computing*. North-Holland, Amsterdam, the Netherlands.

Apkarian, P., Tuan, H. D., and Bernussou, J. (2000). Analysis, eigenstructure assignment and $H_2$ multi-channel synthesis with enhanced LMI characterizations. In *Proceedings of the 39th IEEE Conference on Decision and Control*, Sydney, Australia.

Apkarian, P. C., Tuan, H. D., and Bernussou, J. (2001). Continuous-time analysis, eigenstructure assignment, and $H_2$ synthesis with enhanced linear matrix inequalities (LMI) characterizations. *IEEE Transactions on Automatic Control*, 46(12):1941–1946.

Bachelier, O., Bosche, J., and Mehdi, D. (2006). On pole placement via eigenstructure assignment approach. *IEEE Transactions on Automatic Control*, 51(9):1554–1558.

Bauer, F. L. and Fike, E. (1960). Norms and exclusion theorems. *Numerische Mathematik*, 2:137–144.

Bazaraa, M. S., Sherali, H. D., and Shetty, C. M. (2006). *Nonlinear Programming: Theory and Algorithms*. John Wiley& Sons Inc., Hoboken, NJ.

Bland, R. G., Goldfarb, D., and Tod, M. J. (1981). The ellipsoid method: A survey. *Operations Research*, 29(6):1039–1091.

Bouhtouri, A. E., Hinrichsen, D., and Pritchard, A. J. (1999). $H_\infty$-type control for discrete-time stochastic systems. *International Journal of Robust and Nonlinear Control*, 9(13):923–948.

Boyd, S. and Barratt, C. (1991). *Linear Controller Design: Limits of Performance.* Prentice Hall, Englewood Cliffs, NJ.

Boyd, S. and Ghaoui, L. E. (1993). Method of centers for minimizing generalized eigenvalues. *Linear Algebra and Its Applications*, 188–189:63–111.

Boyd, S., Ghaoui, L. E., Feron, E., and Balakrishnan, V. (1994). *Linear Matrix Inequalities in System and Control Theory.* Society for Industrial and Applied Mathematics, Philadelphia, PA.

Boyd, S. and Vandenberghe, L. (2004). *Convex Optimization.* Cambridge University Press, Cambridge, U.K.

Brooke, A., Kendrick, D., Meeraus, A., and Raman, R. (1998). *GAMS: A User's Guide.* The Scientific Press, Redwood City, CA.

Cai, G. B., Duan, G. R., and Hu, C. H. (2011). A velocity-based LPV modeling and control framework for an airbreathing hypersonic vehicle. *International Journal of Innovative Computing Information and Control*, 7(5A):2269–2281.

Cai, G. B., Duan, G. R., Hu, C. H., and Zhou, B. (2010). Tracking control for air-breathing hypersonic cruise vehicle based on tangent linearization approach. *Journal of Systems Engineering and Electronics*, 21(3):469–475.

Cai, G. B., Hu, C. H., and Duan, G. R. (2012). Efficient LMI-based quadratic stability and stabilization of parameter-dependent interval systems with applications. *International Journal of Innovative Computing Information and Control*, 8(3A):1943–1954.

Cao, J. D., Yuan, K., Ho, D. W. C., and Lam, J. (2006). Global point dissipativity of neural networks with mixed time-varying delays. *Chaos*, 16(1):013105.

Cao, Y. Y. and Lam, J. (2000). Robust $H_\infty$ control of uncertain Markovian jump systems with time-delay. *IEEE Transactions on Automatic Control*, 45(1):77–83.

Cao, Y. Y., Sun, Y. X., and Lam, J. (1998). Delay-dependent robust $H_\infty$ control for uncertain systems with time-varying delays. *IEE Proceedings—Control Theory and Applications*, 145(3):338–344.

Casavola, A., Famularo, D., and Franze, G. (2008). Robust fault detection of uncertain linear systems via quasi-LMIs. *Automatica*, 44(1):289–295.

Chen, J. D. (2007a). Delay-dependent observer-based control design of uncertain time-delay systems: An LMI approach. *Journal of the Chinese Institute of Engineers*, 30(3):401–409.

Chen, J. D. (2007b). Robust output observer-based control of neutral uncertain systems with discrete and distributed time delays: LMI optimization approach. *Chaos Solitons & Fractals*, 34(4):1254–1264.

Chilali, M. and Gahinet, P. (1996). $H_\infty$ design with pole placement constraints: An LMI approach. *IEEE Transactions on Automatic Control*, 41(3):358–367.

Choi, H. H. and Chung, M. J. (1997). An LMI approach to $H_\infty$ controller design for linear time-delay systems. *Automatica*, 33(4):737–739.

Crusius, C. (2002). A parser/solver for convex optimization problems. PhD thesis, Stanford University, Stanford, CA.

Cychowski, M. (2009). *Robust Model Predictive Control.* VDM Publishing, Berlin, Germany.

de Oliveira, M. C., Bernussou, J., and Geromel, J. C. (1999). A new discrete-time robust stability conditions. *Systems & Control Letters*, 37(4):261–265.

Dikin, I. (1967). Iterative solution of problems of linear and quadratic programming. *Soviet Mathematics Doklady*, 8(3):674–675.

Doyle, J., Packard, A., and Zhou, K. M. (1991). Review of LFTs, LMIs, and $\mu[M]$. In *Proceedings of the 30th Conference on Decision and Control*, Brighton, England, pp. 1227–1232.

Doyle, J., Zhou, K. M., Glover, K., and Bodenheimer, B. (1994). Mixed $H_2$ and $H_\infty$ performance-objectives II: Optimal-control. *IEEE Transactions on Automatic Control*, 39(8):1575–1587.

Doyle, J. C., Glover, K., Khargonekar, P. P., and Francis, B. A. (1989). State-space solutions to standard $H_2$ and $H_\infty$ control problems. *IEEE Transactions on Automatic Control*, 34(8):831–847.

Duan, G. R. (1991). Design of control systems with poles located in a robust region (in Chinese). *Information and Control*, 20(3):39–46.

Duan, G. R. (1992). Simple algorithm for robust pole assignment in linear output feedback. *IEE Proceedings, Part D: Control Theory and Applications*, 139(5):465–469.

Duan, G. R. (1993). Robust eigenstructure assignment via dynamical compensators. *Automatica*, 29(2):469–474.

Duan, G. R. (1994). Eigenstructure assignment by decentralized output feedback—A complete parametric approach. *IEEE Transactions on Automatic Control*, 39(5):1009–1014.

Duan, G. R. (1995). Parametric approach for eigenstructure assignment in descriptor systems via output feedback. *IEE Proceedings, Part D: Control Theory and Applications*, 142(6):611–616.

Duan, G. R. (1998). Eigenstructure assignment and response analysis in descriptor linear systems with state feedback control. *International Journal of Control*, 69(5):663–694.

Duan, G. R. (1999). Eigenstructure assignment in descriptor systems via output feedback: A new complete parametric approach. *International Journal of Control*, 72(4):345–364.

Duan, G. R. (2003). Parametric eigenstructure assignment via output feedback based on singular value decompositions. *IEE Proceedings, Part D: Control Theory and Applications*, 150(1):93–100.

Duan, G. R. (2004a). *Linear Systems Theory*, 2nd ed. (in Chinese). Harbin Institute of Technology Press, Harbin, China.

Duan, G. R. (2004b). Parametric eigenstructure assignment in second-order descriptor linear systems. *IEEE Transactions on Automatic Control*, 49(10): 1789–1795.

Duan, G. R. (2008). On numerical reliability of pole assignment algorithms—A case study. In *Proceedings of the 27th Chinese Control Conference*, Kunming, China, pp. 189–194.

Duan, G. R. (2010). *Analysis and Design of Descriptor Linear Systems* (*Advances in Mechanics and Mathematics*, Vol. 23). Springer, New York.

Duan, G. R. (2012). An LMI approach to robust attitude control of BTT missiles. Technical report, *The 5th International Conference on Optimization and Control with Applications*, Beijing, China.

Duan, G. R. and Chen, F. S. (1995). Model matching problem for linear systems (in Chinese). *Heilongjiang Techniques of Automation and Applications*, 14(3):1–3.

Duan, G. R., Gu, D. K., and Li, B. (2010). Optimal control for final approach of rendezvous with non-cooperative target. *Pacific Journal of Optimization*, 6(3):521–532.

Duan, G. R., Howe, D., and Patton, R. J. (2002). Robust fault detection in descriptor linear systems via generalized unknown input observers. *International Journal of Systems Science*, 33(5):369–377.

Duan, G. R. and Huang, W. H. (1990). A stability robustness result for dynamical systems and its application (in Chinese). *Journal of Vibration and Shock*, 3(1):1–5.

Duan, G. R. and Li, Y. J. (2009). Robust passive control for discrete-time T-S fuzzy systems with delays. *Journal of Systems Engineering and Electronics*, 20(5):1045–1051.

Duan, G. R. and Liu, G. P. (2002). Complete parametric approach for eigenstructure assignment in a class of second-order linear systems. *Automatica*, 38(4):725–729.

Duan, G. R., Liu., G. P., and Thompson, S. (2003). Eigenstructure assignment design for proportional-integral observers: The discrete-time case. *International Journal of Systems Science*, 34(5):357–363.

Duan, G. R., Liu, W. Q., and Liu, G. P. (2001). Robust model reference control for multivariable linear systems subject to parameter uncertainties. *Proceedings of the Institution of Mechanical Engineers. Part I: Journal of Systems and Control Engineering*, 215(6):599–610.

Duan, G. R., Lu, L. L., and Wu, A. G. (2009a). Robust dissipative filtering for continuous-time polytopic uncertain neutral systems. *Journal of Systems Engineering and Electronics*, 20(3):598–606.

Duan, G. R. and Patton, R. J. (1998). A note on Hurwitz stability of matrices. *Automatica*, 34(4):509–511.

Duan, G. R. and Patton, R. J. (1999). Eigenstructure assignment in descriptor systems via output feedback: A new complete parametric approach. *International Journal of Control*, 72(13):1193–1203.

Duan, G. R. and Patton, R. J. (2001). Robust fault detection using Luenberger-type unknown input observers—A parametric approach. *International Journal of Systems Science*, 32(4):533–540.

Duan, G. R., Qiang, W. Y., Feng, W. J., and Sun, L. J. (1994). A complete parametric approach for model reference control system design (in Chinese). *Journal of Astronautics*, (2):7–13.

Duan, G. R. and Wang, Q. C. (1992). Modes decoupling control for linear systems (in Chinese). *Journal of Astronautics*, (2):7–13.

Duan, G. R., Wu, A. G., and Hou, W. N. (2007). Parametric approach for Luenberger observers for descriptor linear systems. *Bulletin of the Polish Academy of Sciences: Technical Sciences*, 55(1):15–18.

Duan, G. R., Wu, G. Y., and Huang, W. H. (1990). Robust state feedback controllers for linear systems (in Chinese). *Information and Control*, (1):31–34.

Duan, G. R. and Wu, Y. L. (2005). Robust pole assignment in matrix descriptor second-order linear systems. *Transactions of the Institute of Measurement and Control*, 27(4):279–295.

Duan, G. R., Wu, Z. Y., Bingham, C., and Howe, D. (2000). Robust magnetic bearing control using stabilizing dynamical compensators. *IEEE Transactions on Industry Applications*, 36(6):1654–1660.

Duan, G. R., Xu, S. J., and Huang, W. H. (1989). Generalized positive definite matrix and its application in stability analysis (in Chinese). *Acta Mechanica Sinica*, 21(6):754–757.

Duan, G. R. and Yu, H. H. (2008). Robust pole assignment in high-order descriptor linear systems via proportional plus derivative state feedback. *IET Control Theory and Applications*, 2(4):277–287.

Duan, G. R., Yu, H. H., and Tan, F. (2009b). Parametric control systems design with application in missile control. *Science in China Series F—Information Sciences*, 52(11):2190–2200.

Dumitrescu, B. (2008). LMI stability tests for the Fornasini–Marchesini model. *IEEE Transactions on Signal Processing*, 56(8):4091–4095.

Fan, L. Y., Liu, H. L., Duan, G. R., and Feng, W. J. (2010). Delay-dependent stabilization for delay stochastic differential systems with Markov jumping parameters based on SLQ controllers. In *Proceedings of the 8th World Congress on Intelligent Control and Automation (WCICA)*, Jinan, China, pp. 1528–1532.

Feng, D. X. (1995). *Fundamental of Convex Analysis (in Chinese)*. Science Press, Beijing, China.

Fiacco, A. and McCormick, G. (1968). *Nonlinear Programming: Sequential Unconstrained Minimization Techniques*. Wiley, New York. Reprinted 1990 in the SIAM Classics in Applied Mathematics series.

Fourer, R., Gay, D., and Kernighan, B. (1999). *AMPL: A Modeling Language for Mathematical Programming*. Duxbury Press, Belmont, CA.

Freund, R. W. and Jarre, F. (1994). An interior-point method for fractional programs with convex constrains. *Mathematical Programming*, 67(3):407–440.

Fridman, E. and Shaked, U. (2002). A descriptor system approach to $H_\infty$ control of linear time-delay systems. *IEEE Transactions on Automatic Control*, 47(2): 253–270.

Fu, Y. M. and Duan, G. R. (2004). Stochastic stabilizability and passive control of linear time-delay systems with jumping parameters. In *Proceedings of the Eighth International Conference on Control, Automation, Robotics and Vision*, December 6–9, Kunming, China, pp. 1757–1761.

Fu, Y. M. and Duan, G. R. (2005). Robust guaranteed cost observer for uncertain descriptor time-delay systems with Markovian jumping parameters. *Acta Automatica Sinica*, 31(3):479–483.

Fu, Y. M., Duan, G. R., and Song, S. M. (2004). Design of unknown input observer for linear time-delay systems. *International Journal of Control, Automation and Systems*, 2(4):530–535.

Fu, Y. M., Wu, D., Zhang, P., and Duan, G. R. (2006a). Design of unknown input observer with $H_\infty$ performance for linear time-delay systems. *Journal of Systems Engineering and Electronics*, 17(3):606–610.

Fu, Y. X., Zhao, Y., and Duan, G. R. (2006b). Robust dissipative filtering for linear time-delay systems with Markovian jumping parameters. In *Proceedings of International Conference on Impulsive Dynamical Systems and Applications*, Qingdao, China, pp. 417–421.

Gács, P. and Lovász, L. (1981). Khachiyan's algorithm for linear programming. *Mathematical Programming Studies*, 14:61–68.

Gahinet, P. and Apkarian, P. (1993). An LMI-based parametrization of all $H_\infty$ controllers with applications. *In Proceedings of 32nd IEEE Conference on Decision and Control, San Antonio*, TX, pp. 656–661.

Gahinet, P. and Apkarian, P. (1994). A linear matrix inequality approach to $H_\infty$ control. *International Journal of Robust and Nonlinear Control*, 4:421–448.

Gahinet, P., Apkarian, P., and Chilali, M. (1996). Affine parameter-dependent Lyapunov functions and real parametric uncertainty. *IEEE Transactions on Automatic Control*, 41(3):436–442.

Gahinet, P., Nemirovskii, A., Laub, A. J., and Chiluli, M. (1994). LMI control toolbox. In *Proceedings of the 33rd Conference on Decision and Control*, Lake Buena Vista, Fl, pp. 2038–2041.

Gahinet, P., Nemirovski, A., Laub, A. J., and Chilali, M. (1995). *LMI Control Toolbox–For Use with Matlab*. The MathWorks, Natick, MA.

Gao, H. J., Chen, T. W., and Lam, J. (2008). A new delay system approach to network-based control. *Automatica*, 44(1):39–52.

Gao, H. J., Lam, J., Wang, C. H., and Wang, Y. (2004). Delay-dependent output-feedback stabilisation of discrete-time systems with time-varying state delay. *IEE Proceedings, Part D: Control Theory and Applications*, 151(6):691–698.

Gao, H. J., Lam, J., Xie, L. H., and Wang, C. H. (2005). New approach to mixed $H_2/H_\infty$ filtering for polytopic discrete-time systems. *IEEE Transactions on Signal Processing*, 53(8):3183–3192.

Gao, H. J. and Wang, C. H. (2003a). Delay-dependent robust $H_\infty$ and $L_2 - L_\infty$ filtering for a class of uncertain nonlinear time-delay systems. *IEEE Transactions on Automatic Control*, 48(9):1661–1666.

Gao, H. J. and Wang, C. H. (2003b). Robust $L_2 - L_\infty$ filtering for uncertain systems with multiple time-varying state delays. *IEEE Transactions on Circuits and Systems 1-Fundamental Theory and Applications*, 50(4):594–599.

Gao, H. J. and Wang, C. H. (2004). A delay-dependent approach to robust $H_\infty$ filtering for uncertain discrete-time state-delayed systems. *IEEE Transactions on Signal Processing*, 52(6):1631–1640.

Gao, X. Y., Teo, K. L., and Duan, G. R. (2011a). Non-fragile guaranteed cost control for robust spacecraft orbit transfer with small thrust. *IMA Journal of Mathematical Control and Information*, 28(4):507–524.

Gao, X. Y., Teo, K. L., and Duan, G. R. (2012a). Non-fragile robust $H_\infty$ control for uncertain spacecraft rendezvous system with pole and input constraints. *International Journal of Control*, 85(7):933–941.

Gao, X. Y., Teo, K. L., Duan, G. R., and Wang, N. (2012b). Robust reliable $H_\infty$ control for uncertain systems with pole constraints. *International Journal of Innovative Computing Information and Control*, 8(5A):3071–3079.

Gao, X. Y., Teo, K. L., Duan, G. R., and Zhang, X. (2011b). Necessary and sufficient condition for robust stability of discrete-time descriptor polytopic systems. *IET Control Theory and Applications*, 5(5):713–720.

Geromel, J. C. (1999). Optimal linear filtering under parameter uncertainty. *IEEE Transactions on Signal Processing*, 47(1):168–175.

Geromel, J. C., de Oliveira, M. C., and Hsu, L. (1998). LMI characterization of structural and robust stability. *Linear Algebra and Its Applications*, 285(1–3):69–80.

Ghaoui, L. E. and Niculescu, S. I. (2000). *Advances in Matrix Inequality Methods in Control (Advances in Design and Control)*. SIAM, Philadelphia, PA.

Grace, A., Laub, A. J., and Thompson, C. M. (1992). *Control System Toolbox for Use with MATLAB*. The Math Works Inc., Natick, MA.

Grant, M., Boyd, S., and Ye., Y. (2006). *Disciplined convex programming*. In *Global Optimization: From Theory to Implementation* (Nonconvex Optimization and Its Applications). Springer, New York, pp. 155–210.

Grigoriadis, K. M. and Watson, J. T. (1997). Reduced-order $H_\infty$ and $L_2 - L_\infty$ filtering via linear matrix inequalities. *IEEE Transactions on Aerospace and Electronic Systems*, 33(4):1326–1338.

Grötschel, M., Lovász, L., and Schrijver, A. (1988). *Geometric Algorithm and Combinatorial Optimization (Algorithm and Combinatorics*, Vol. 2). Springer, Berlin, Germany.

Gu, K. Q. (1997). Discretized LMI set in the stability problem of linear uncertain time-delay systems. *International Journal of Control*, 68(4):923–934.

Guan, X. P., Long, C. N., Hua, C. C., and Duan, G. R. (2002). Decentralized stabilization of discrete time-delay large-scale systems with structured uncertainties (in Chinese). *Control Theory and Applications*, 19(4):537–540.

Guo, L. and Wang, H. (2010). *Stochastic Distribution Control System Design: A Convex Optimization Approach (Advances in Industrial Control)*. Springer, Berlin, Germany.

Han, Q. L. (2005). Absolute stability of time-delay systems with sector-bounded nonlinearity. *Automatica*, 41(12):2171–2176.

He, L. and Duan, G. R. (2005). Multiobjective control synthesis for a class of uncertain fuzzy systems. In *Proceedings of International Conference on Machine Learning and Cybernetics*, Guangzhou, China, pp. 2563–2567.

He, L. and Duan, G. R. (2006). Robust $H_\infty$ control with pole placement constraints for T-S fuzzy systems. In *Proceedings of International Conference on Machine Learning and Cybernetics*, Guangzhou, China, pp. 338–346.

He, L., Duan, G. R., and Wu, A. G. (2006a). Robust $L_1$ filtering with pole constraint in a disk via parameter-dependent Lyapunov functions. In *Proceedings of SICE-ICASE International Joint Conference*, Busan, South Korea, pp. 833–836.

He, L., Fu, Y. M., and Duan, G. R. (2004a). Multiobjective control synthesis based on parametric eigenstructure assignment. In *Proceedings of International Conference on Control, Automation, Robotics and Vision*, Kunming, China, pp. 1838–1841.

He, L., Wu, A. G., and Duan, G. R. (2006b). Further results on delay-dependent robust stabilization for uncertain continuous-time delayed fuzzy systems. *Dynamics of Continuous Discrete and Impulsive Systems-Series B-Applications & Algorithms*, 13(Part 2, S):593–597.

He, L., Wu, A. G., and Duan, G. R. (2006c). Improved robust mixed $H_2/H_\infty$ filtering for uncertain continuous-time systems. In *Proceedings of International Conference on Sensing, Computing and Automation*, Dalian, China, pp. 2412–2416.

He, Y., Wu, M., She, J. H., and Liu, G. P. (2004b). Delay-dependent robust stability criteria for uncertain neutral systems with mixed delays. *Systems & Control Letters*, 51(1):57–65.

He, Y., Wu, M., She, J. H., and Liu, G. P. (2004c). Parameter-dependent Lyapunov functional for stability of time-delay systems with polytopic-type uncertainties. *IEEE Transactions on Automatic Control*, 49(5):828–832.

Horisberger, H. and Belanger, P. (1976). Regulators for linear, time invariant plants with uncertain parameters Horisberger and Belanger. *Automatic Control*, 21(5):705–708.

Hou, M. Z. and Duan, G. R. (2008). Integrated guidance and control of homing missiles against ground fixed targets. *Chinese Journal of Aeronautics*, 21(2):162–168.

Hu, W. Y. (1997). *Research on robust stabilization of uncertain linear systems and system families* (in Chinese). PhD thesis, Harbin Institute of Technology, Harbin, China.

Huard, P. (1967). *Resolution of Mathematical Programming with Nonlinear Constraints by the Method of Centers*. North Holland, Amsterdam, the Netherlands.

Jarre, F. (1993). Optimal ellipsoidal approximations around the analytic center. Technical report, Intitüt für Angewandte Mathematik, University of Würzburg, Würzburg, Germany.

Jiang, Y. X. (1987). Decoupling of a kind of missile control system modes (in Chinese). *Journal of Astronautics*, (3):8–15.

Jin, S. H. and Park, J. B. (2001). Robust H∞ filtering for polytopic uncertain systems via convex optimisation. *IEE Proceedings, Part D: Control Theory and Applications*, 148(1):55–59.

Jing, X. J., Tan, D. L., and Wang, Y. C. (2004). An LMI approach to stability of systems with severe time-delay. *IEEE Transactions on Automatic Control*, 49(7):1192–1195.

Johnson, E. A. and Erkus, B. (2007). Dissipativity and performance analysis of smart dampers via LMI synthesis. *Structural Control & Health Monitoring*, 14(3): 471–496.

Kaczorek, T. (2009). LMI approach to stability of 2D positive systems. *Multidimensional Systems and Signal Processing*, 20(1):39–54.

Kautsky, J., Nichols, N. K., and Van Dooren, P. (1985). Robust pole assignment in linear state feedback. *International Journal of Control*, 41:1129–1155.

Khachiyan, L. (1979). A polynomial algorithm in linear programming. *Soviet Mathematic Doklady*, 20:191–194.

Kim, J. E. and Park, H. B. (1999). H∞ state feedback control for generalized continuous/discrete time-delay system. *Automatica*, 35(8):1443–1451.

Kim, J. H. (2001). Delay and its time-derivative dependent robust stability of time-delayed linear systems with uncertainty. *IEEE Transactions on Automatic Control*, 46(5):789–792.

Kim, J. H., Ahn, S. J., and Ahn, S. (2005). Guaranteed cost and H∞ filtering for discrete-time polytopic uncertain systems with time delay. *Journal of the Franklin Institute-Engineering and Applied Mathematics*, 342(4):365–378.

Kimura, H. (1977). A further result on the problem of pole assignment by output feedback. *IEEE Transactions on Automatic Control*, 22:458–463.

Kwon, O. M., Park, J. H., and Lee, S. M. (2008). Exponential stability for uncertain dynamic systems with time-varying delays: LMI optimization approach. *Journal of Optimization Theory and Applications*, 137(3):521–532.

Lam, H. K. and Leung, F. H. F. (2011). *Stability Analysis of Fuzzy-Model-Based Control Systems: Linear-Matrix-Inequality Approach*. Springer, Berlin, Germany.

Li, H. Z. and Fu, M. Y. (1997). A linear matrix inequality approach to robust H∞ filtering. *IEEE Transactions on Signal Processing*, 45(9):2338–2350.

Li, X. and deSouza, C. E. (1997). Delay-dependent robust stability and stabilization of uncertain linear delay systems: A linear matrix inequality approach. *IEEE Transactions on Automatic Control*, 42(8):1144–1148.

Li, Y. J. and Duan, G. R. (2008). Robust dissipative control for discrete-time T-S fuzzy systems with time delays. In *Proceedings of the 27th Chinese Control Conference*, Kunming, China, pp. 276–280.

Li, Y. J., Fu, Y. M., and Duan, G. R. (2006a). Robust dissipative control for T-S fuzzy systems with time-varying delays. In *Proceedings of IEEE International Symposium on Industrial Electronics*, Montreal, Canada, pp. 97–101.

Li, Y. J., Fu, Y. M., and Duan, G. R. (2006b). Robust passive control for T-S fuzzy systems. In *Proceedings of Computational Intelligence International Conference on Intelligent Computing*, Kunming, China, pp. 146–151.

Li, Z. Y., Wang, Y., Zhou, B., and Duan, G. R. (2009). Detectability and observability of discrete-time stochastic systems and their applications. *Automatica*, 45(5):1340–1346.

Li, Z. Y., Wang, Y., Zhou, B., and Duan, G. R. (2010). On unified concepts of detectability and observability for continuous-time stochastic systems. *Applied Mathematics and Computing*, 217(2):521–536.

Lieu, B. and Huard, P. (1966). La méthode des centres dans un espace topologique. *Numerische Mathematik*, 8:56–67.

Lin, C., Wang, Q.-G., Lee, T. H., and He, Y. (2007). *LMI Approach to Analysis and Control of Takagi-Sugeno Fuzzy Systems with Time Delay*. Springer, Berlin, Germany.

Liu, G. P., Duan, G. R., and Daley, S. (2000a). Stable observer-based controller design for robust state-feedback pole assignment. *Proceedings of the Institution of Mechanical Engineers. Part I: Journal of Systems and Control Engineering*, 214(4):313–318.

Liu, G. P., Duan, G. R., and Patton, R. (2000b). Mixed time- and frequency-domain robust eigenstructure assignment. *International Journal of Systems Science*, 31(1):63–71.

Liu, H. L., Ding, B. C., and Duan, G. R. (2006). Non-fragile generalized $H_2$ control for linear time-delay systems. In *Proceedings of Chinese Control Conference*, Harbin, China, pp. 2050–2055.

Liu, H. L. and Duan, G. R. (2006a). Generalized $H_2$ control of uncertain linear neutral systems with time-varying delay. *Dynamics of Continuous Discrete and Impulsive Systems-Series B-Applications & Algorithms*, 13E(Part 6 Suppl. S): 2660–2665.

Liu, H. L., Duan, G. R., and Fan, L. Y. (2010). Delay-dependent passive control of stochastic differential system with time delay. In *Proceedings of the 8th World Congress on Intelligent Control and Automation (WCICA)*, Jinan, China, pp. 963–968.

Liu, M., Zhang, H. S., and Duan, G. R. (2007). $H_\infty$ measurement-feedback control for discrete-time systems with multiple delayed measurements. *Control Theory and Applications*, 24(1):46–52.

Liu, T. and Zhao, J. (2003). *Dynamics of Spacecraft* (in Chinese). Harbin Institute of Technology Press, Harbin, China.

Liu, X. L. and Duan, G. R. (2006b). Robust $H_\infty$ filtering for discrete-time switched systems. *Dynamics of Continuous Discrete and Impulsive Systems-Series A-Mathematical Analysis*, 13:410–417.

Liu, X. L. and Duan, G. R. (2006c). Robust stabilization of switched systems with time-delay. *Dynamics of Continuous Discrete and Impulsive Systems-Series A-Mathematical Analysis*, 13:819–826.

Liu, X. L., Duan, G. R., and Teo, K. L. (2008). Optimal soft landing control for moon lander. *Automatica*, 44(4):1079–1103.

Lu, P. L. and Yang, Y. (2009). On delay-dependent global asymptotic stability for pendulum-like systems. *Journal of Optimization Theory and Applications*, 143(2):295–308.

Lur'e, A. I. (1951). *Some Nonlinear Problems in the Theory of Automatic Control* (in Russian). HMSO, London.

Mahmoud, M. S., Terro, M. J., and Abdel-Rohman, M. (1998). An LMI approach to $H_\infty$ control of time-delay systems for the benchmark problem. *Earthquake Engineering & Structural Dynamics*, 27(9):957–976.

Montagner, V. F., Oliveira, R. C. L. F., Peres, P. L. D., and Bliman, P. A. (2007). Linear matrix inequality characterisation for $H_\infty$ and $H_2$ guaranteed cost gain-scheduling quadratic stabilisation of linear time-varying polytopic systems. *IET Control Theory and Applications*, 1(6):1726–1735.

Nemirovskii, A. and Yudin, D. (1983). *Problem Complexity and Method Efficiency in Optimization*. John Wiley & Sons, Chichester, U.K.

Nesterov, Y. and Nemirovsky, A. (1988). A general approach to polynomial-time algorithms design for convex programming. Technical report, Central Economics and Mathematics Institute, USSR Academy of Sciences, Moscow, USSR.

Nesterov, Y. and Nemirovsky, A. (1990a). Optimization over positive semidefinite matrices: Mathematical background and user's manual. Technical report, Central Economical & Mathematical Institute, USSR Academy of Sciences, Moscow, USSR.

Nesterov, Y. and Nemirovsky, A. (1990b). Self-concordant functions and polynomial time methods in convex programming. Technical report, Central Economical & Mathematical Institute, USSR Academy of Sciences, Moscow, USSR.

Nesterov, Y. and Nemirovsky, A. (1991). Conic formulation of a convex programming problem and duality. Technical report, Central Economical & Mathematical Institute, USSR Academy of Sciences, Moscow, USSR.

Nesterov, Y. and Nemirovsky, A. (1993). *Acceleration of the path-following method for optimization over the cone of positive semidefinite matrices*. Technical report, Information and Automation Research Institute of the French National.

Nesterov, Y. and Nemirovsky, A. (1994). *Interior-point Polynomial Methods in Convex Programming* (*Studies in Applied Mathematics*, Vol. 13). SIAM, Philadelphia, PA.

Nesterov, Y. and Nemirovsky, A. (1995). An interior point method for generalized linear-fractional programming. *Mathematical Programming*, 69(1):177–204.

Oliveira, M. C. D., Geromel, J. C., and Bernussou, J. (2002). Extended $H_2$ and $H_\infty$ norm characterizations and controller parametrizations for discrete-time systems. *International Journal of Control*, 75(9):666–679.

Ostertag, E. (2011). *Mono- and Multivariable Control and Estimation: Linear, Quadratic and LMI Methods*. Springer, Berlin, Germany.

Palhares, R. M. and Peres, P. L. D. (1999). Mixed $L_2$-$L_\infty$/$H_\infty$ filtering for uncertain linear systems: An LMI approach. *IEEE International Symposium on Industrial Electronics*, (3):1070–1075.

Park, P. G. (1999). A delay-dependent stability criterion for systems with uncertain time-invariant delays. *IEEE Transactions on Automatic Control*, 44(4): 876–877.

Paszke, W. (2006). *Analysis and Synthesis of Multidimensional Systems Classes Using Linear Matrix Inequality Methods*. University of Zielona Gora Press, Zielona Gora, Poland.

Peaucelle, D., Arzelier, D., Bachelier, O., and Bernussou, J. (2000). A new robust D-stability condition for real convex polytopic uncertainty. *Systems & Control Letters*, 40(1):21–30.

Peterson, I. R., Anderson, B. D. O., and Jonckheere, E. A. (1991). A first principles solution to the non-singular $H_\infty$ control problem. *International Journal of Robust and Nonlinear Control*, 1(3):171–185.

Petres, Z., Baranyi, N., Korondi, P., and Hashimoto, H. (2007). Trajectory tracking by TP model transformation: Case study of a benchmark problem. *IEEE Transactions on Industrial Electronics*, 54(3):1654–1663.

Pyatnitskii, E. S. and Skorodinskii, V. I. (1982). Numerical methods of Lyapunov function construction and their application to the absolute stability problem. *Systems and Control Letters*, (2):130–135.

Qin, L. and Duan, G. R. (2006). Robust dissipative control for uncertain descriptor linear systems with time delay. In *Proceedings of World Congress on Intelligent Control and Automation*, Dalian, China, pp. 2327–2333.

Ramos, D. C. W. and Peres, P. L. D. (2002). An LMI condition for the robust stability of uncertain continuous-time linear systems. *IEEE Transactions on Automatic Control*, 47(4):675–678.

Rendl, F., Vanderbei, R., and H.Wolkowocz (1993). A primal-dual interior-point method for the max-min eigenvalue problem. Technical report, Department of Combinatorics and Optimization, University of Waterloo, Ontario, Canada.

Ruszczynski, A. P. (2006). *Nonlinear Optimization*. Princeton University Press, Princeton, NJ.

Saif, A. W., Mahmoud, M. S., and Shi, Y. (2009). A parameterized delay-dependent control of switched discrete-time systems with time-delay. *International Journal of Innovative Computing Information and Control*, 5(9):2893–2906.

Scherer, C., Gahinet, P., and Chilali, M. (1997). Multi-objective output-feedback control via LMI optimization. *IEEE Transactions on Automatic Control*, 42(7):896–911.

Scherer, C. and Weiland, S. (2000). *Linear Matrix Inequalities in Control*. DISC course lecture notes.

Seiler, P. and Sengupta, R. (2005). An $H_\infty$ approach to networked control. *IEEE Transactions on Automatic Control*, 50(3):356–364.

Shor, N. Z. (1985). *Minimization Methods for Non-differentiable Functions*. Springer, Berlin, Germany.

Sidi, M. J. (1997). *Spacecraft Dynamics and Control—A Practical Engineering Approach*. Cambridge University Press.

Skelton and Iwasaki (1995). Increased roles of linear algebra in control education. *IEEE Control Systems Magazine*, 15(4):76–90.

Song, B. and Hedrick, J. K. (2011). *Dynamic Surface Control of Uncertain Nonlinear Systems: An LMI Approach*. Springer, New York.

Stilwell, D. J. and Rugh, W. J. (1999). Interpolation of observer state feedback controllers for gain scheduling. *IEEE Transactions on Automatic Control*, 44(6):1225–1229.

Suplin, V., Fridman, E., and Shaked, U. (2006). H$_\infty$ control of linear uncertain time-delay systems—A projection approach. *IEEE Transactions on Automatic Control*, 51(4):680–685.

Syrmos, V. L. and Lewis, F. L. (1993). Output feedback eigenstructure assignment using two Sylvester equations. *IEEE Transactions on Automatic Control*, 38(3):495–499.

Tan, F. and Duan, G. R. (2008). Global stabilizing controller design for linear time-varying systems and its application on BTT missiles. *Journal of Systems Engineering and Electronics*, 19(6):1178–1184.

Tanaka, K. and Wang, H. O. (2001). *Fuzzy Control Systems Design and Analysis: A Linear Matrix Inequality Approach*. John Wiley & Sons, Inc, New York.

Toker, O. and Sunar, M. (2005). Substructural multiobjective H$_\infty$ controller design for large flexible structures: A divide-and-conquer approach based on linear matrix inequalities. *Proceedings of Institution of Mechanical Engineers Part I-Journal of Systems and Control Engineering*, 219(I5):319–334.

Torn, A. and Zilinskas, A. (1989). *Global Optimization*. Springer, Berlin, Germany.

VanAntwerp, J. G. and Braatz, R. D. (2000). A tutorial on linear and bilinear matrix inequalities. *Journal of Process Control*, 10:363–385.

Vandenberghe, L. and Boyd, S. (1995). A primal-dual potential reduction method for problems involving matrix inequalities. *Mathematical Programming*, 69(1):205–236.

Wang, G. S., Liang, B., and Duan, G. R. (2005). Reconfiguring second-order dynamic systems via P-D feedback eigenstructure assignment: A parametric method. *International Journal of Control, Automation and Systems*, 3(1): 109–116.

Wang, N. and Zhao, K. Y. (2007). Parameter-dependent Lyapunov function approach to stability analysis for discrete-time LPV systems. In *Proceedings of the IEEE International Conference on Automation and Logistics*, Jinan, China, pp. 724–728.

Willems, J. C. (1971b). Least squares stationary optimal control and the algebraic Riccati equation. *IEEE Transactions on Automatic Control*, 16(6):621–634.

Wilson, R. F., Cloutier, J. R., and Yedavalli, R. K. (1992). Control design for robust eigenstructure assignment in linear uncertain systems. *IEEE Control System Magazine*, 12:29–34.

Wu, A. G., Dong, H. F., and Duan, G. R. (2007). Improved robust H$_\infty$ estimation for uncertain continuous-time systems. *Journal of Systems Science and Complexity*, 20(3):362–369.

Wu, A. G., Dong, J., and Duan, G. R. (2009). Robust H$_\infty$ estimation for linear time-delay systems: An improved LMI approach. *International Journal of Control, Automation and Systems*, 7(4):668–673.

Wu, M., He, Y., and She, J. H. (2004a). New delay-dependent stability criteria and stabilizing method for neutral systems. *IEEE Transactions on Automatic Control*, 49(12):2266–2271.

Wu, M., He, Y., She, J. H., and Liu, G. P. (2004b). Delay-dependent criteria for robust stability of time-varying delay systems. *Automatica*, 40(8):1435–1439.

Wu, S. and Boyd, S. (2000). *SDPSOL: A parser/solver for semidefinite programs with matrix structure*. In *Recent Advances in LMI Methods for Control*. SIAM, Philadelphia, PA, Chapter 4, pp. 79–91.

Wu, Y. L. and Duan, G. R. (2005). Unified parametric approaches for observer design in matrix second-order linear systems. *International Journal of Control, Automation and Systems*, 3(2):159–165.

Xie, L., Shishkin, S., and Fu, M. Y. (1997). Piecewise Lyapunov functions for robust stability of linear time-varying systems. *Systems & Control Letters*, 31(3): 165–171.

Xu, S. and Lam, J. (2006). *Robust Control and Filtering of Singular Systems*. Springer, Berlin, Germany.

Xu, S. Y. and Chen, T. W. (2002). Robust H$_\infty$ control for uncertain stochastic systems with state delay. *IEEE Transactions on Automatic Control*, 47(12):2089–2094.

Xu, S. Y. and Lam, J. (2005). Improved delay-dependent stability criteria for time-delay systems. *IEEE Transactions on Automatic Control*, 50(3):384–387.

Xu, S. Y. and Van Dooren, P. (2002). Robust H$_\infty$ filtering for a class of nonlinear systems with state delay and parameter uncertainty. *International Journal of Control*, 75(10):766–774.

Xu, S. Y., Van Dooren, P., Stefan, R., and Lam, J. (2002). Robust stability and stabilization for singular systems with state delay and parameter uncertainty. *IEEE Transactions on Automatic Control*, 47(7):1122–1128.

Xu, S. Y. and Yang, C. W. (2000). An algebraic approach to the robust stability analysis and robust stabilization of uncertain singular systems. *International Journal of Systems and Science*, 31(1):55–61.

Yakubovich, V. A. (1962). The solution of certain matrix inequalities in nonlinear control theory. *Soviet Mathematics Doklady*, (3):620–623.

Yakubovich, V. A. (1964). Solution of certain matrix inequalities in automatic control theory. *Soviet Mathematics Doklady*, (5):652–656.

Yakubovich, V. A. (1967). The method of matrix inequalities in the stability theory of nonlinear control systems, I, II, III. *Automation and Remote Control*, 25–26(4):905–917, 577–592, 753–763.

Yang, C. Y., Zhang, Q. L., Lin, Y. P., and Zhou, L. N. (2007). Positive realness and absolute stability problem of descriptor systems. *IEEE Transactions on Circuits and Systems I-Regular Papers*, 54(5):1142–1149.

Yang, X. B., Gao, H. J., Shi, P., and Duan, G. R. (2010). Robust $H_\infty$ control for a class of uncertain mechanical systems. *International Journal of Control*, 83(7):1303–1324.

Yoshise, A. (1994). An optimization method for convex programs: Interior-point method and analytical center. *Systems, Control and Information*, 38(3):155–160. Special issue on Numerical Approaches in Control Theory. [In Japanese.]

Yu, L. (2002). *An LMI Approach to Robust Control* (in Chinese). Tsinghua University Press, Beijing, China.

Yue, D. and Han, Q. L. (2005). Delay-dependent exponential stability of stochastic systems with time-varying delay, nonlinearity, and Markovian switching. *IEEE Transactions on Automatic Control*, 50(2):217–222.

Yue, D. and Won, S. (2001). Delay-dependent robust stability of stochastic systems with time delay and nonlinear uncertainties. *Electronics Letters*, 37(15):992–993.

Zhang, F. and Duan, G. R. (2012a). Fuel near-optimal guidance law for the powered descending phase of a lunar module. *Proceedings of the Institution of Mechanical Engineers Part G-Journal of Aerospace Engineering*, 226(G1):108–120.

Zhang, F. and Duan, G. R. (2012b). Integrated translational and rotational finite-time maneuver of a rigid spacecraft with actuator misalignment. *IET Control Theory and Applications*, 6(9):1192–1204.

Zhang, F., Duan, G. R., and Hou, M. Z. (2012). Integrated relative position and attitude control of spacecraft in proximity operation missions with control saturation. *International Journal of Innovative Computing Information and Control*, 8(5B):3537–3551.

Zhang, H. S., Feng, G., Duan, G. R., and Lu, X. (2006a). $H_\infty$ filtering for multiple-time-delay measurements. *IEEE Transactions on Signal Processing*, 54(5):1681–1688.

Zhang, H. S., Xie, L. H., and Duan, G. R. (2007). $H_\infty$ control of discrete-time systems with multiple input delays. *IEEE Transactions on Automatic Control*, 52(2):271–283.

Zhang, P., Fu, Y. M., and Duan, G. R. (2005a). Robust reliable $H_\infty$ control for uncertain stochastic systems with time-varying delays. In *Proceedings of International Conference on Control and Automation*, Hungarian Academy of Sciences, Budapest, Hungary, pp. 181–184.

Zhang, P., Fu, Y. M., and Duan, G. R. (2006b). Design of robust $L_2-L_\infty$ filter for linear uncertain time-delay systems. In *Proceedings of World Congress on Intelligent Control and Automation*, Dalian, China, pp. 606–610.

Zhang, Q. L. and Yang, D. M. (2003). *Analysis and Synthesis of Uncertain Descriptor Linear Systems (in Chinese)*. The Northeastern University Press, Shenyang, China.

Zhang, R. W. (1998). *Satellite Orbit and Attitude Dynamics and Control* (in Chinese). Beihang University Press, Beijing, China.

Zhang, X. M., Wu, M., She, J. H., and He, Y. (2005b). Delay-dependent stabilization of linear systems with time-varying state and input delays. *Automatica*, 41(8):1405–1412.

Zhang, Y. and Duan, G. R. (2005a). Guaranteed cost observer for uncertain discrete-time switched systems with time-delay. In *Proceedings of Chinese Control and Decision Conference*, Guangzhou, China, pp. 912–916.

Zhang, Y. and Duan, G. R. (2005b). $H_\infty$ control for uncertain discrete-time switched systems with time-delay based on LMI. In *Proceedings of International Multi-Conference on Automation, Control, and Information Technology*, Novosibirsk, Russia, pp. 309–313.

Zhang, Y., Duan, G. R., and Zhang, X. Q. (2008). $H_\infty$-performance analysis for discrete-time switched systems with time-delay. In *Proceedings of IEEE International Conference on Automation and Logistics*, Qingdao, China, pp 1698–1702.

Zhang, Y., Wu, A. G., and Duan, G. R. (2009). Filtering for continuous-time state-delayed systems. *International Journal of Automation and Computing*, 6(2):159–162.

Zhang, Y., Wu, A. G., and Duan, G. R. (2010). Improved $L_2$-$L_\infty$ filtering for stochastic time-delay systems. *International Journal of Control Automation and Systems*, 8(4):741–747.

Zhao, Y., Fu, Y. M., and Duan, G. R. (2007). Robust passive filtering for switched systems with time-varying delays. In *Proceedings of IEEE Conference on Industrial Electronics and Applications*, Harbin, China, pp. 1350–1354.

Zheng, D. Z. (2002). *Linear Systems Theory (in Chinese)*. Tsinghua University Press, second edition.

Zhong, Z. and Duan, G. R. (2007). The analysis of finite-time stability for time varying polytopic systems. In *Proceedings of 26th Chinese Control Conference*, Zhangjiajie, China, pp. 210–221.

Zhong, Z. and Duan, G. R. (2009). Analysis of finite-time practical stability for time-varying polytopic systems. *Journal of Systems Engineering and Electronics*, 20(1):112–119.

Zhou, B., Cui, N. G., and Duan, G. R. (2012a). Circular orbital rendezvous with actuator saturation and delay: A parametric Lyapunov equation approach. *IET Control Theory and Applications*, 6(9):1281–1287.

Zhou, B., Duan, G. R., and Lam, J. (2010a). On the absolute stability approach to quantized feedback control. *Automatica*, 46(2):337–346.

Zhou, B., Hu, J., and Duan, G. R. (2010b). Strict linear matrix inequality characterisation of positive realness for linear discrete-time descriptor systems. *IET Control Theory and Applications*, 4(7):1277–1281.

Zhou, B., Lam, J., and Duan, G. R. (2009). An ARE approach to semi-global stabilization of discrete-time descriptor linear systems with input saturation. *Systems and Control Letters*, 58(8):609–616.

Zhou, B., Lin, Z. L., and Duan, G. R. (2011a). Lyapunov differential equation approach to elliptical orbital rendezvous with constrained controls. *Journal of Guidance Control and Dynamics*, 34(2):345–358.

Zhou, B., Lin, Z. L., and Duan, G. R. (2012b). A Lyapunov inequality characterization of and a Riccati inequality approach to $H_\infty$ and $L_2$ low gain feedback. *SIAM Journal on Control and Optimization*, 50(1):1–22.

Zhou, B., Zheng, W. X., and Duan, G. R. (2011b). Stability and stabilization of discrete-time periodic linear systems with actuator saturation. *Automatica*, 47(8):1813–1820.

Zhou, B., Zheng, W. X., Fu, Y. M., and Duan, G. R. (2011c). $H_\infty$ filtering for linear continuous-time systems subject to sensor non-linearities. *IET Control Theory and Applications*, 5(16):1925–1937.

Zhou, K., Khargonekar, P. P., Stoustrup, J., and Niemann, H. H. (1995). Robust performance of systems with structured uncertainties in state-space. *Automatica*, 31(2):249–255.

Zhou, K. M., Glover, K., Bodenheimer, B., and Doyle, J. (1994). Mixed $H_2$ and $H_\infty$ performance-objectives I: Robust performance analysis. *IEEE Transactions on Automatic Control*, 39(8):1564–1574.

Zhu, X. L., Yang, G. H., Li, T., Lin, C., and Guo, L. (2009). LMI stability criterion with less variables for time-delay systems. *International Journal of Control Automation and Systems*, 7(4):530–535.

# Index